U.S. Marine Corps Physical Readiness Training for Combat

Plus

Physical Fitness Training

Department of Defense

Pentagon
Publishing

UNITED STATES MARINE CORPS
Marine Corps Combat Development Command
Quantico, Virginia 22134-5001

29 January 1988

FLEET MARINE FORCE REFERENCE PUBLICATION............0-1B

MARINE PHYSICAL READINESS TRAINING FOR COMBAT

1. PURPOSE

Fleet Marine Force Reference Publication (FMFRP) 0-1Marine Physical Readiness Training for Combat is intended for use by all Marines. It provides the information and references necessary to establish and conduct physical conditioning programs to prepare Marines for the physical demands of combat.

2. APPLICABILITY

Although not directive, this handbook is intended for use as a reference by all Marines in developing unit physical readiness training for combat. This FMFRP does not establish policy for the conduct of physical fitness training. Refer to Marine Corps Order 6300.3_ for applicable Marine Corps policy on physical fitness. It may also be used as a basis for instruction conducted by unit leaders or as a reference for instruction or training.

3. RECOMMENDATIONS

To improve this manual, the MAGTF Warfighting Center seeks comments and recommendations from the user. Officers and noncommissioned officers with constructive comments are encouraged to respond. Lack of response will be considered as concurrence. The User Suggestion Form which follows can be detached and forwarded to - -

Commanding General
Marine Corps Combat Development Command (WF 06)
Quantico, Virginia 22134-5001

4. DISTRIBUTION

Initial distribution of this FMFRP 0-1B is shown below. Addressees on this list should ensure that appropriate subordinate organizations are provided copies.

JOHN I. HOPKINS
Brigadier General, U.S. Marine Corps
Assistant Chief of Staff, Warfighting
Marine Corps Combat Development Command
Quantico, Virginia

DISTRIBUTION: TDJ

From:

To: Commanding General, Marine Corps Combat Development Command (WF 06), Quantico, Virginia 22134-5001

Subj: FMFRP 0-1B, MARINE PHYSICAL READINESS TRAINING FOR COMBAT, RECOMMENDATIONS CONCERNING

1. In accordance with the Promulgation Page to FMFRP 0-1B, which invites individuals to submit suggestions concerning this FMFRP directly to the above addressee, the following unclassified recommendation is forwarded:

Page________Article/Para No.________Line No.________Figure/Table No.________

Nature of Change:_____Add_____Delete_____Change_____Correct

2. Proposed New Verbatim Text: (Verbatim, double spaced; continue on additional pages as necessary.)

3. Justification/Source: (Need not be double spaced.)

NOTE: Only one recommendation per page.

PREFACE

MARINE PHYSICAL READINESS TRAINING FOR COMBAT

Chapter 1. Physical Readiness Leadership
Section I. Role of Physical Fitness In Combat

Chapter 1
PHYSICAL READINESS LEADERSHIP
Section I. Role of Physical Fitness in Combat

1101. PURPOSE OF PHYSICAL FITNESS TRAINING

Physical fitness training in the Marine Corps has one purpose: to prepare Marines to physically withstand the rigors of combat. All other goals of physical fitness training are subordinate to and must support attainment of this goal. The idea that only infantry or reconnaissance units and their attachments normally face physically demanding combat is wrong. This error must not influence the priority commanders of combat support, combat service support, aviation, and headquarters units give to physical fitness for combat. Physical fitness for combat has a high priority for all Marines. A sound, effective unit program, requiring limited time and material, offers a greater payoff in combat than many more expensive and time-consuming training programs. This handbook describes the concept and provides the ingredients of a program for commanders to use in carrying out one of their most serious responsibilities: ensuring that their Marines are physically ready for combat.

1102. SCOPE OF THE MANUAL

This handbook provides guidance for all leaders, trainers, and planners of physical training programs. It describes unit physical fitness for combat training in the following chapters:

a. **Physical Readiness Leadership.** Chapter 1 provides guidance to leaders in the conduct of physical combat readiness training.

b. **Physical Readiness Training Programs.** Chapter 2 discusses how to structure programs to meet physical readiness goals in a variety of combat training situations.

c. **Physical Conditioning Activities.** Chapter 3 discusses the primary physical conditioning activities which commanders may use to attain readiness for combat goals. Sections in this chapter address foot marches under load, strength-building activities, and activities which build confidence and the aggressive spirit needed in combat.

d. **Combat Water Survival.** As an amphibious force, Marines cannot ignore the importance of combat water survival training. Chapter 4 describes a program for commanders to conduct this training.

e. **Competitive Conditioning Activities.** Chapter 5 describes competitive activities which stress the relationship to combat actions, featuring team-building types of competition.

f. **Evaluation of Performance During Training.** Chapter 6 describes tests which measure individual and unit physical fitness for combat.

g. **The Human Body.** Chapter 7 covers the structure and functioning of the body.

1103. COMMANDER'S ROLE

Major General Commandant John A. Lejeune in the 1921 edition of Marine Corps Manual expressed our philosophy of leadership. Among other things, General Lejeune stated that "it will be necessary for officers ... to endeavor to enlist the interest of their men in building up and maintaining their bodies in the finest physical condition." It is significant that the Commandant prefaced these writings in 1921 with the statement that his thoughts were greatly influenced by the experience of World War I. We have never varied as a Corps from this belief that one of a commander's most serious responsibilities is to have Marines physically prepared for combat. Senior commanders and their staffs carry out this role primarily by providing command emphasis, including planning, support, and supervision. Commanders at the battalion/squadron and company/battery level execute a program of physical fitness for combat by close supervision and, most importantly, leadership by example.

1104. HISTORY OF MILITARY PHYSICAL READINESS

Every war has revealed our military physical deficiencies during the initial periods of mobilization. This realization followed the Civil War and has recurred regularly with each national emergency.

a. Training programs in each war were geared to the physical need of the era. Success was dependent upon the amount of time available during training to physically prepare Marines for battle conditions. Frequently, casualties in initial engagements were attributed to the inability of our Marines to physically withstand the rigors of combat over rugged terrain and under unfavorable climatic conditions. With adequate preparation, Marines have always handled the test of battle.

b. During World War II, the first physical conditioning doctrine that could be scientifically justified by testing was introduced. As the war progressed, this program was effective in the physical conditioning of millions of Marines for combat.

c. Postwar periods have traditionally been a time of consolidation. Unfortunately, some leaders considered the conditioning phase of training to be a wartime tool. With such a philosophy prevalent between wars, physical readiness was relegated to secondary importance resulting in a serious lowering of combat effectiveness. The initial commitment of Marines in Korea dramatically displayed this failure to recognize the extreme physical demands of warfare. Call-up of large numbers of reserve Marines and pressing them into the war within weeks after call-up mainly caused the loss in effectiveness.

d. Costly lessons learned from our military experiences over a period of years and the course of several wars led to an increasing interest in the physical conditioning of the individual Marine. No longer can we afford to emphasize physical fitness during wartime and de-emphasize it during peacetime. It is evident that, in spite of increased mechanization and modern weapons, physical readiness retains a vital place in the life of each individual Marine and in every unit within the Marine Corps regardless of mission.

e. Commanders are well aware of the need for rugged and well conditioned Marines. The daily demands of housekeeping, maintenance, support, training, operations, and other time-consuming tasks make it necessary for commanders to set aside time for frequent, regular, and vigorous exercise periods.

1105. PHYSICAL DEMANDS OF COMBAT

There are three primary elements of effective physical fitness for combat: lower body strength and stamina; upper body strength and stamina; and a competitive, combative spirit. Cardiovascular functioning is not treated as a separate element since it is a necessary foundation for the other three. Exercises and activities designed to improve Marines' condition in the three primary elements will foster good cardiovascular and respiratory condition.

a. Lower Body Strength and Stamina. Some common demands that combat places on lower body strength and endurance are marching long distances under load and functioning effectively at the destination; moving quickly and evasively under fire; and carrying wounded Marines to safety.

b. Upper Body Strength and Stamina. Some common demands that combat places on upper body strength and stamina are rapidly emplacing crew-served weapons; handling large-caliber ammunition for extended periods; climbing walls, cliffs, and other high obstacles; and performing field maintenance on air craft or heavy machinery.

c. Competitive, Combative Spirit. While part of this critical aspect of combat readiness is mental and emotional, robust physical condition and a training program which promotes physical aggressiveness greatly enhance a competitive and combative spirit. Activities which force Marines to overcome natural physical fear, which directly leads to fatigue, are particularly valuable.

1106. PHYSICAL EFFECTS OF COMBAT STRESS

It has been thoroughly documented that the added uncertainty and stress of combat have a major physical effect on Marines as well as the acknowledged psychological effect. In other words, in combat, fear equals fatigue. In training, we cannot easily reproduce this combat stress which reduces the effectiveness of individuals and units. However, we can produce fatigue and psychological doubt in training situations by developing a challenging physical training program which stresses the participants. This compels Marines to exercise their ability to continue to function under difficult and trying conditions. Such activities as long foot marches under load and difficult confidence courses train Marines to overcome their own fear and fatigue. Rugged competitive activities such as martial arts and pugil stick fighting prepare Marines psychologically to overcome an opponent as well as their own fear and fatigue. This sort of physically demanding training builds self-confidence and unit morale. The adage that "the unit that sweats more in peacetime will bleed less in battle" certainly applies to its physical training program. Athletic coaches have long followed this approach by such practices as having teams shoot foul shots after wind sprints in order to strengthen their mental conditioning and toughness. Although pushing Marines to their limit is necessary, commanders often set lower standards initially than ultimately desired. The program should then call for steadily increasing the difficulty. A good guide for a starting point is a standard that will be physically and psychologically demanding for no less than three-fourths of the unit. Expressed another way, no more than one-fourth of the unit should be working at less than their limit. The program should progress quickly, augmented, if needed, by remedial training for those unable to meet the unit standards, until the training is challenging for all but the most exceptional Marines in the command. Commanders must never forget that Marines expect to be challenged and thrive on being pushed to their limit.

1107. LIMITS OF PHYSICAL READINESS

A well-conditioned Marine unit has a significant advantage in combat, but this does not mean that physical conditioning can substitute for technical knowledge or good planning. For instance, units must, as discussed in chapter 3, pursue a continuing and challenging program of marches under loads. This does not disprove the axiom that fighting men, if required to carry over one-third of body weight, will generally become ineffective in battle. Leaders and planners must decide what is essential to be carried and must use an effective logistic distribution system, such as unit trains, to make remaining items available when they are needed. In a similar vein, the valid requirement to foster a competitive and combative spirit in Marines does not excuse leaders from the responsibility to practice tactics which will not needlessly risk lives. Physical readiness for combat is a vital goal, but it is only one part of a unit's overall readiness for the lest of combat.

Section II. Fundamentals of Physical Fitness

1201. FITNESS FOR MARINES

Total combat readiness includes technical proficiency and mental, emotional, and physical fitness. If any one of these attributes is lacking, combat effectiveness suffers proportionately. Without technical fitness, Marines lack the knowledge and skill to fight; without mental and emotional fitness, they lack the incentive and will to fight; and without physical fitness, they lack the physical ability and confidence to fight. Physical fitness in a Marine means a whole and healthy body, the capacity for skillful and sustained performance, the ability to recover from exertion rapidly, the desire to complete a designated task, and the confidence to face any eventuality.

1202. COMPONENTS OF PHYSICAL FITNESS

A sound body, free of disease and defect, does not in itself constitute physical fitness. Before an untrained Marine can be considered physically fit for combat, the following traits must be developed:

a. Strength. Every Marine needs enough strength to perform the heaviest task encountered in routine and emergency activities. The basic areas where strength is required are in the arm and shoulder girdle, abdomen, back, and legs. Muscles increase in size, strength, and firmness with regular and strenuous exercises. Without work, they grow flabby and weak.

b. Endurance. Each Marine needs sustaining power to maintain the maximum ability without undue fatigue. There are two types of endurance:

(1) Muscular Endurance. The Marine needs muscular endurance to fight the enemy under the most tiring combat conditions. Muscular endurance is the ability to perform continuous work over long periods of time. Endurance depends on the bloodstream's ability to deliver large amounts of oxygen and nutrition to the muscle masses and then carry away the waste products quickly.

(2) Cardiovascular-Respiratory Endurance. The development of cardiovascular-respiratory endurance ("wind") is necessary to maintain muscular endurance. Cardiovascular-respiratory endurance depends on the efficiency of the blood vessels, lungs, and heart. The maximum effort a Marine can exert over a period of time is limited by the capacity to absorb oxygen and expel carbon dioxide. The average Marine's cardiovascular-respiratory capacity can be greatly increased by exercise.

c. Agility. A Marine must be able to change direction quickly and as faultlessly as possible. The ability to react instantly and to maintain orientation during rapid changes of body position is important to survival. This important ability may be developed by conditioning exercises that require varied and rapid changes of body position on the ground and in the air, such as obstacle courses.

d. Coordination. Coordination is the ability to move all parts of the body in a smooth, efficient, and concerted effort (commonly called timing). A well-coordinated individual does not make useless movements. An individual who moves with precision and accuracy saves energy. Coordination is best developed by practicing diversified muscular activities and skills affecting all body parts.

1203. TYPES OF EXERCISE

Basically, there are two forms of exercise: isotonic and isometric. Both forms are contained in the activities found in this manual.

a. Isotonic. Isotonic exercises are those in which the expenditure of energy is regulated and released during concentrated efforts. The regulated expenditure of energy is controlled by both the mode of exercise and the individual's effort. This type of effort is common to most exercise and sports. To develop endurance, coordination, and flexibility, isotonic exercise must be used. Strength can also be developed or increased through this type of exercise.

b. Isometric. Isometric exercises are those in which maximum effort is applied and held until the engaged muscle energy is depleted during a single contraction effort. The individual exerts full force against an immovable object for a relatively short period of time (6 to 10 seconds) and then repeats the contraction several times with a short period of rest between each contraction. This type of exercise develops only strength; therefore, it has limited application.

1204. PRINCIPLES OF PHYSICAL CONDITIONING

Men and women vary in their physical makeup. Physiological (body) function and reaction also vary in proportion to heavy demands placed upon the body. To attain the maximum benefit without sacrificing Marines' welfare, fitness must grow with a careful program of conditioning. To allow for adjustments in body functioning as the conditioning program progresses and to ensure attainment of objectives, the following principles of physical conditioning must be applied:

a. Overload. As strength and endurance increase, the physical load must be increased until the desired level of fitness is reached.

b. Progression. In beginning stages, the load must be moderate. Gradual progression from this low state of fitness to a higher state is possible through application of a progressive program.

c. Balance. An effective program utilizes various types of activities and provides for the concurrent development of strength, endurance, and coordination as well as basic physical skills.

d. Variety. Some programs fail because the routine becomes boring. The most successful programs always include conditioning activities, competitive events, and military physical skill development.

e. Regularity. There is no easy or occasional way to develop physical fitness. Regularity of exercise is a must, with daily exercise preferred.

1205. THREE STAGES OF PHYSICAL CONDITIONING

Unconditioned or poorly conditioned Marines pass through the following stages in reaching the desired state of physical condition.

a. Toughening Stage. This stage is approximately 2 weeks long and is usually characterized by muscular stiffness and soreness followed by recovery.

b. Slow Improvement Stage. This stage is approximately 6 to 10 weeks long and is characterized by slow and steady improvement until the desired level, or a high level, of fitness is attained.

c. Sustaining Stage. This stage goes on indefinitely in order to maintain the level of conditioning achieved by passage through the previous stages.

1206. EFFECTS OF CLIMATIC CONDITIONS

Temperature, both atmospheric and body, affects the physical performance of Marines. Proper maintenance of body temperature through warm up exercise, proper dress in cold weather, and removal or adjustment of clothing in hot weather is necessary for effective performance and health. Climatic factors to be considered are the following:

a. Exercise In High Temperatures and High Humidity. Marines can endure strenuous physical activity in extremely hot temperatures if they are given an opportunity to become acclimated and if they take enough salt and water. It is essential to continue physical training programs in hot climates. Marines can better withstand high temperatures when they are well-conditioned. High humidity combined with high temperatures presents a serious danger. These conditions prevent the natural cooling of the body by the evaporation of perspiration. Training schedules should conform to the provisions of the base commander in respect to wet bulb conditions. Those who conduct training under these conditions should monitor weight loss and be careful to make fluid replacement adjustments.

b. Exercise at High Altitudes. Certain problems are encountered in conditioning Marines stationed in high altitudes because the heart undergoes greater exertion during exercise. It is particularly important that only light exercise be given initially at such altitudes. A Marine's body gradually adjusts to high altitudes within a few weeks. After this adjustment, progressively greater amounts of exercise are possible.

c. Exercise in Arctic Regions. Military duty in the arctic is so arduous that a high level of physical conditioning is essential. Because of the difficulties of carrying on physical conditioning exercises in extreme cold, Marines should be conditioned to the highest level possible before they arrive. A sustaining program will then maintain that level. When exercising in cold weather, Marines should be required to remove excess clothing to prevent them from becoming damp with perspiration.

1207. WARMING UP AND COOLING OFF

It is a fundamental physiological principle that Marines should warm up gradually before taking strenuous exercise. Such action speeds up the circulation to prepare the body to take an overload and helps to prevent injury to muscles and joints after exercising. Marines should be kept mildly active, walking, stretching, or performing some other mild muscular activity until breathing and temperature have returned to normal. Marines should never be allowed to cool off too rapidly; in cool or cold weather, they should put on additional clothing during the cooling-off period.

1208. PHYSICAL ACTIVITY AS AGE INCREASES

Combat makes severe physical demands on Marines. All Marines, regardless of age, must be physically ready to meet these demands. There is no physiological reason to cease exercise or exertion with age. Increased age usually brings increased responsibility which, in many instances, leads to a routine that can become almost devoid of physical activity. The key to fitness with increased age is to continue exercising at a reasonable level and to include exercise of a vigorous type in the daily routine. Older persons who have not regularly maintained a reasonable state of physical fitness require more time than younger persons to become fit. Such individuals usually require a longer period of time to recover from physical effort than younger Marines. If general health is good, evidence shows that older Marines can develop and maintain a degree of fitness which permits vigorous activity and proper performance of duties. It is both the individual's and the commander's responsibility that all Marines continue a daily sustaining exercise program.

Section III. Goals of Physical Readiness Training

1301. NECESSITY FOR PHYSICAL READINESS TRAINING

A very important objective of training is attainment and maintenance of operational readiness. Marines must be physically ready for operations at any time and under all conditions of climate and environment. A combination of training to develop proficiency in physical skills and conditioning to improve strength and endurance results in physical readiness for combat. The degree of physical fitness required of Marines can be acquired through physical exercises in a normal well-coordinated and closely monitored training routine. The performance of purely military duties, such as drills and marching, is not enough to build all the desired areas of fitness. Few recruits are physically fit for the arduous duties ahead of them. The softening influences of our mechanized civilization and the difficulties of conditioning Marines make physical fitness more important than ever before. If Marines are to be fully developed to and maintained at the desired standard of physical fitness, a well-conceived plan of physical readiness training must be a basic part of every training program. Marines cannot be adequately prepared in any other way for the hard work and arduous demands required on the battlefield.

1302. OBJECTIVE OF PHYSICAL READINESS TRAINING

The overall objective of the physical readiness training program is to develop individuals and units who are physically able and ready to perform their duty assignments or missions during training and in combat. To attain the objective of physical readiness, exercise activities must develop the following areas:

a. Strength and Endurance. Developing adequate strength to perform required duties and adequate endurance to sustain activity over a long period of time.

b. Muscle Tone. Developing muscle tone adequate to maintain proper posture and reasonable weight control.

c. Skills. Developing military physical skills which are essential to personal safety and effective combat performance. As skill is developed, agility and coordination are attained. The essential skills are-

(1) Marching Under Load. Marching with individual and unit weapons and equipment.

(2) Running. Distance and sprint running on roads and cross-country.

(3) Jumping. Broad jumping and vertical jumping downward from a height.

(4) Dodging. Changing body direction rapidly while running.

(5) Climbing and Traversing. Vertical climbing of ropes, poles, walls, and cargo nets. Traversing horizontal objects such as ropes, pipes, and ladders.

(6) Crawling. High crawling and low crawling for speed and stealth.

(7) Throwing. Propelling objects, such as grenades, for a distance with accuracy.

(8) Vaulting. Surmounting low objects, such as fences and barriers, by use of hand assists.

(9) Carrying. Carrying objects and employing man-carries.

(10) Balancing. Maintaining proper body balance on narrow walkways and at heights above normal.

(11) Falling. Contacting the ground from standing, running, and jumping postures.

(12) Surviving in Water (Or Other Specialized Situations) Using water survival techniques.

d. Character Traits. Jnstilling character traits which help accomplish military missions to include-

(1) Confidence. Confidence develops through achieving progressively more difficult tasks as physical ability grows.

(2) Aggressiveness. Participation in combative activities and contests develops desire and willingness to overcome an opponent.

(3) Reaction Under Pressure. Competitive contests and game situations train Marines to think and to act quickly while under pressure.

(4) Teamwork. Teamwork develops through competitive events in which a number of Marines must coordinate their efforts to accomplish a physical task.

1303. BENEFITS OF EXERCISE

The benefits of exercise are not always understood. Some of the more important results of exercise are listed below:

a. Improved Muscle Tone. Muscular tone improves and, at the same time, muscular strength and endurance are built up.

b. Cardiovascular-respiratory Endurance. Cardiovascular-respiratory endurance, or wind, improves through a process of opening up dormant lung capacity to absorb greater amounts of oxygen.

c. Circulation. Circulation of the blood speeds up and extends to a greater portion of the body as exercise forces the blood to service all parts of the body. The efficiency and effectiveness of the heart, lungs, and blood vessels improve.

d. Flexibility. A wider range of muscular movement is possible and rapidity in physical skills grows.

e. Elimination of Body Waste. Bending and twisting the body and the general speedup of body processes caused by exercise regulate and help eliminate body wastes.

f. Tension. Working off excess nervous energy and relief from daily worries and cares relieve tension. Participation in exercises leaves little time for worry.

g. Sleep. Sleep improves because muscles are healthfully tired after about of exercise. A by-product of sound sleep is relief of tension.

h. Obesity Control.. Control of obesity (fat) is made possible by using up excessive amounts of fat-producing food elements.

i. Injury Susceptibility. Susceptibility to injury is reduced through exercise. Muscles, tendons, and joints are strengthened. Injuries such as hernia, back strain, and joint sprains are less likely to occur if muscles are maintained in proper tone.

Section IV. Leadership Roles

1401. PSYCHOLOGICAL LEADERSHIP

The full development of a Marine's resources is not all physical. To be effective in developing physical readiness, leaders must realize that mind and attitude are also important to success. The more important psychological considerations are to—

a. Promote Understanding of the Value of Physical Readiness. A desire to be physically ready should be created in all Marines. Motivation is increased and Marines take greater interest in their individual physical fitness if they understand the value and benefits of vigorous exercise. When Marines realize their efforts are an investment in their own personal welfare, it should not be difficult to obtain their cooperation. Marines should understand the objectives, the benefits, and the value of each type of exercise activity in their program. They should also understand the relation of physical readiness to survival in combat.

b. Maintain a Positive Approach. Physical readiness training for combat is strenuous and demanding. It is a responsibility of leadership to create an atmosphere where all desire to participate fully. This attitude should be fostered. A negative approach must not be identified with physical readiness training, even with those having difficulty. Only in unusual cases should fear of punishment be the motivating factor behind good performance. For those few who cannot keep up or attempt to malinger, an effective remedial program is essential.

c. Seek Cooperation and Develop Morale. In a program placing maximum physical stress upon individuals, it is necessary to gain their cooperation. Favorable reaction is enhanced by proper planning and organization, challenging requirements, use of competition, and application of a progressive program resulting in physical fitness. As physical fitness grows, morale also grows.

1402. COMMAND AND SUPERVISORY FUNCTIONS

a. Command Functions. Commanders should take the following actions to support physical readiness training:

(1) Lead by personal example.

(2) Instill command interest and indicate to subordinate personnel the importance of this training to the welfare of the organization.

(3) Allot sufficient time for the achievement of objectives and monitor the use of such allotted time. The substitution of other training or routine duties for scheduled physical readiness training is unsound and unwise.

(4) Assign and properly utilize qualified personnel to supervise and conduct physical readiness training. If leaders are not competent, take action to ensure they become competent quickly. Reassign those who do not meet this standard to other duties.

(5) Make necessary facilities and funds available to support a program to develop physical readiness within all personnel.

(6) Measure the physical fitness of individuals and units in order to evaluate progress and to determine if the program is successful.

b. Supervisory Functions. Leaders responsible for planning, conducting, and supervising physical readiness training should take the following actions:

(1) Prepare physical readiness training schedules which apply the principles of physical conditioning and which aim for a particular type of program plan.

(2) Provide for wide participation of as many Marines as possible. All Marines, regardless of position or age, will benefit from regular exercise. In some instances, special efforts are necessary to overcome obstacles to regular and frequent training. Special effort is also necessary to ensure remedial conditioning. Such conditioning should occur for those who are physically substandard and after extended absence due to leave, sickness, injury, and travel.

(3) Prevent waste or unwise use of time allotted for physical readiness training. Time-wasters include unprepared instructors; assignment of one instructor to a group larger than a platoon; progression which does not keep pace with the physical development of the Marines; extreme formality; inadequate equipment or facilities which require waiting turns to exercise; and lengthy rest periods between exercises which interfere with the application of overload.

(4) Ensure that the program contains vigorous physical activity. Such activity places progressively greater demands upon the body during each exercise session and also over the duration of the training program. To be of benefit, exercise must tire the muscles and cause the heart to increase its rate of beat.

(5) Set an overall objective for each physical fitness program.Observe the training as necessary to ensure that the established objectives are being achieved.

(6) Observe physical readiness training to insure the use of a positive approach. To implement a positive attitude, small-unit leaders and instructors should personally set the example; have an understanding, fair, and sympathetic attitude; recognize individual differences; and motivate Marines toward their best effort.

(7) Guide and inform small-unit leaders and instructors concerning approved techniques, directives, and literature. As necessary, arrange for local training of instructors to include clinics, conferences, schools, and demonstrations.

(8) Determine the effectiveness of physical readiness training by personal participation in and observation of training, analysis of field inspection reports, and analysis of individual physical fitness test scores. Scores may be combined to reflect the fitness of the unit.

1403. SMALL-UNIT LEADERS AND INSTRUCTORS

a. Responsibility. The instruction and conduct of physical readiness activities are the function of company/battery commanders, platoonleaders and persons assigned as instructors. Experience has proved the effectiveness of physical fitness development when conducted in company-and platoon-size units under direct control of the leader with overall supervision by the parent-unit commander. For example, all the platoons of a company may exercise at the same time under the general supervision of the company commander, with each platoon conducting the assignment separately and under its own leadership.

b. Leader's Assignment. A small-unit leader or an instructor in a school or training activity is assigned to a combat unit or to a support unit. In this assignment, the leader is responsible for all training to include physical readiness training. In a different situation, the noncommissioned officer (NCO) or officer may beassigned as a full-time physical readiness instructor. This contrasts to a unit leader assignment where only part of the time is devoted to such training. In either case, the leader will hold an important and vital position for the physical fitness of Marines.

c. Leader's Training. Leaders may come to the assignment either fully or partially trained. It may be their first responsibility for the development of physical fitness. If they have had previous training through experience, make certain that their information is supplemented with study of this manual. If they have had professional training in physical education during civilian life, but no military experience, they should also use this manual to learn the methods used by the Marine Corps. A new leader should take advantage of various ways to learn including attendance at leader training courses, self-study, practice, and discussion with more experienced leaders.

d.Leader's Objective. As a physical readiness training instructor, the leader has two general objectives. The first is to motivate Marines to want to be physically fit. The second is to conduct a program that will develop a high degree of physical fitnss. Motivated Marines will react enthusiastically to such a program. It aids greatly in achieving local program objectives.

e. Leader's Personal Fitnes. A unit leader who must instruct and demonstrate physical activities must be in physical condition to do the job without undue physical stress.The leader should be able to do those things that must be demonstrated. The leader's strength, endurance, posture, and skill should set the example. This does not mean that the leader must excel, as other Marines do not expect championship performance. However, they do expect, and deserve, a creditable showing of fitness for the job.

f. Leader's Knowledge. The leader must have three types of knowledge to properly administer physical readiness training. Theyare- -

(1) Knowledge of Marines. The leader must understand Marines, know how to lead and motivate them, understand how they learn, and apply this knowledge wisely in the day-to-day training situation.

(2) Understanding of Body Functioning. A more intelligent exercise program results from understanding and applying the principles which govern physical conditioning of the body. The leader with such knowledge can better prescribe, adjust, and regulate exercise types, amounts, and progression to attain fitness.

(3) Understanding Exercise Activities. The leader needs to understand the contribution each type of physical activity makes to physical fitness, and how to use each activity to develop fitness. Skill to demonstrate and lead the various activities is a necessary part of technique and is invaluable to the instructor or small-unit leader.

Chapter 2
PHYSICAL READINESS TRAINING PROGRAMS
Section I. Development of a Program

2101.GUIDANCE FOR PLANNERS

This chapter instructs planners on physical readiness training procedures. It contains program planning guidance including factors to consider when developing programs; steps in assembling a program; definitions of activity packages and systems of exercise; selection of activity packages; and selection of systems used in implementing those packages.

2102. MARINE CORPS MISSION

The mission of the Marine Corps is to seize and defend advanced naval bases and to perform other missions as the President of the United States may direct. Traditionally, to perform this mission, Marines have been projected into the area of operations by Navy vessels and have attacked hostile beaches over the shore in landing craft, assault amphibian vehicles, and helicopters. Recently, performance of this task has been somewhat complicated by the introduction of the concepts of maritime prepositioning and the air-landed Marine expeditionary-brigade. Under these concepts, Marines, landed at airfields in foreign countries, will join their equipment at commercial ports. The distance between these airfields and ports is often 10 or more miles. In time of international crisis, it can be expected that transportation assets will not be adequate. Marines from all elements of a Marine Air-Ground Task Force may have to march from the airfield to the site of their equipment. All Marines must be physically conditioned to rigors of conducting foot marches with individual weapons and equipment.

2103. EVALUATION OF FITNESS

The Marine Corps mission is such that all Marines must constantly and consistently achieve a high level of fitness that prepares them for the demands of combat. While Marines have many different MOS's, all Marines must be prepared for the demands of marching under load and performing basic infantry tasks, such as rear security and patrolling. While the physical fitness test remains our universal measure of individual fitness, it is the commander's responsibility to observe and evaluate the unit's ability to perform effectively in combat. Often the commander's most effective evaluation tool is weekly physical training sessions. The commander should participate in these sessions and see that they are fully integrated into the unit physical conditioning program.

2104. INTEGRATION OF TRAINING

Training time must be used efficiently and wisely. Every opportunity to integrate physical training into other training activities should be seized. For example, foot marches under load can be integrated into the normal daily activities as a way to move from one training or work site to another. This integration reflects realism in training and should be used to maintain overall proficiency. To accomplish this requires imagination and enthusiasm on the part of the commander. The area of operations in which the training is conducted determines what can be integrated and how. If possible, training should also consider the terrain and climate of the area in which the unit will subsequently conduct training or operations. Training should include familiarization with special equipment and the application of specialized techniques to tactical principles. Activities such as gun drills and command post emplacement drills also have physical conditioning value, particularly if supervised and done competitively or measured against time standards.

2105. ACTIVITIES AT THE UNIT LEVEL

Physical readiness training is a command responsibility and is generally conducted in Marine units at the company and platoon levels. Battalion/squadron commanders must constantly supervise, making sure that all Marines are physically prepared for combat.

a. Physical readiness training should usually be conducted weekly by each battalion/squadron-level unit, using events designed for all personnel (e.g., marches under load, battalion/squadron runs, military field days, etc.).These events enable the commanders to demonstrate personal leadership, and observe and evaluate the physical condition and combat physical readiness of their Marines. This exercise period should be designed to elevate the unit's morale and emphasize unit identity through the wearing of uniforms and the carrying of battalion and company guidons. The battalion/squadron commander should make every effort to be a visible participant in the unit's physical readiness program on a weekly basis. These activities could be conditioning marches under load or a weekly calisthenics/unit run.The suggested times for these activities are early on Monday morning or late on Friday afternoon after the training week has been completed. The uniform may be physical training gear orutilities and boots.

b. Commanders should be cautious of long runs in boots and utilities. This is not to suggest that units should not run in utilities, but to remind commanders that boots are designed for marching, not running. If commanders choose boots and utilities, they should select off-road routes which will provide a cushioning effect.

c. Units should be made well aware of the commander's intent before the unit falls out for exercise. If the exercise period is conducted early in the morning, it need not necessitate an extensive shower/clean-up period afterward, so long as all Marines from the commander on down remain dirty. Commanders should inform their Marines not to fall out for these sessions in clean, pressed uniforms.

d. The physical readiness training of headquarters units is often difficult to manage. Often physical training is left for individuals or sections to develop and manage on their own. This approach can be effective, but if not supervised, it can result in a "paper" program. Headquarters unit commanders should endeavor to conduct weekly integrated training which will enable them to evaluate the physical readiness of their personnel. Often this training has to be scheduled before or after normal working hours to maximize participation of the entire unit.

2106. STEPS IN PLANNING

To implement workable and effective programs (as directed by CMC ALMAR 261/87) the planner must-

a. Determine the Type of Program Needed. Marine units are inherently different in organization and mission. The physical readiness program must be tailored to the mission and to the current physical condition of most unit personnel. Programs to meet this need are of the following types:

(1) **Developmental Programs.** Marines in a beginning or poor state of physical readiness need a program which will develop strength, endurance, physical skills, and character traits which are beneficial to successful accomplishment of military missions. Such programs should be applied progressively to rise gradually to a peak of fitness and skill.

(2) Maintenance Programs. Once Marines reach the sustaining stage of conditioning, their goal is then to maintain this level by participation in a maintenance program.

(3) Remedial Programs. The term "remedial" is usually applied to those individuals or groups who possess substandard physical fitness. For example, a remedial physical conditioning program could be applied to persons who are overweight, who fail to reach physical fitness test standards, or who have missed extended periods of conditioning due to illness, injury, extended hospitalization, or other absence. Saturday training sessions are designed as remedial sessions for all individuals who either miss the daily training or have failed to meet adequate standards. Note that personnel in light duty or no-duty status should be expectedto attend all training sessions, observing, supervising, or assisting as needed.

b. Determine the Time Required. The amount of time for training operations varies considerably. However, every unit can find time to conduct physical readiness training. Frequent (i.e.,daily) physical training of short duration-30 to 60 minutes-is preferred over occasional longer periods. Other demands for training time are so urgent that every minute of time allotted for physical readinesstraining should be used. Determine the time required per week anddivide it into daily blocks.

c. Organize for Various Group Size. It is essential to stress exercise rather than formality. Marines must complete the program where they are- -on the training field, in the motor pool, on the range, next to the classroom, in the office area, in the shop, aboard ship, or elsewhere. Although desirable, it is not always possible to assemble company-size units for physical training. Platoon-size groups are appropriate for the proper conduct of physical conditioning activities. Certain situations may require exercise programs for section- or squad-size units. This manual outlines programs for all situations and types of organizations.

d. Allow for Weather and Exercise Area. In programing and scheduling, the climate and terrain often govern the selection of activities.

(1) Weather changes cause differences in temperature, rainfall, wind chill, and snow. These changes should be anticipated as they dictate the type of program. Alternate plans should be part of the schedule. Nevertheless, normal weather changes and rain should not drive training inside.
(2) Local terrain and available exercise areas may also influence the selection of activities and the type of program which it is possible to support. Some activities can be completed in nearly any area.
e. Plan for Seasonal Change. As most physical readiness training is conducted outdoors, it is necessary to recognize seasonal change. A program should be divided into fall, winter, spring, and summer parts. Seasonal change also causes change in light. For example, an early morning program started in the summer will have ideal lightconditions, yet in the fall or winter, darkness will occur at that same hour and interfere with the conduct of the program. Develop programs in seasonal blocks, and make provisions for anticipated changes.

f. Consider Needed Facilities. An excellent program can be conducted with practically no facilities since there are exercises which require no equipment. However, a better program can be developed when supported by certain facilities and items of equipment. Proper command support, plus ingenuity, will solve this problem. (Items of equipment, when necessary to support the recommended exercises, are included in the chapters on exercise activities.)

g. Specify Appropriate Uniform. The uniform worn for exercising depends upon the season of the year, the state of the weather, and local regulations. Whenever possible, Marines should be dressed alike. Undershirts are preferred as the upper garment when the weather permits. A uniform that restricts the free movement of the body should not be worn when exercising.

h. Consider Availability of Instructors. Leaders who can lead and direct the scheduled activity must be available. Organizational units should train junior officers and noncommissioned officers down to squad or section leaders to instruct and lead the various activities.

i. Select Activity and System. With the type of program needed and the objectives in mind, select an activity package or an integrated training package for each day's scheduled physical training. At this same time, the system to be used in employment of the selected activity or activities must be determined. This selection affects equipment, areas, instructors, transportation, and other support requirements.

j. Secure Command Participation and Support. Prepare and brief the commander to assure full understanding of the objectives and administration of the program. The full participation and support of the commander will greatly improve the level of success.

k. Supervise Execution. Determine needs, publish the program, and supervise its execution as a necessary part of developing the unit's training schedule.

2107. SELECTION OF ACTIVITIES AND SYSTEMS

a. Exercise Activities. The planner must determine exercise activities which will be appropriate to include in the program. An exercise activity is a single means of exercise usually identified by the name applied; for example, running, log exercises, and obstacle course. Many exercise activi ties can become part of activity packages.

b. Activity Packages. Many of the physical activities described in this manual are arranged in prescribed sequences and are known as activity packages. An activity package is a number of exercises of the same type, assembled as a group or a set, and arranged in a specific sequence. Exercise packages are organized in such manner that not more than 15 minutes will be required to complete the execution of any package. Each type of activity is explained in laterchapters. The number of available drills, tables, or circuits; the manner of organization; and the contribution each makes to the total program are covered. Full understanding of this information will greatly assist in developing effective programs. Various designations are used to identify exercise packages; for example, conditioning exercises when arranged in a set order are known as drills, and other packaged activities are designated as tables or circuits. The following are activity packages:

(1) Activity packages for groups:

o Conditioning Drill Two
o Conditioning Drill Three
o Rifle Drill
o Log Drill
o Grass Drill
o Running Tables
o Guerrilla Tables
o Circuit-Interval Table
o Combatives Tables
o Relay Tables

(2) Strength Circuits:

o Fixed Circuit
o Movable Circuit
o Simplified Circuit (Circuit Interval Table)

(3) Activity packages for individuals:

o The 6-12 Plan
o Weight Training
o Isometric Exercise

c. Advantages of Using Packages. These of exercise packages simplifies scheduling and conducting of exercise and results in the following benefits:

(1) Schedule development is simplified as the planner assembles packages which will satisfy the training objective. There is no need to deal with selection of individual activities or to be concerned about the amount of time to be expended on each.

(2) Any 15-minute period, and in some cases less time, can be scheduled or used to perform an activity.

(3) If longer periods of time are available or if the objective demands, several packages can be assembled to provide a more complete period of activity.

(4) Marines are assured a balanced set of exercises or activities as each package is carefully arranged to reach all muscle groups.

(5) The instructor can concentrate on the conduct of a vigorous workout as the type and duration of the activity have already been determined.

d. Nonpackaged Activities. Several types of activity are not packaged. Activities in this category are conditioning marches, unit runs, obstacle courses, combat water survival swimming, teamcontests, and team sports. These activities can be scheduled in combination with packaged activities, or they may be scheduled separately. Often they require a longer period of time. Most nonpackaged activities require a 50-minute period to satisfactorily complete their objective. For conditioning marches, a few hours are required. The benefits of these activities should not be overlooked as some desirable objectives cannot be attained without them.

e. Systems of Exercise. Several methods or systems of organizing exercise and activity packages can be used. Each system is based upon a specific organization as follows:

(1) Single Activity System. The unit leader immediately assumes command of the unit at the beginning of the exercise period. The leader moves the unit to a predesignated exercise site at double time, forms the unit in a circle around the leader, groundsclothing and equipment as appropriate, and quickly moves into the exercise routine. There is usually no time to teach; therefore,the Marines must know the activity to be used. At the conclusion of a 5- to 15-minute period, the unit leader returns the unit to the instructional area at double time and releases the unit for the next scheduled activity.

(2) Progressive Activity System. All Marines (company or platoon) complete activities in the same order during the period. For example, Drill One is followed by dual combatives, and finally a 1-mile run. This system is usually progressive from a warm up activity, such as Drill One, to an activity which contributes in a major way to one of the objectives, such as aggressiveness development through combatives or cardiovascular development, such as running.

(3) Rotating Activity System. Set up the same number of activities or stations as there are platoons in the company. Each platoon rotates through each station in turn. With four platoons in a 50-minute period, about 10 minutes can be devoted to each station. With three platoons, approximately 15 minutes may be spent at each station. Activities must be of a type that can be covered in the time allotted. For example, with three platoons, Station 1 could be Conditioning Drill One; Station 2, running; and Station 3, a team contest.

(4) Circuit System. Set up a number of stations to provide various types of exercise equipment and/or items of apparatus. The idea is to keep all Marines busy and exercising vigorously for a short period of time at each station. The fixed strength circuits and the movable strength circuits are examples of this system. Station changes must be rapid, and the exercise must be started quickly after each change. Since the objective is to exercise at top speed, the motivation comes from frequent change of activity by moving to another station. Rotation by station groups continues until all Marines have covered all stations.

(5) Interval System. This system stresses the development of strength and endurance. It involves heavy work for a given distance within a specified time, alternated with lighter work and recovery, but never stopping during the workout. This procedure is repeated, and the intensity is increased gradually as physical condition improves, but always with adequate recovery. The important concept is stress, recover, stress, recover, and so on. This system is often applied through running, but other activities of a continuous nature may also be used. An example is the Circuit-Interval Table.

f. Steps in Selecting Activities and Systems. The planner should follow these steps.

(1) Evaluate the needs of one unit with emphasis upon unit mission, objective, and time available.

(2) For each day, select an activity package or a combination of activities which will contribute to the objective.

(3) Then determine the system to be used in implementing the selected activities. There are several possible choices and much opportunity for flexibility in program development. For example, the weekly program may contain various exercise packages, systems, and time periods. Additional guidance is contained in following chapters.

2108. SAMPLE WEEKLY PROGRAMS

The variety of weekly programs given shows available alternatives. Programs and schedules other than those illustrated may be assembled to provide schedules to fit any situation. Many other scheduling combinations are possible through use of the packaged activities, varied time periods, and different types of assigned duty.

a. Sample A. Figure 2-1 illustrates a 60-minute time allotment in which the single activity system is used with activity packages that vary from day to day. In this program, 5 hours of training are scheduled for the week. This program needs some type of integrated unit training to supplement it each week. A bimonthly conditioning march of 10 miles in 3 hours is suggested to round out this program. The Saturday session is designed for remedial physical training.

NOTE: A physical training activity must be conducted at a minimum of once every 72 hours. After 72 hours, muscles begin to atrophy.

Figure 2-1 is available in the illustration section at the end of the book

b. Sample B. A more comprehensive weekly program (fig. 2-2) illustrates the scheduling of various-length periods, a variety of activities, and the use of three systems during the week. In this schedule, 3.5 hours are included for the week. The following should be noted concerning this schedule:

Figure 2-2 is available in the illustration section at the end of the book

(1) On Monday, a 60-minute period is available for physical readiness training. Three 20-minute packages are scheduled. Each platoon, under its own leadership, will progress through each scheduled activity in turn.

(2) On Tuesday, Wednesday, and Thursday, only 30-minute periods are available. Here each platoon leader supervises each platoon in a single activity system.

(3) On Friday, the company commander directs training to exhibit personal leadership example and supervision. In this example, only one strength circuit and one cross-country course are available, and each will accommodate only one platoon. For this reason, Conditioning Drill One station is included. A platoon rotates to each station within the 60-minute period.

(4) The Saturday session once again is designed for remedial physical training.
(5) Bimonthly, a 10-mile/3-hour march under load will be necessary to round out this program.
NOTE: Cardiovascular/respiratory exercise to improve endurance must be of long enough duration (20 minutes or more) to induce and maintain oxygen debt.

c. Sample C. A third sample schedule (fig. 2-3) illustrates the use of the single activity system for 5.25 hours of training. This schedule includes sustaining-type activities for a unit that has passed through both the toughening and slow improvement stages of conditioning. The use of competitive activities is featured to holdinterest and provide self-motivation. A bimonthly 10-mile/3-hour march under load will be necessary to round out this program. The Saturday session once again is designed forremedial physical training.

Figure 2-3 is available in the illustration section at the end of the book

NOTE:Thc body will respond to hard work or physical training without soreness and stiffness provided the work or physical training is conducted regularly.
NOTE:Excrciscs of short duration increase strength and bulk but do little to effect endurance.
NOTE:Thc times shown in the charts arc participation times.

Section II. Remedial Physical Training

2201. NEED FOR ATTENTION

Remedial physical training is a process by which physically substandard individuals are conditioned to meet standard levels of performance. To achieve a full degree of operational readiness, it is necessary to bring all Marines up to the prescribed standard. Experience has demonstrated that some Marines have difficulty due to a poor state of fitness, obesity, or lack of motivation. Attention to these individual deficiencies will help improve unit combat readiness.

2202. IDENTIFICATION OF INDIVIDUALS

The company or battery commander identifies persons who cannot achieve the prescribed level of proficiency through the physical fitness test or as a result of observation during training. The commander notes particular weakness of body parts as indicated by failure of certain test events. These Marines are then placed in a special remedial program either at platoon, company/battery, or battalion/squadron level. Extra time is devoted to overcoming the weakness. These extra periods of conditioning may be during or after duty hours, as determined by the commanders. Other persons need reconditioning as result of hospitalization. Restoring physical fitness to damaged areas of the body is accomplished through progressively graded physical activities under professional supervision, not by the physical fitness trainers.

a. Types of Deficiencies. Physical deficiencies which can be corrected by exercise fall into several categories.

(1) Lack of Strength in One or More Body Parts. The major muscle areas concerned are the arms and shoulder girdle, back, abdomen, and legs.

(2) Lack of Overall Endurance. Usually, there is deficient muscular and cardiovascular endurance.

(3) Deficiency in Coordination and Agility. In these cases, physical skill is not developed to a satisfactory degree in activities such as crawling, running, jumping, climbing, traversing, vaulting, pushing, pulling, lifting, and carrying.

(4) Overweight or Underweight. Either condition may interfere with physical fitness and mission accomplishment. Lack of exercise is not always the cause. The cause may be malfunction of normal physiological functions or it may be poor health habits such as over- or under-eating, lack of adequate rest, or overconsumption of alcohol.

(5) Lack of Motivation. Not all persons are motivated to attain or maintain **a** desirable state of fitness. Some Marines do not understand the importance of physical fitness, some find proper exercise too difficult, and others find it inconvenient.

b. Causes of Deficiencies. There are several causes for Marines being deficient in physical fitness:

(1) Absence of exercise.

(2) Exercise which fails to develop all muscle groups and components of fitness.

(3) Exercise which is not vigorous enough or which lacks progression.
(4) Injury or illness which depletes fitness.
(5) Inadequate amounts of sleep or rest.

c. Methods of Detection. The commander has several means by which to detect physical deficiencies:

(1)

from those who pass. Make an analysis to determine the cause of failure as revealed by test scores.

(2) Observe Marines as they perform physical tasks-

o Marines who have difficulty during training or in physical types of work,

o Marines who have become obese and therefore experience difficulty.

(3) Be alert to those Marines who are often on sick call, returnees from hospitalization, or newly assigned.

2203. ADMINISTRATION OF REMEDIAL ACTION

a. Group. Attitude Marines who are singled out as being physically deficient are self-conscious and are not always convinced they need extra help. Within a deficient group, motivation may be low. These Marines must be convinced that a remedial program which is tailored to their needs will help them and will eliminate their deficiencies. Encouragement is often needed and desirable.

b. Leadership. The leader of this group must study individuals and know their deficiencies. The unit leader should counsel Marines individually, observe them closely as they progress through the remedial program, maintain records, and adjust the program as required.

c. Measurement.. Whenever improvement in performance is noted, Marines should be measured by the physical fitness test either individually or by group. When testing reveals an individual to be satisfactory, that individual should be released from the remedial group. There may be exceptions to this policy in the case of Marines who are overweight or in the execution of an exercise program prescribed for some special purpose.

d. Organization of a Remedial Group

(1) A remedial group is usually a small group of Marines within a company or battalion. In some situations, it may be a larger group numbering 50 to 75 Marines. In the case of a larger group, divide the Marines initially into subgroups according to ability. Prescribe exercise loads commensurate with their ability. General conditioning is sufficient in the early part of the program to qualify Marines who are on the borderline. Qualification will reduce the size of the group, permitting more individual and specialized attention to those who remain.

(2) As the program progresses, regroup individuals who have like deficiencies in order that they can concentrate on their weakness. For example—

(a) One group may be weak in the arms and shoulders as revealed by failure in the chin up event. This group, with an assistant instruction charge, can work on pull-ups, rope climb, pushups, rifle or logexercises, horizontal ladder, and similar types of exercise.

(b) Another group may be weak in general endurance as revealed by the3-mile run. This group could profit from participation in conditioning drills, running, grass drill, and strength circuit.

(3) It may be necessary to form some groups to overcome weaknesses in skills, such as an inability to throw, to quickly change direction while running, to crawl rapidly, or to carry a load. Lack of coordination or lack of practice may cause these deficiencies. In this instance, instructors must provide an opportunity to practice and correct poor form and other errors as they are noted.

Chapter 3
PHYSICAL CONDITIONING ACTIVITIES
Section I. Marching Under Load

3101. GENERAL

Few physical fitness activities are as directly related to readiness for combat as foot marches under load. In addition to the obvious physical conditioning and unit cohesiveness benefits, marches under load prepare most Marines for numerous foreseeable tasks in combat. The idea that only infantry and reconnaissance units actually require training marches to prepare them for combat is clearly a mistake. In addition to the very real possibility of Marines from all types of units being used to fill combat-depleted infantry units, it is likely in maritime prepositioning force operations that aviation, combat service support, and command element personnel will have to march with weapons and equipment, from the arrival airfield to the equipment marshalling site. In American coastal cities, the harbor is nearly always 10 or more miles from the airport. It is difficult to imagine any different condition in a lesser developed country .A further combat-related benefit is that foot marching under load is the most effective way to develop leg strength in the context of a unit training program. In his classic study of troop performance in World War II, Men Against Fire, S.L.A. Marshall noted that the greatest single weakness of replacements of all occupational fields was leg strength. Truck drivers must push trucks which are stuck in mud, and headquarters communicators must climb hills to emplace antennas. The demands on the lower body in combat apply to all Marines, and foot marching is one activity which can help build these muscles while being conveniently integrated with the rest of the training program.

3102.TRAINING GOALS

The standard for success of a foot march is very simple to measure: did the unit arrive at the destination at the prescribed time with Marines in condition and required equipment present to accomplish the mission?A progressive program can increase a unit's readiness. It can instill pride by increasing distances and rates of march and by selecting routes over increasingly challenging terrain as the Marines become better conditioned. However, the ability to execute the mission at the conclusion of the march must remain the standard of success.

3103.MOVEMENT PLANNING

The basic considerations in planning a foot march are the mission, tactical situation, terrain and weather, and the units to participate. The success of the march will depend largely upon the thoroughness with which it is planned. A successful march is characterized by adherence to prescribed routes and time schedules, the efficient employment of the means available, and the ability of the unit to accomplish its assigned mission upon arrival at the destination.

a. Movement Orders. Movement planning culminates in the preparation and issuance of an operation order prepared in the standard, five-paragraph format. Necessary annexes are attached to furnish detailed information required for the movement. Written movement orders are rarely prepared at company level. A discussion of the annexes is contained in this chapter.

b.March Planning. March planning, as discussed here, is the planning conducted at battalion level. March planning may be organized into the following steps:

(1) **Preparation and Issuance of the Warning Order.** In order to afford subordinate units the maximum possible time to prepare for a pending move, a warning order containing all available information about the march is issued. The amount of planning time available will determine the time of issuance and the content of the warning order.

(2) Estimate of the Situation. In his estimate, the commander considers the mission, terrain, weather, time, and space factors, available routes, available transportation for the movement of equipment and/or shuttling of Marines, enemy capabilities, disposition of own forces, physical condition and training of Marines, and courses of action available to the command.

(3) Organization and Dispatch of a Reconnaissance. Party very march plan is based on as thorough and complete a ground reconnaissance as time and the situation will permit. Map and aerial reconnaissance are valuable in formulating a plan, but are not a substitute for ground reconnaissance. Route reconnaissance is accomplished by a reconnaissance party which usually consists of a reconnaissance element, an engineer element from the attached or supporting engineer unit, and a traffic control element. Unit standing operating procedures generally establish the basic composition of the reconnaissance party. lt is modified as necessary to meet the requirements of a particular march. A recommended method of reporting information obtained by the route reconnaissance party is shown in figure 3-1.The minimum information required from the reconnaissance party is—

(a) Available routes and conditions.(Routes may be specified by higher headquarters.)

(b) Recommended rate of march.

(c) Selection of start point and release point, or confirmation of the suitability of start point and release point previously selected by map reconnaissance.

(d) Confirmation of location of the assembly or bivouac area.

(e) Location of critical points on the route.

(f) Distance between critical points on the route and total distance from start point to release point.

Figure 3-1 is available in the illustration section at the end of the book

(g) Location of obstacles and estimation of necessary Marines and equip- ment needed to repair and maintain routes.

(h) Number of guides required and their location on the route.

(4) Development of Detailed Movement Plan. Some of the elements included in a detailed movement plan are-

(a) Organizing the Column. To facilitate control and scheduling, units will be organized into serials and march units and given an order of march. In determining the order of march, the march planner must consider the enemy situation and the desirable order of arrival of the units at the destination. Where dispersion is required, a unit may be organized into two or more columns, each assigned a different route.

(b) Using Reconnaissance Information. Results of the route reconnaissance will be used to select the route(s); determine the start point, critical points along the route(s), and the release point; and select the rate of march.

(c) Determining March Computations. March unit pass time is based on the strength, formation, and rate of march. The pass time of the marching columns, plus necessary time distance computations, will be used to determine the completion time of the march.

(d) Drafting of Road Movement Table. Using the completed march computations, a draft road movement tabic is compiled.

(e) Checking the Plan. Using the draft road movement table and a road movement graph, the movement plan is checked to ensure that it conforms to the directive of the higher headquarters and the battalion commander's instructions.

(5) Preparation and Issuance of the Road Movement Order. After the plan has been checked and approved by the commander, an operation order is prepared and issued. The operation order may be in written form or issued orally, and is accompanied by a road movement table, overlay and/or strip map, and appropriate administrative details.

(6) Road Movement Table. A road movement table, prepared as an annex to the operation order, provides serial commanders with arrival and clearance times at critical points along the route of march. It also provides the column commander with information as to the proposed location of elements of the column at various times.

(7) Overlay/Strip Map. An overlay serves the normal purpose and should show, as a minimum, the present location of units, route of march, critical points, and the new location of units at the destination. A strip map is a schematic diagram of the route of march and shows landmarks and critical points with the distances between them. A strip map may be issued as an annex to the road movement order, in addition to or in lieu of an overlay.

c. March Computations. Prior to issuance of the operation order, the S-3 must verify time and space computations as they provide him with the necessary data for the preparation of a road movement table.

(1) Time-Distance. Time distance (TD) is determined by dividing the distance to be traveled (D) by the rate of inarch (R):

TD = D/RWhere TD = hours D = distance in kilometers R = kilometers per hour

(2) Length of ColumnThe length of column (LC) is used to determine the pass time of a column. The sum total of the following two parts determines the length of column. (See pars, (a) and (b)).The two parts are the space occupied by the Marines alone (including the distance between Marines) and the sum of the distances between the units of the column (column gap).

(a) The length of column of Marines alone is determined by multiplying the number of Marines by the appropriate factor selected from the table below. The length of column does not include distances between units. LC Marines = No. of Marines x factor.

(b) The total distance in meters between units is determined as follows:

Step One. Determine the number of gaps between serials (total serials minus one).

Step Two. Multiply the number of serial gaps from Step One by the length (in meters) between respective units.

Step Three. Determine the number of gaps between march units (total march units minus one, minus the number of serial distances).

Step Four. Multiply the number of march unit gaps from Step Three by the length (in meters) between respective units.

Step Five. Add the totals from Step Two and Step Four to get the total meters for the column gap.

EXAMPLE:A battalion foot column is organized into 12 platoon-sized march units and 3 company-sized serials. REQUIRED: total column gap distances when there are 100 meters between serials and 50 meters between march units.

(3) Pass-Time.Pass-time (PT) is the time a unit takes to pass a specified point. For foot columns, the pass-time is determined by applying the following formula:PT (minutes) = LC x FACTOR (for appropriate rate of march).

PASS-TIME FACTORS - FOOT TROOPS

.0150 for 4.0 km/h

.0187 for 3.2 km/h

.0250 for 2.4 km/h

.0375 for 1.6 km/h

EXAMPLE: Determine the pass-time of a unit whose length of column is 1,500meters and is marching at a rate of 4 km/h.PT (min) = 1,500 x .0150 (the factor for 4.0 km/h) = 22.5 mm.

(4) Completion Time. Completion time is the time of day that a march will be completed. Completion time is determined by using the following formula: Completion Time = SP time + TD + PT + Scheduled Halts (other than normal hourly halts).

EXAMPLE: A column's starting point (SP) time is 0700.The time-distance is6 hours and 40 minutes. Pass-time of the column is 30 minutes. A 35-minutelunch halt has been scheduled. What is the completion time of the march?Employing the 24-hour clock system, the formula can be applied as follows for simplified addition of the times:

HrMin SP time................0700

TD.....................0640

PT.....................0030

Lunch halt.............0035

Completion time.........13 105or 1445

The march will be completed at1445.

(5) Experience Tables. Based on previous movements made by a unit, data is accumulated to facilitate march planning. Such data includes approximate pass-times lor various elements of the battalion. The S-3 can utilize these data rather than computing them each time a march is scheduled. Such experience tables serve to reduce the time required to complete the computation phase of march planning. Matter appropriate to the unit standing operating procedure should be integrated therein.

d. Road Movement Graph

(1) A road movement graph is a time-distance diagram used in planning, preparing, or checking road movement tables, and for controlling marches. The graph provides a visual representation of a march plan so that conflicts and discrepancies may be prevented in the planning stage, before congestion occurs on the route. It is not usually issued as a part of the order. Road movement graphs may be applied to small units, to movements of a single column, or to a large organization scheduling separate elements, moving by various means, with different rates of march, over one or more routes.

(2) To construct a road movement graph, use the following steps: (See fig. 3-2.)

Figure 3-2 is available in the illustration section at the end of the book

Step One. Determine the number of hours available for the march. Designate the lower left corner of a sheet of graph paper as the SP time or the earliest even hour before the march is to begin. Select a scale and plot the hours available in sequence from the left on the horizontal axis.

Step Two. Determine the distance to be moved in kilometers. Indicate the SP at the lower left corner of the graph sheet and, using an appropriate scale, plot the number of kilometers on the vertical scale from the SP to the release point. Indicate the location of critical points on the vertical scale.

Step Three. At the proper distance from the SP, draw a horizontal line indicating the location of the RP. Indicate by a vertical line the hour when the movement must be completed. Plot lines representing route restriction, if any, at the proper distances and times on the graph.

Step Four. Determine the pass-times of foot and motor elements of the column.

Step Five. Starting at the SP at the specified hour, plot the movement of the head of the leading element. If the rate of march is 3 kilometers per hour, the lead element will have moved 1.5 kilometers in 30 minutes, 3 kilometers in 1 hour, etc. Plot the trace of the lead man to the RP. The last Marine will cross the SP in pass-times after the lead man. Measure this time on the graph and plot the trace of the last Marine of the column. The lines describing the head and tail of a march serial are parallel. Indicate the time subsequent serials reach the SP and plot the trace of the head and tail. Check to see that the plan complies with all restrictions and orders. If there are any violations or conflicts, the plan may be made to conform by changing the order of march, the starting time of the serials, the planned rate of march, or the organization of the column.

e. Conversion to the Metric System. The adoption of the metric system for expressing linear distances requires a simple means of conversion.

(1) To accomplish this, use the following tables:

o Multiply inches x 2.54 to obtain centimeters. Multiply centimeters x .39 to obtain inches,

o Multiply yards x .91 to obtain meters. Multiply meters x 1.1 to obtain yards,

o Multiply miles x 1.6 (or 8/5) to obtain kilometers. Multiply kilometers x .62 (or 5/8) to obtain miles.

(2) Linear distances can be accurately measured on maps which have been converted to the metric system by using a straightedge graduated in centimeters (cm).The centimeter scale can be used in lieu of the graphic scale when the scale of the map is shown, as follows:

Map Scale:	One cm Equals:
1:500,000	5,000 meters
1:250,000	2,500 meters
1:100,000	1,000 meters
1:50,000	500 meters
1:25,000	250 meters

3104. INDIVIDUAL LOAD

Since the backbone of Marine ground combat forces is the infantry, tactical mobility is largely dependent on the marching ability of the individual Marine. The load that each Marine must carry is the greatest impediment to mobility. In general, the commander must make every effort to reduce the individual load of his Marines to the absolute minimum. While the load has the greatest effect on Marines marching on foot, the value of carrying the minimum essential supplies and equipment applies for all Marine forces.

a. Fighting Load. The fighting load consists of items of clothing, equipment, weapons, and ammunition that are carried by, and essential to, the effectiveness of the combat Marine and the accomplishment of the immediate mission. The fighting load should not weigh more than 40 pounds. The commander must ensure that his Marines' loads- -and his own- -be stripped to the bare minimum. The addition of small, light-weight items in excess of the limit should not be tolerated, since cumulatively, these items will take a toll in energy. Every extra pound a Marine must carry decreases his combat effectiveness.

b. Existence Load. The existence load consists of items other than those in the fighting load that are required to sustain or protect the combat Marine, that may be necessary for increased personal and environmental protection, and that are not normally carried by the individual. The existence load is normally carried by the Marine's unit. Marching, but not engaged in combat, a Marine may be expected to carry a maximum of 50 pounds of supplies and equipment.

c. Considerations

(1) The primary consideration is not how much a Marine can carry, but how much he can carry without impaired combat effectiveness-moral or physical. The combat strength of a unit is not counted simply in numbers of Marines, but in the numbers of willing and physically able Marines.

(2) It is generally better to risk temporary inconvenience from lack of health and comfort items than to exhaust Marines due to overloading.

(3) It is a fundamental truth that men become physically exhausted more quickly when under the stress of combat. Marines must be required to carry less into battle than they are conditioned to carry in training. Marines should be conditioned for carrying weight, but should be equipped in combat for fleetness of foot.

(4) A common mistake is to base the fighting load on the gear and supplies necessary to meet every contingency. The commander cannot reasonably expect the Marines to carry enough gear for every possible eventuality. The items to be carried must be based on reasonable expectations.
(5) It is the commander's responsibility to procure the transport to carry any additional gear. As a rule of thumb, a rifle company, or unit of similar size, requires one 5-ton truck and trailer in support to carry additional gear. In cold weather, or other conditions in which the necessary personal gear increases, this requirement will increase.
(6) The commander must ensure that the supply system provides, on a dependable and timely basis, the balance of essential supplies and equipment not carried by the unit. Marines must feel confident that they will be supported with the necessary supplies and equipment.
(7) In training, commanders must implant in their Marines the pride in operating under austere conditions. To effectively demonstrate the need for this spartan mentality, the commander must personally set the standard. Marines should be trained in field craft, foraging techniques, and the use of caches and field expedients. Maximum use should be made of captured stores.

3105. MARCH TECHNIQUES AND PROCEDURES

a. Organization. A command executing a march is organized into march units, march serials, and march columns, as necessary for control.
(1) March Units. A march unit is a unit of command which moves and halts at the command of a single commander. The march unit normally corresponds to one of the smaller Marine units such as a platoon or company.
(2) March Serials. A march serial consists of one or more march units organized under the senior officer and given a numerical or alphabetical designation to facilitate control. A serial is usually a battalion, but may be a company if the battalion is marching alone.
(3) March Columns. A march column is composed of elements of a command moving over the same route. It may be composed of one or more serials. To facilitate control, a column commander is designated. A column is normally a regiment or larger unit, but may be a battalion if marching alone.

b. Order of March. The order of march is determined by the mission, terrain, and the commander's desired order of commitment of units into action.

c. Control and Coordination Measures. The commander establishes initial control of the march by designating control measures in his march order. The most frequently used measures include:

o Start point and release point.
o Time at which head or tail of column passes the start point.
o Rate of march.
o Order of march.
o Route of march.
o Location of command post.
o March communications.

d. The Pacesetter. The pace-setter is an experienced individual carrying the same load as the majority of the Marines and marching from 4 to 10 meters at the head of the column. (See fig. 3-3.)The pace-setter's primary duty is to maintain the rate of march ordered by the column commander.

(l) Setting the Pace. The pacesetter does this by establishing the pace (length of step) and cadence (steps per minute) to obtain the prescribed rate of march. He should be of medium height so normal strides will be taken. Overstriding or understriding tends to tire leg muscles quickly and affects the combat efficiency of marching Marines. The officer marching at the head of the column supervises the pacesetter to ensure that he takes normal strides and maintains a uniform cadence.

Figure 3-3 is available in the illustration section at the end of the book

(2) Reducing Accordion Effect. Accordion effect occurs when the column alternately lengthens and shortens, causing the rear elements of the column to double time to maintain distance. It is caused by changes in the rate of march by the lead elements, usually after passing a slowing obstacle. To reduce this effect as much as possible, lead elements should slow their rate of march for a sufficient time after crossing obstacles to allow the rear elements to maintain the prescribed distance without running. The rear elements can signal the head when the last man has cleared the obstacle and the rate can be resumed. Since some accordion effect is nearly inevitable, the order of march should be periodically rotated to prevent units in the rear from becoming physically exhausted at the conclusion of the march.

e. Length, Rate and Pace

(1) **Length of March.** The length of march varies depending on terrain and weather, enemy situation, and physical and mental condition of Marines. The normal length of march for a 24-hour period is from 20 to 32 kilometers (12 to 20 miles) marching from 5 to 8 hours at a rate of 4 kilometers (2.5 miles) per hour (km/hmi/h).A march in excess of 32 kilometers in a 24-hour period is considered a forced march. Well-trained units that have been progressively trained will be able to achieve a unit goal of 60 miles in a 3-day period.

(2) Rate of March. The same factors of terrain, weather, load to be carried, and condition of Marines affect the rate of march. The commander considers these and selects a rate which will place his unit at the destination in the shortest possible time in condition to accomplish the mission. Unit standing operating procedure should state the standard rate of march for that unit over normal terrain for both road and cross-country movement and in both day and night. The commander then modifies this rate if the situation requires. Normally prescribed standards are as follows:

Figure is available in the illustration section at the end of the book

(3) Pace and Cadence. The normal pace is 30 inches. A pace of 30 inches and a cadence of 106 steps per minute result in a speed of 4.8 kilometers per hour or 3 miles per hour and a rate of 4 kilometers per hour or 2.5 miles per hour if a 10-minute rest halt per hour is taken. Since the pace of each man may vary, the cadence may have to be adjusted to maintain the prescribed rate. Slope of ground and surface footing affect pace. A normal cadence is relatively simple to maintain on level or gently rolling terrain on firm footing. Mud, sand, loose gravel, and hills all greatly affect pace.

f. Halts. Halts during a day's march are taken at regular intervals to rest personnel and adjust loads.Halts are specified by standing operating procedures or march order. Such factors as midday heat or enemy action may force the adoption of long midday halts or night marches. At long halts, each unit disperses to a previously selected location near the route of march.

(1) Normal Time and Frequency. Under normal conditions, a 15-minute halt is made after the first 45 minutes of marching. Following the first halt, a 10-minute halt is made after each 50 minutes of marching. Variations of this schedule are made when a scheduled halt time occurs when passing through a built-up area or when cover and concealment are required by the tactical situation and none is available. Observation posts may be established, if required for security of the unit during halts.

(2) Actions at Halts. All units in the column should be halted at the same time. At the halt signal, Marines should move to the side of the road, staying within the immediate vicinity of their unit. Marines should remove or loosen gear and sit or lie with feet elevated. Commanders inspect Marines and equipment, and corpsmen administer medical treatment, as required.

g. Communications. The four primary means of communications used in controlling foot marches are radio, visual, sound, and messenger.

(1) Radio. Radios are normally used for communications between platoons and higher headquarters in the march column. Radio transmissions should be held to the absolute minimum necessary for control and should be at minimum power required.

(2) Visual.. Visual communications may include flashlights, luminous markers, panel sets, pyrotechnics, and hand and arm signals. When contact with the enemy is not expected, the loss of light discipline from some of these signals may be a lesser security risk than the risk of disclosure from radio transmissions.

(3) Sound. Sound communications include voice and such signaling devices as horns and whistles. Some of their uses include assembling Marines at the conclusions of halts and providing warnings of air or chemical attacks.

(4) Messengers. Messengers are particularly effective during periods of radio silence and reduced visibility. Messages should be simple and brief to preclude mistakes by the messenger.

h. Security. Both active and passive measures are used against attack by enemy aircraft and indirect fire weapons during movement. Active measures include the use of air sentries and organic and attached weapons in accordance with the unit air defense standing operating procedure. Passive measures include the use of concealed routes and assembly areas, night marches, and extended interval between elements of the column. With imagination and planning, all of these measures can be practiced in training without detracting from the physical conditioning benefit of the march. In fact, such drills as rapidly taking cover during simulated attacks and then quickly resuming the march can prove physically challenging.

i. Reconnaissance. Reconnaissance determines in advance critical points along the march route such as bridges, fords, and obstacles in order that congestion or delay may be held to a minimum and local security provided. Paragraph 3103.C describes organization of the reconnaissance party.

3106.TACTICAL MARCHES

These marches are movements of Marines and equipment not in direct ground contact with the enemy but expecting early ground contact either en route or upon arrival at the objective. Some characteristics of this type of movement include security elements to the front, flanks, and, if necessary, to the rear of the march column. Additional considerations include the selection of concealed areas on defendable terrain for start and release points and rest stops. In training, infantry units of regimental size or smaller will normally derive the greatest benefit from tactical marches. The security requirements usually dictate a slower rate of march and, therefore, reduce the physical conditioning value for other types of units which would seldom move in combat with the expectation of early ground contact with the enemy. Two key points should be remembered regarding tactical marches in a training program.

a. Security. Although security requirements often slow the rate of march, it is a valid and important training objective for infantry units to work toward reducing march times while still practicing sound tactical security. March standing operating procedures, good land navigation skills, and simple but complete march orders contribute as much to this goal as physical conditioning. The idea that proceeding slowly enhances security is seldom true. Keeping the enemy off balance by rapid movement frequently is the best guarantee of security.

b. Training. In peacetime training, it is critical that support personnel who normally are located with infantry units in combat, participate completely with infantry units in forced march training. Not only must they not slow the infantry rate of march because of inferior conditioning, but they must be thoroughly familiar with the infantry unit's standing operating procedure for security on the march. Personnel from within the ground combat element such as artillery forward observer teams and combat engineer units, can be easily identified for this training and should be habitually associated with the same infantry units. Personnel from other elements, such as low-altitude air defense teams and radio battalion detachments, are more difficult to identify and earmark for support of the same infantry unit. Their normal mode of employment is in vehicles and their units frequently provide general support vice direct support. This does not, however, lessen their requirement to be prepared to provide support to dismounted infantry over terrain or in a tactical situation not conducive to vehicles. Such training needs to take place, and planning and cooperation among the elements of the MAGTF can overcome the obstacles.

3107.ADMINISTRATIVE MARCHES

These marches are movements of Marines and equipment made when no enemy interference is expected except from aircraft or indirect fire. During these marches, units are administratively grouped for ease of control and speed of movement. The term administrative foot marches should not be taken to imply that tactical considerations are disregarded. There are simply less stringent security measures used than when ground contact with the enemy is a distinct possibility. The normal formation is the route column with one file moving on either side of the road and with negligible security. Air sentries are used and dispersion between individuals and units is practiced. Although roads usually provide the fastest route, there may be reasons to conduct these marches cross-country (particularly as part of a training program). Administrative marches are the type of foot marches most practical for training programs for non-infantry units. These marches may also play a key role in the physical conditioning program of infantry units, because of the greater speeds possible due to lessened security requirements.

3108. FORCED MARCHES

A forced march is a march which covers a greater distance than 32 kilometer (20 miles) in a 24-hour period. Normally the extra required distance should be achieved by increasing the number of hours marched in a day beyond the norm (8 hours) rather than increasing the rate of march beyond the norm (4 kilometers per hour/2.5 miles per hour). However, there will be occasions when the situation or mission demands an increase in the rate of march.

a. Maximum Recommended Distances. The maximum recommended distances for forced marches are-56 kilometers (35 miles) in 24 hours; 96 kilometers (60 miles) in 48 hours; or 128 kilometers (80 miles) in 72 hours.

b. Sample Time Breakdown. A sample time breakdown for a forced march of 52 kilometers, beginning at first light is as follows:

Figure is available in the illustration section at the end of the book

3109. NIGHT MARCHES

Night marches are characterized by closed formations, more difficult control and reconnaissance, and a slower rate of march than day marches.

a. Control. Control is increased by reducing the distance between individuals and units, and by using connecting files to maintain contact between platoons and companies. Connecting files are normally constituted from personnel from the rear most march unit.

b. Safety on Roads. Because of reduced visibility, night marches on roads used by vehicles require attention to safety procedures to prevent accidents. Commanders should not think these safety measures apply only to peacetime training. In combat, nearly all vehicles operate with blackout lights at night, making visibility even more difficult than in garrison driving. If consistent with the tactical situation, the following measures can reduce the possibility of marchers being struck by vehicles:

(1) Use off-road trails and routes as much as possible.

(2) Guards to the front and rear of march units should be marked with strips of luminous tape and carry red-filtered flashlights. These measures are generally consistent with light discipline, while still affording reasonable warning to approaching drivers.

Section II. Conditioning Drills (One, Two, and Three)

3201. GENERAL

a. Description and Objective. Conditioning drills are calisthenic exercises. Each drill contains seven exercises organized and numbered in a set pattern. Each drill takes 15 minutes to complete. The objective of conditioning drills is to exercise all major muscle areas in order to develop strength, endurance, coordination, and flexibility.

b. Area and Equipment. Any level area is satisfactory for conduct of the drills. Drills One and Two contain ground exercises. If ground conditions are unsatisfactory, Drill Three should be used as it contains no ground positions. Usually, no equipment is required;however, if the group exceeds a platoon in size, an instructor's stand is necessary.

c. Formation. The extended rectangular formation is used in this drill. (See app. A.)

d. Starting Level and Progression. The starting level is six repetitions of each exercise. An increase of one repetition for each three periods of exercise in which the drill is performed is an acceptable rate of progression. This rate is continued until 12 repetitions can be completed. To maintain, continue the drill at 12 repetitions. To progress, move to 6 repetitions at a more difficult drill which exercises the same muscle group. Progression can also be gained by moving from Drill One to Drill Two, as Drill Two is more demanding.

e. Starting Positions. Starting positions vary with the exercise and are explained in each exercise.

f. Leadership. A principal instructor demonstrates and leads the drills. The instructor must be familiar with leadership techniques peculiar to conditioning drills to include the exercises, commands, counting cadence, cumulative count, formation, method of teaching the exercises, and utilization of assistant leaders.

g. Use With Other Programs. Conditioning Drills One, Two, and Three reach all major muscles of the body. They are easy to learn and to perform, and they are simple to administer and supervise. These features, coupled with the short time required for completion, the fact that no equipment is necessary, and adaptability to most areas of execution, make these drills possible in any programs.

3202. CONDITIONING DRILL ONE

This conditioning drill is similar to the calisthenics drill commonly known throughout the Marine Corps as the "Daily Seven."The "Daily Seven" may be used as a substitute for Conditioning Drill One.

a. Exercise 1:High Jumper

(1) Starting Position. Feet separated shoulder width, knees flexed, body bent forward at the waist, arms aligned with the trunk and hips, elbows locked, palms facing, fingers extended and joined, head and eyes to the front.(See fig. 3-4, A.)(Elbows remain locked throughout the exercise.)

Figure 3-4 is available in the illustration section at the end of the book.

(2) Cadence. Moderate.

(3) Movement. A four-count exercise: at the count of—

(a) ONE -- Take a slight jump into the air, swinging the arms forward and up to shoulder level.

(b) TWO -- Take a slight jump into the air and swing the arms downward and back, returning to the starting position.

(c) THREE -- Take a vigorous leap into the air, swinging the arms forward and up to an overhead position, momentarily looking skyward, on returning to the ground the knees are flexed, head and eyes return to the front.

(d) FOUR--Repeat the action of count two.

b. Exercise 2: Bend and Reach

(1) Starting Position. Feet spread more than shoulder width, arms overhead, elbows locked, palms facing, fingers extended and joined, head and eyes to the front.(See fig. 3-4, B.)

(2) Cadence. Moderate.

(3) Movement. A four-count exercise: at the count of—

(a) ONE-Bend at the knees and waist, swing the arms straight downward and reach between the legs. Touch the ground as far to the rear as possible and look to the rear. (Elbows remain locked throughout the exercise).

(b) TWO-- Recover sharply to the starting position.

(c) THREE--Repeat the action of count ONE.

(d) FOUR-- Repeat the action of count TWO.

c. Exercise 3: Pushup

(1) Starting Position. Front leaning rest position. To assume this position there is a silent one-two count:on the silent count of one, assume the squatting position, heels together, elbows locked inside the knees, hands flat on the ground directly beneath the shoulders; on the silent count of two, thrust the legs to the rear, toes and heels together, body straight from head to heels.(Seefig. 3-4, C.)

(2) Cadence. Moderate.

(3) Movement. A four-count exercise: at the count of—

(a) ONE -- Flex the elbows lowering the body until the thick portion of chest touches the ground.

(b) TWO -- Raise the body until elbows are straight and locked.

(c) THREE-- Repeat the action of count ONE.

(d) FOUR -- Repeat the action of count TWO. (On returning to position of attention, the silent one-two count is used in reverse).

d. Exercise 4: Trunk Twister

(1) Starting Position. Feet are spread more than shoulder width apart, fingers placed behind neck, thumbs pointing downward, elbows back. (See fig. 3-4, D.)(Elbows remain well back throughout the exercise).

(2) Cadence. Slow.

(3) Movement. A four-count exercise: at the count of-

(a) ONE-- Keeping the knees locked and the back straight, bend forward at the waist sharply, with a slight recovery.

(b) TWO-- Twist the trunk to the left vigorously at the waist, keeping the elbow back. The left elbow is higher than the right.

(c) THREE-- Twist vigorously to the right, so the left elbow comes under the right.

(d) FOUR-- Straighten sharply to the starting position. NOTE: Do not attempt to touch the elbows to the knees on counts two and three.

e. Exercise 5: Squat Bender

(1) Starting Position. Feet are spread less than shoulder width apart, hands on hips, thumbs in small of back, elbows back. (See fig. 3-4, E.)

(2) Cadence. Moderate.

(3) Movement. A four-count exercise: at the count of—

(a) ONE-- Assume the squatting position, maintain balance on the balls of the feet, with trunk erect thrust arms forward to shoulder level, elbows locked, palms down.

(b) TWO-- Recover to starting position. Elbows are well back.

(c) THREE-- Keeping the knees locked, bend forward at the waist, touching the ground in front of the toes.

(d) FOUR-- Vigorously recover to the starting position.

f. Exercise 6: Body Twist

(1) Starting Position. On the back, arms extended sideward on the ground, palms down. The legs are raised to a near vertical position, feet together, knees locked.(See fig. 3-4, F.)

(2) Cadence. Slow-fast.

(3) Movement. A four-count exercise: at the count of-

(a) ONE-- Lower legs slowly to the left until they touch the ground near the left hand, keeping the knees straight and shoulders on the ground.

(b) TWO-- Recover the starting position by quickly raising the legs, keep knees straight and feet together.

(c) THREE-- Repeat movement of count ONE, except the movement is to the right side.

(d) FOUR-- Recover sharply to the starting position.

g. Exercise 7: Stationary Run

(1) Starting Position. Position of attention. (See fig, 3-4, G.)

(2) Cadence. Fast.

(3) Movement.

(a) At the command of execution, start running in place at double-time cadence, lifting the left foot first time cadence. Follow the instructor as he counts two repetitions of cadence;

e.g., 1, 2, 3, 4--1, 2, 3, 4. The instructor then gives informal commands such as FOLLOW ME, running on the toes and balls of the feet, keeping the back straight, speeding up the cadence to a sprint, raising the knees high, leaning forward at the waist, and pumping the arms vigorously.

(b) To halt the exercise, the instructor will count two repetitions of cadence as the left foot strikes the ground: 1, 2, 3, 4 - 1 , 2, 3, HALT.

NOTE: When counting cadence, the instructor counts only as the left foot strikes the ground. The duration of the exercise is approximately 1 1/2 minutes.

3203. CONDITIONING DRILL TWO

a. Exercise 1: Jumping Jack

(1) Starting Position. Feet separated more than shoulder width, arms overhead. (See fig. 3-5, A.)

(2) Cadence. Moderate.

(3) Movement. A four-count exercise: at the count of~

(a) ONE-- Jump to position with the feet together and assume the squatting position, swinging the arms sideward and downward, placing the hands palms down on the ground, elbows locked inside the knees.

(b) TWO-- Recover to the starting position by jumping to the side straddle and swinging the arms sideward overhead.

(c) THREE-- Repeat the action of count ONE.

(d) FOUR-- Recover to the starting position.

b. Exercise 2: Turn and Bend

(1) Starting Position. Side straddle, arms overhead. (See fig. 3-5, B.)

(2) Cadence. Moderate.

(3) Movement. A four-count exercise: at the count of-

(a) ONE-- Turn the trunk to the left and bend forward over the left thigh, attempting to touch the fingertips to the ground outside the left foot. Keep the left knee straight. On successive repetitions attempt to touch farther and farther to the side.

(b) TWO-- Recover to the starting position.

(c) THREE-- Turn the trunk to the right and bend forward over the right thigh, trying to touch the hands to the ground outside the right foot. Keep the right knee straight.

Figure 3-5 is available in the illustration section at the end of the book.

(d) FOUR-- Recover to the starting position.

c. Exercise 3: Eight-Count Pushup

(1) Starting Position. Position of attention. (See fig. 3-5, C.)

(2) Cadence. Moderate.

(3) Movement. An eight-count exercise: at the count of—

(a) ONE-- Assume the squatting position, palms on the ground directly beneath the shoulders, elbows locked inside the knees.

(b) TWO-- Thrust the legs to the rear assuming the front leaning rest position.

(c) THREE-- Flex the elbows until the thick portion of the chest touches the ground.

(d) FOUR-- Raise the body on a straight plane until the elbows are locked.

(e) FIVE-- Repeat the action of count THREE.

(f) SIX-- Repeat the action of count FOUR.

(g) SEVEN-- Recover to the squatting position as in count ONE (elbows locked inside the knees).

(h) EIGHT—Return sharply to the position of attention.

d. Exercise 4: Turn and Bounce

(1) Starting Position. Feet spread more than shoulder width apart, arms sideward at shoulder level, palms up. (See fig. 3-5, D.)

(2) Cadence. Slow.

(3) Movement. An eight-count exercise: at the count of—

(a) ONE-- Turn sharply to the left as far as possible, then recover slightly.

(b) TWO-- Again turn to the left as far as possible and recover as in ONE.

(c) THREE-- Repeat the action of count TWO.

(d) FOUR-- Recover sharply to the starting position.

(e) FIVE—Turn sharply to the right as far as possible, then recover slightly.

(f) SIX—Again turn to the right as far as possible and recover as in FIVE.

(g) SEVEN-Repeat the action of count SIX.

(h) EIGHT-Return to the starting position.

NOTE: The head and hips remain to the front throughout the exercise and the knees and elbows are locked at all times.

e. Exercise 5: Squat Stretch

(1) Starting Position. Attention. (See fig. 3-5, E.)

(2) Cadence. Moderate.

(3) Movement. A four-count exercise: at the count of—

(a) ONE-- Squat, placing the hands on the ground about 12 inches in front of the feet.

(b) TWO— Keeping the fingertips on the ground, straighten the knees completely and raise the hips.

(c) THREE-- Recover to position ONE.

(d) FOUR-- Recover to the starting position.

f. Exercise 6: Leg Circular

(1) Starting Position. On the back, arms stretched sideward, palms down, feet raised foot from ground, knees straight. (See fig 3-5,F.)

(2) Cadence. Slow.

(3) Movement A four-count exercise: at the count of—

(a) ONE-- Swing the legs as far as possible to the left, keeping the knees straight and the legs together.

(b) TWO-- Swing the extended legs over head with the thighs as close as possible to the trunk.

(c) THREE-- Swing the legs as far as possible to the right.

(d) FOUR-- Recover to the starting position.

g. Exercise 7: Stationary Run

(1) Starting Position. Position of attention. (See fig. 3-5, G.)

(2) Cadence. Fast.

(3) Movement

(a) At the command of execution, start running in place at double-time cadence, lifting the left foot first. Follow the instructor as he counts two repetitions of cadence; for example: 1, 2, 3, 4--1, 2, 3, 4. The instructor then gives informal commands such as FOLLOW ME. Running on the toes and balls of the feet, keeping the back straight, speeding up the cadence to a sprint, raising the knees high, leaning forward at the waist, and pumping the arm vigorously.

(b) To halt the exercise, the instructor will count two repetitions of cadence as the foot strikes the ground: 1, 2, 3, 4-1,2, 3, HALT.

NOTE: When counting cadence, the instructor counts only as the left foot strikes the ground. The duration of the exercise is approximately 1 1/2 minutes.

3204. CONDITIONING DRILL THREE

a. Exercise 1: Side Straddle Hop

(1) Starting Position. Position of attention. (See fig. 3-6, A.)

(2) Cadence. Moderate.

(3) Movement. A four-count exercise: at count of-

(a) ONE-- Take a slight jump into the air, moving the legs sideward (more than shoulder width apart); at the same time, swing the arms overhead (to an overhead position) clapping the palms together.

Figure 3-6 is available in the illustration section at the end of the book.

(b) TWO-- Take a slight jump into the air, swing the arms sideward and downward returning to the starting position.

(c) THREE-- Repeat the action of count ONE.

(d) FOUR-- Repeat the action of count TWO.

b. Exercise 2: Back Bender

(1) Starting Position. Standing, feet 12 inches apart, fingers placed behind the head. (See fig. 3-6, B.)

(2)Cadence. Slow.

(3) Movement. A four-count exercise: at the count of-

(a) ONE-- Bend the upper trunk backward, raising the chest high, pulling the elbows back, and looking upward. Keep the knees straight.

(b) TWO-- Recover to the starting position.

(c) THREE-- Repeat the action of count ONE.

(d) FOUR-- Recover to the starting position.

c. Exercise 3: Squat Thrust

(1) Starting Position. Position of attention. (See fig. 3-6, C.)

(2) Cadence Moderate.

(3) Movement. A four-count exercise: at the count of—

(a) ONE-- Assume the squatting position; heels together, placing the hands flat on the ground, shoulder width apart, elbows locked and inside the knees.

(b) TWO-- Thrust the legs to the rear, assuming the front leaning rest position, body in line from head to toe, heels and toes together.

(c) THREE-- Return to the squatting position as in ONE.

(d) FOUR-- Return to position of attention.

d. Exercise 4: Side Bender

(1) Starting Position. Feet are spread more than shoulder width apart, arms are raised sideward and overhead, thumbs interlocked palms to front, fingers extended and joined, elbows locked. (See fig. 3-6, D.)

(2) Cadence. Slow.

(3) Movement. An eight-count exercise: at the count of—

(a) ONE-- Bend to left as far as possible, then recover slightly.

(b) TWO-- Again bend to the left as far possible, then recover slightly.

(c) THREE-Repeat the action of count TWO.

(d) FOUR-- Recover sharply to the starting position.

(e) FIVE-- Bend to the right as far as possible, then recover slightly.

(f) SIX-- Again bend to the right as far as possible, then recover slightly.

(g) SEVEN-- Repeat the action of count SIX.

(h) EIGHT-- Recover sharply to the starting position.

NOTE:Keep the elbows and knees locked throughout the exercise. The bend should occur to the side and not the front.

e. Exercise 5: Knee Bender

(1) Starting Position. Feet are spread less than shoulder-width apart, hands on hips, thumbs in small of back, elbows back. (See fig 3-6, E.)

(2) Cadence. Moderate.

(3) Movement. A four-count exercise: on the count of—

(a) ONE-Do a knee bend, lean trunk forward at the waist, thrust arms between legs until the extended fingers touch the ground palms to the ground, hands 6 inches apart.

(b) TWO~Recover sharply to the starting position.

(c) THREE-Repeat the action of count ONE.

(d) FOUR-Repeat the action of count TWO.

f. Exercise 6: Bottoms Up

(1) Starting Position. Front leaning rest position. A silent one-two count is used as in the pushups. (See fig. 3-6, F.)

(2) Cadence. Moderate.

(3) Movement A four-count exercise: at the count of-

(a) ONE-- With the weight on the hands, and knees locked, jump forward bringing the feet as close to the hands as possible; look to the rear.

(b) TWO--Keeping the knees locked, thrust the legs backward assuming the front leaning rest position.

(c) THREE-- Repeat the action of count ONE.

(d) FOUR-- Repeat the action of count TWO.

g. Exercise 7: Stationary Run

(1) Starting Position. Position of attention. (Sec fig. 3-6, G.)

(2) Cadence. Fast.

(3) Movement

(a) At the command of execution, start running in place at double time, lifting the left foot first. Follow the instructor as he counts two repetitions of cadence; for example: 1, 2, 3, 4--1, 2,3, 4. The instructor then gives informal commands such as FOLLOW ME, running on the toes and balls of the feet, keeping the back straight, speeding up the cadence to a sprint, raising the knees high, leaning forward at the waist, and pumping the arms vigorously.

(b) To halt the exercise, the instructor will count two repetitions of cadence as the left foot strikes the ground: 1, 2, 3, 4 - 1, 2, 3, HALT.

NOTE: When counting cadence, the instructor counts only as the left foot strikes the ground. The duration of the exercise is approximately 1 1/2 minutes.

Section III. Rifle and Log Drills

3301. RIFLE DRILL

a. Description and Objective. Rifle exercises are conditioning exercises performed with a rifle. Each drill contains six exercises (fig. 3-7) and they are numbered in a set pattern. The drill takes 15 minutes to complete. The objective of rifle drills is to exercise the arms, shoulders, and back muscles in order to develop strength and endurance, particularly in the upper body. In units without rifles, log drills may be substituted.

b. Area and Equipment. Any level area is satisfactory for conducting this drill. Each Marine completes these exercises from a standing position and no ground contact is required. Each Marine will need a rifle and, if the group exceeds a platoon in size, then the instructor will need an instructor's stand.

c. Formation. The extended rectangular formation is used in this drill. (See app. A.)

d. Starting Positions. Starting positions vary with the exercises and are explained in each exercise. As in all set conditioning drills, the command used to start the exercise is STARTING POSITION, MOVE. The following directions apply to rifle drill.

(1) In those exercises which start from the rifle downward position, on the command MOVE, execute port arms as prescribed in NAVMC 2691, Drill and Ceremonies Manual, and then assume the starting position. The command to return to the men to the position of attention at the conclusion of the exercise is POSITION OF ATTENTION, MOVE.

(2) In exercises which terminate in the rifle downward position, on the command MOVE, execute the position followed by order arms as prescribed in NAVMC 2691.

(3) In the exercises which terminate in a position other than the rifle downward position, Marines first assume the rifle downward position before executing port arms and order arms.

(4) These movements are executed without command. This procedure promotes uniformity, but precision is not expected. To be effective, rifle exercises must be strenuous enough to tire the arms, but not to the point where the arms cannot move with precision.

e. Leadership. A principal instructor demonstrates and leads the drill. He must be familiar with leadership techniques for conditioning exercises and the peculiar techniques for rifle drill.

3302. EXERCISES PERFORMED WITH RIFLES

The exercises of rifle drill are outlined in the following paragraphs,

a. Exercise 1: Foreup, Behind Back

(1) Starting Position. Rifle downward, feet together.(See fig. 3-7, A.)

Figure 3-7 is available in the illustration section at the end of the book.

(3) Movement. A four-count exercise: at the count of—

(a) ONE-- Swing the arms forward and upward to the overhead position. Inhale.

(b) **TWO**-- Lower the rifle to the back of the shoulders. Exhale.
(c) **THREE**-- Recover to position ONE and inhale.
(d) **FOUR**-- Recover to the starting position and exhale.

b. Exercise 2: Lunge Side, Turn and Bend

(1) Starting Position. Rifle downward, feet together.(See fig. 3-7, B.)
(2) Cadence. Moderate.
(3) Movement. An eight-count exercise:at the count of—

(a) **ONE**-- Lunge sidewards to the left, swing the rifle forward and upward to the overhead position.

(b) **TWO**-- Turn the trunk to the left and bend forward over the left hip. At the same time, swing the rifle to a **low** horizontal in front of the left ankle.

(c) **THREE**-- Recover to position ONE.
(d) **FOUR**-- Recover to the starting position.
(e) **FIVE, SIX, SEVEN, and EIGHT**-- Repeat on the right side.

c. Exercise 3: Foreup, Back Bend

(1) Starting Position. Rifle downward, feet together.(See fig. 3-7, C.)
(2) Cadence. Moderate.
(3) Movement. A four-count exercise:at the count of—

(a) **ONE**-- Swing the arms forward and upward to the overhead position.

(b) **TWO**-- Bend backward, emphasizing the bend in the upper back. The face is up.Keep the knees straight.

(c) **THREE**-- Recover to position ONE.
(d) **FOUR**-- Recover to the starting position.

d. Exercise 4:Up and Forward

(1) Starting PositionRifle downward, feet together.(See fig. 3-7, D.)
(2) Cadence. Fast.
(3) Movement. A four-count exercise:at the count of—

(a) **ONE**-Swing the arms forward and upward to the overhead position.
(b) **TWO**- Swing the arms forward to shoulder level.
(c) **THREE**-Recover to position ONE.
(d) **FOUR**-Recover to the starting position.

e. Exercise 5: Foreup, Full Squat

(1) Starting Position. Rifle downward, feet in narrow stance. (See **fig.** 3-7, E.)
(2) Cadence. Moderate.
(3) Movement. A four-count exercise: at the count of—

(a) **ONE**-- Swing the arms forward and upward to the overhead position.
(b) **TWO**-- Swing the arms down to shoulder level and assume the squatting position.
(c) **THREE**-- Recover to position ONE.
(d) **FOUR**-- Recover to the starting position.

f. Exercise 6: Arms Forward, Side Bend

(1) **Starting Position.** Sidestraddle, regular stance, rifle forward.(See fig. 3-7, F.)
(2) **Cadence.** Moderate.
(3) **Movement.** A four-count exercise:at the count of-
(a) **ONE--** Bend the trunk to the left. Keep the knees straight.
(b) **TWO--** Recover to the starting position.
(c) **THREE--** Bend the trunk to the right. Keep the knees straight.
(d) **FOUR--** Recover to the starting position.
NOTE:Keep the rifle on the same level as the shoulders throughout the exercise.

3303. LOG DRILL

a. Description and Objective. Log exercises are conditioning exercises performed with a log. Each drill contains six exercises and they are numbered in a set pattern. The drill takes 15 minutes to complete. The objective of log drills is to develop strength and muscular endurance and, in this instance, under maximum loads. Log exercises also develop teamwork. Log may be used in lieu of conditioning drills after the Marines have become somewhat conditioned.

b. Area and Equipment. Any level area is satisfactory for conducting this drill. Each Marine completes these exercises from a standing position and no ground contact is required. If the group exceeds a platoon in size, then the instructor will need instructor's stand. Each six-person group or eight-person group will need a log. The logs should be from 6 to 8 inches in diameter. They may vary in length from 14 feet (for 6 people) to 18 feet (for 8 people).The logs should be skinned, smoother, and dried. The 14-foot logs should weigh approximately 300 pounds and the 18-foot logs, approximately 400 pounds. Rings should be painted on the logs to indicate each person's position. When not in use, the logs should be stored on a rack to keep them off the ground.

c. Formation. All the Marines assigned to the same log team should be about the same height at the shoulders. The recommended method of dividing the platoon is to have the Marines form a single file or column with short people to the front and tall people to the rear. Have the Marine assume their positions in the column according to shoulder height, not head height. When they are in position, they are given the command COUNT OFF BY SIXES (OR EIGHTS), COUNT OFF, to divide them into six- or eight-person log teams. Each team in turn, can then proceed to the log rack, shoulder a log, and carry it to the designated exercise area. The log teams form in columns in front of the instructor. With the Marine holding the log in the chest position, have them face the instructor and ground the log at least 10 yards between columns and 10 yards between log teams within the columns.

d. Starting Positions. The Marines fall in, facing the log, with their toes about 4 inches from it. The basic starting positions and commands are as follows:(See fig. 3-8.)
(1) **RIGHT-HAND STARTING POSITION, MOVE.** It the command MOVE, move the left foot 12 inches to the left, and lower the body into a flat foot squat. Keep the back straight, head up, and arms between the legs. Encircle the far side of the log with the left hand. Place the right hand underneath the log.(See fig. 3-8, A.)

(2) LEFT-HANDED STARTING POSITION, MOVE. These commands are executed in the same manner as in paragraph a. except that the left hand is underneath the log and the right hand encircles its far side. (See fig. 3-8, B.)

Figure 3-8 is available in the illustration section at the end of the book.

(3) RIGHT SHOULDER POSITION, MOVE. At the command MOVE, pull the log upward in one continuous motion to the right shoulder. At the same time, move the left foot to the rear and stand up, facing left. Balance the log on the right shoulder with both hands.(See fig. 3-8, C.)This movement cannot be performed from the left-hand starting position because of the position of the hands.

(4) LEFT SHOULDER POSITION, MOVE. These commands should be given from the left hand starting position. At the command MOVE, pull the log upward in one continuous motion, to the left shoulder. At the same time, move the right foot to the rear and stand facing right. Balance the log on the left shoulder with both hands.(See fig. 3-8, D.)This movement cannot be performed from the right-hand starting position.

(5) WAIST POSITION, MOVE. From the right hand position pull the log waist high. Keep the arms straight and fingers laced underneath the log. The body is inclined slightly to the rear and the chest is lifted and arched. (See fig. 3-8, E.)

(6) CHEST POSITION, MOVE. This command should be given after the waist position has been assumed .On the command MOVE, shift the log to a position high on the chest, bring the left arm under the log and hold the log in the bend of the arms.(See fig. 3-8, F.)Keep the upper arms parallel to the ground.

(7) To move the log from the right shoulder to the left shoulder, the command is: LEFT SHOULDER POSITION, MOVE. On the command MOVE, push the log overhead and lower it to the opposite shoulder.

(8) To return the log to the ground from any of the above positions, the command is: STARTING POSITION, MOVE. At the command MOVE, slowly lower the log to the ground. The hands and fingers must be kept from under the log.

e. Leadership. A principal instructor demonstrates and leads the drill. He must be familiar with the leadership techniques for conditioning exercises and the peculiar techniques for log drill.

3304. EXERCISE PERFORMED WITH LOGS

The exercises of log drill are outlined in the following paragraphs. Figure 3-9 graphically explains log drill.

a. Exercise l:Two-Arm Pushup

(1) Starting Position. Right or left shoulder position. Regular stance. (See fig. 3-9, A.)

(2) Cadence. Moderate.

(3) Movement. A four-count exercise:at the count of—

- **(a) ONE--** Push the log overhead until the elbows lock.
- **(b) TWO--** Lower the log to the opposite shoulder.
- **(c) THREE--** Repeat the action of count ONE.
- **(d) FOUR--** Recover to the starting position.

b. Exercise 2: Forward Bender

(1) Starting Position. Chest position. Regular stance.(See fig. 3-9, B.)

(2) Cadence. Moderate.

(3) Movement. A four-count exercise:at the count of-

(a) ONE-- Bend forward at the waist, keeping the back and legs straight.

(b) TWO-- Recover to the starting position.

(c) THREE-- Repeat the action of count ONE.

(d) FOUR-- Recover to the starting position.

c. Exercise 3: Straddle Jump

(1) Starting Position. Right or left shoulder position, feet together, fingers interlaced on top of the log.(See fig. 3-9, **C.)**

(2) Cadence. Moderate.

(3) Movement. A four-count exercise:at the count of—

(a) ONE-- Jump to a side-straddle. Pull down on the log with both hands to keep it from bouncing on the shoulder.

(b) TWO-- Recover to the starting position.

(c) THREE-- Repeat the action of count ONE.

(d) FOUR-- Recover to the starting position.

Figure 3-9 is available in the illustration section at the end of the book.

d. Exercise 4: Side Bender

(1) Starting Position. Right shoulder position, feet regular stance.(See fig. 3-9, D.)

(2) Cadence. Moderate.

(3) Movement. A four-count exercise:at the count of—

(a) ONE-- Bend sideward to the left as far as possible, bending the left knee.

(b) TWO-- Recover to the starting position.

(c) THREE-- Repeat the action of count ONE.

(d) FOUR-- Recover to the starting position.

(4) After completing the required number of repetitions, change shoulders and execute an equal number of repetitions to the other side.

e. Exercise 5: Knee Bend

(1) Starting Position. Right or left shoulder position. Narrow stance. Fingers interlocked on top of the log.(See fig. 3-9, E.)

(2) Cadence. Slow.

(3) Movement. A four-count exercise:at the count of—

(a) ONE-- Flex the knees to a quarter-squat position.

(b) TWO-- Flex the knees to a half-squat position.

(c) THREE-- Lower the body to a three-quarter squat position.(Lean slightly forward.)

(d) FOUR-- Recover to the starting position.

NOTE:Pull forward and downward on the log throughout the exercise.

f. Exercise 6: Overhead Toss

(1) Starting Position. Right or left shoulder position, regular stance. The knees are bent to a quarter-squat.(See fig. 3-9, F.)

(2) Cadence. Moderate.

(3) Movement. A four-count exercisc:at the count of-

(a) ONE-Straighten the knees and toss the log into the air approximately 12 inches overhead. Catch the log with both hands and lower it toward the opposite shoulder. As the log is caught, lower the body into a quarter-squat.

(b) TWO-- Again toss the log into the air and when caught return it to the original shoulder.

(c) THREE-- Repeat the action of count ONE.

Section IV. Grass Drills

3401. GENERAL

a. Description and Objective. Grass drills are extremely strenuous exercises and are performed at top speed for only short periods of time. No cadence is counted but the Marines continue to execute the multiple repetitions of the command until the next command is given. The grass drill consists of two drills:Drill One and Drill Two. Each drill contains six exercises. The objective of the drills is to decrease reaction time, to develop cardiovascular endurance, and to provide a vigorous workout for all major muscles.

b. Area and Equipment. Any level area suitable for ground contact and of a size to accommodate the group is adequate. No equipment is needed.

c. Formation. All movements are executed in place. The extended, rectangular formation is recommended for a platoon- or company-size unit. The circle formation is suitable for groups of squad or section size. At the beginning of an exercise program, 2 to 3 minutes of grass drills will insure a good workout.

d. Progression. Progression is gained by gradually increasing the length of time devoted to the drills. As the physical condition of the Marines improves, the periods should be gradually lengthened to 5 minutes. As the second drill is more difficult than the first, some progression can be attained by initially executing Drill One; then as the program and the Marines progress, introduce Drill Two. To extend the duration of the drill, it may be necessary to repeat the drill.

e. Starting Position

(1) The drills are started from the GO position. Other basic positions are FRONT, BACK, and STOP. (See fig. 3-10, A.)

(a) GO. Running in place (top speed): on the toes and balls of feet, knees raised high, arms pumping, body bent forward at waist.

(b) FRONT. Prone position: elbows bent (along body), palms flat on ground directly under the shoulders, legs together and straight.

(c) BACK. Supine position (flat on back): arms extended near side on ground with palms down, legs together and straight, feet toward the stand or instructor.

(d) STOP. Football lineman stance: feet spread and staggered, left arm across left thigh, right arm straight, knuckles on ground, head up, back parallel with ground.

(2) To assume the FRONT or BACK position from the STANDING, GO, or STOP position, vigorously get into the prescribed position as quickly as possible. (See fig. 3-10, B.)

(3) To change from the FRONT to the BACK position, quickly do a pushup, move the feet several short steps to the right or left, lift the arm on the side toward which the feet move, and thrust the legs vigorously to the front. (See fig. 3-10, C.)

Figure 3-10 is available in the illustration section at the end of the book.

(4) To move from the BACK to the FRONT position, sit up quickly, place both hands on the ground to the right or the left of the legs. Move the feet several short steps to the rear on the side opposite the hands. When the feet are opposite the hands, thrust the legs vigorously to the rear and lower the body to the ground.(See fig. 3-10, D.)

f. Leadership. A warm up activity of lesser intensity should proceed grass drill. During the instructional phase and conduct of these drills, the following points should be applied.

(1) The instructor executes only GO and STOP with the Marines. This allows the instructor to supervise the drill.

(2) The commands for grass drills are given in rapid succession without the usual preparatory command.

(3) To prevent confusion, the instructor should give the commands sharply to distinguish them from comments or encouragement.

(4) As soon as the Marines know the drill, they should respond to the instructor's commands and perform all exercises vigorously and as rapidly as possible. All exercises are executed continuously until the next command is given. Insist on top speed performance; anything less is not effective.

(5) The commands peculiar to each exercise are identical to the name of the exercise.

(6) Marines are not to be required to assume the position of attention once the drills are started. To halt the drill for instructions or for rest, the command UP is used. At this command, the Marines assume a relaxed standing position. Do not demand formality. At the conclusion of a fast and vigorous 5-minute grassdrill, it is physically impossible for people to stand at attention.

(7) The sequence of commands for the execution of grass drills should occur as follows:Drill One. GO, FRONT, Bouncing Ball; GO, BACK, Bicycle; GO, Full Squatter; GO, BACK, Situps; GO, FRONT, Mountain Climber; GO, FRONT, Roll Left; GO, STOP, UP.

g. Use With Other Programs. Since grass drills can be executed in a short period of time, they may be executed where only a few minutes are available for exercise or in conjunction with another type of activity. Grass drills are an excellent substitute for running when time is a factor.

3402. GRASS DRILL ONE AND TWO

a.Grass Drill One

(1) Bouncing Ball. From the FRONT position, push up, supporting the body on the hands (shoulder-width apart) and feet. (See fig. 3-11.)Keep the back and legs in line and the knees straight. Bounce up and down by a series of short, upward springs from the hands, hips, and feet simultaneously,

(2) Bicycle. From the BACK position, raise the legs and hips. Keep the elbows on the ground and support the hips with the hands. Move the legs vigorously as if pedaling a bicycle.

(3) Full Squatter. From the STOP position, assume a full knee bend, the feet on line, hands on hips. Bounce up and down in place by short, bouncing jumps.

Figure 3-11 is available in the illustration section at the end of the book.

(4) Situps. From the BACK position and with arms stretched overhead, sit up, reach forward, and touch toes. Return to the supine position.

(5) Mountain Climber. From the STOP position, place both hands on the ground directly under the shoulders. Thrust the right leg to the rear, knee straight. The left foot should be close to the left hand, the left knee outside the left arm. Shift the weight to the hands, thrust off with the rear (right) foot and bring that foot up close to the right hand, the right knee outside the right arm. At the same time, thrust the left leg vigorously to the rear, knee straight. Continue at a fast cadence, alternating the legs.

(6) Roll Left. From the BACK or FRONT position, make one complete roll in the direction commanded. On completing the roll, return to the FRONT or BACK position.

b. Grass Drill Two

(1) Legs Over. .From the BACK position and with arms stretched overhead, palms up, raise the legs upward and then swing them backward over the head until the toes touch the ground behind the head. Return legs to the starting position. (See fig. 3-12.)

(2) V-Up and Touch Toes. .From the BACK position, raise the legs with the knees straight, sit up until the trunk and legs form a V, and touch the toes with the hands. Return to the BACK position.

(3) Rocker. .In the FRONT position, clasp the hands behind the back, arch the body, holding the head back. Start rocking, using the front part of the trunk as a rocker.

(4) Bounce and **Clap Hands**. The procedure is the same as for bouncing ball, but while in the air, clap the hands. This requires a more vigorous bounce or spring. (See fig. 3-12.)

(5) Leg Spreader. .From the BACK position, raise the legs so that the heels are 10 to 12 inches from the ground, spread them apart as far as possible, then close them together. Open and close legs as rapidly as possible.

(6) Forward Roll. .For forward roll from the STOP position, place both hands on the ground, tuck the head, and do one complete forward roll, keeping the legs tucked as you roll, and come back to the STOP position.

Figure 3-12 is available in the illustration section at the end of the book.

Section V.Guerrilla Exercise

3501.GENERAL

a. Description and Objective. Guerrilla exercises are individual exercises of various basic skills that are performed rapidly while moving forward in a circle formation. There are two tables of guerrilla exercises. Each table takes 15 minutes to complete. The objective of this exercise is to increase strength and endurance, aid flexibility, and develop coordination.

b. Area and Equipment. Any level area is suitable for conducting guerrilla exercises. No ground contact is required other than the hands. There is no equipment requirement.

c. Formation
(1) The circle formation (app. A) is used for guerrilla exercises. Each platoon forms its own circle and engages in guerrilla exercises under a platoon instructor. If the platoon exceeds 30 Marines, double or concentric circles may be used.
(2) When the circle is formed, the instructor steps into the center of the circle and moves clockwise in a small circle. He commands:QUICK TIME, MARCH, 1-2-3-4. (Rapid cadence of approximately 130 counts per minute. Cadence and step are maintained between exercises.)
(3) To reform the platoon after completing guerrila exercises, the instructor halts the Marines and places the base man or platoon guide where he wishes and commands:
(a) BASE MAN (or platoon guide),POST.
(b) FALL OUT AND FALL IN ON THE BASE MAN (or platoon guide).

d. Progression. Progression may be attained by moving from table I to table II or by shortening the quick time marching periods between exercises and performing all exercises a second time.

e. Leadership
(1) To execute the exercises, the Marines continue at quick time while the instructor simultaneously explains and demonstrates the exercise to be performed, and then commands the Marines accordingly. In each instance, the preparatory command will be the name of the exercise and, in all instances, the command of execution will ***MARCH***.. *To* terminate each exercise, the command *is* ***QUICK TIME, MARCH.*** *The* Marines immediately pick up the step as the instructor counts cadence.
(2) Unless specified differently, each exercise should be continued for 20 to 40 seconds depending upon the vigor of the exercise. The leader can determine the duration of each exercise by observing its effect upon the Marines.
(3) To form for double guerrillas, the commands for pairing the Marines (who are in circle formation) are-
(a) PLATOON, HALT.
(b) FROM (designate an individual), BY TWO S, COUNT OFF(.Examplel-2; 1-2; 1-2; etc.)
(c) EVEN NUMBERS MOVE UP BEHIND ODD NUMBERS(At this time, adjust pairs according to height and weight.)

(d) YOU ARE NOW PAIRED UP FOR DOUBLE GUERRILLAS. To change the Marines position, merely command CHANGE.)
(e) FORWARD, MARCH.

f. Place in the Program. Many Marines have not had the opportunity to perform the simple skills involved in guerrilla exercises. The conduct of these exercises is a simple matter since they can be performed easily and quickly in almost any situation. The tables of exercise are applicable to all personnel. The tables can constitute a station within a 1-hour period or be completed within a separate 15-minute period.

3502.GUERRILLA TABLES

a. Table I

(1) Double Time.(See fig. 3-13, A.)Hold arms in the thrust position. Execute a double-time run, maintaining the circle formation and the prescribed distance between your fellow Marines. Duration-1 minute.

(2) All Fours.(See fig. 3-13, B.) Face downward. Support the body with the hands and feet. Walk forward hands first.

(3) Crab Walk.(See fig. 3-13, C.) Get in the sitting position, face upward, and lift the hips. Support the body with the hands and feet. Walk forward feet first.

(4) Squat Walk.(See fig. 3-13, D.) Assume a full knee bend position. Grasp the ankles (left ankle with the left hand, right ankle with the right hand). Walk forward.

(5)Broad Jump.(See fig. 3-13, E.) Jump forward on both feet in a series of broad jumps. Swing the arms vigorously to assist the jumps.

(6)Toe-Touch Walk.(See fig. 3-13. F.) Walk forward, bending at the waist and touching one hand to the toe of the opposite foot while it is on the ground. Raise the trunk to the vertical position between steps. Keep the knees straight.

(7)Bottoms-Up Walk.(See fig. 3-13, G.) Assume the front leaning rest position and move the feet toward the hands in short steps, keeping the knees locked. When the feet are as close to the hands as possible, walk forward on the hands to the front leaning rest position.

(8) Straddle Run.(See fig. 3-13, H.) Run forward, leaping to the right from the left foot and to the left from the right foot.

(9) Fireman's Carry .(See fig. 3-13,1.)

(lO)Single-Shoulder Carry(See fig. 3-13, J.) Two men execute the carries as indicated by the diagram. No. 1 man executes one type; No. 2 man executes the other.

Figure 3-13 is available in the illustration section at the end of the book.

b. Table II

(1) Double Time.(See fig. 3-14, A.)Hold arms at the thrust position. Execute a double-time run, maintaining the circle formation and the prescribed distance between your fellow Marine.Duration-1 minute.

(2) Toe-Grasp Walk.(See fig. 3-14, B.) Bend forward and grasp toes. With knees slightly bent, walk forward.

(3) Hand-Kick Walk..(See fig. 3-14, C.) Walk forward, kicking the moving foot upward on every step. At the same time, lean forward and touch the elevated toe with the hand of the opposite arm.

(4) Pike Juinping.(See fig. 3-14, D.) Jump forward and upward from both feet, keeping the knees straight, and at the same time, swing the legs forward and touch the toes with the hands at the top of each jump.

(5) Squat Jump.(See fig. 3-14, E.) Leap forward from the squatting position, with the hands on the ground and the arms between the legs. Land on the ground with hands and legs extended. Bring up the legs to the squatting position.

(6) Steam Engine..(See fig. 3-14, F.) Lace the fingers behind the neck and walk forward in the following manner: as the left leg moves forward, raise the knee high, bend the trunk forward, and touch the outside of the right elbow to the outside of the knee. Then lower the left leg and step forward on the left foot and raise the right leg. Repeat with the right leg and left elbow.

(7) Knee-Touch Walk.(Scc fig. 3-14, G.) Walk forward, bending the knees and touching the ground on each step. The knees are bent and straightened on each step.

(8)Hobble Hopping.(See fig. 3-14, H.) Hold foot behind back with opposite hand and hop forward. On the command CHARGE, grasp the opposite foot with opposite hand and hop **forward.**

(9) Cross Carry.(See fig. 3-14,1.)(10)Saddle Back Carry. (See fig. 3-14, J.) Two Marines execute the carries as indicated in the diagram. No. 1 Marine executes onetype;No. 2 Marine executes the other.

Figure 3-14 is available in the illustration section at the end of the book.

Section VI. Running and Orienteering

3601. RUNNING

a. General

(l)Description. The general form and technique for all types of running are fairly constant. (See fig. 3-15.)The head is erect, body slightly forward without bending at the waist, and the arms are at a loose thrust position alternating from front to rear in straight planes. A cross-body arm movement wastes energy .The movement of the legs and feet will be discussed in subsequent paragraphs dealing with the different types of running. Of primary importance is the fact that in all types of running, the toes should be pointed straight ahead. Toeing out is a common error in both running and walking and should be an item of individual correction.

(2) Objective. The objective of running is to develop cardiovascular endurance. Despite the fact that Marines have developed their muscle structure and the strength of their muscle tissue, unless they have developed cardiovascular endurance to a satisfactory degree, they are not entirely physically fit or combat ready. Cardiovascular endurance (wind) depends on the efficiency of the lungs and heart. The efficiency of the lungs and heart depend on the amount of oxygen the lungs can absorb with each breath inhaled and the amount of carbon dioxide the lungs can expel. The process of absorbing oxygen and expelling carbon dioxide (cardiopulmonary process) is performed by the blood that circulates through the lungs. The condition of this process will determine the amount of effort a person can exert over a period of time. Running is one of the best activities to develop this vitally important endurance.

b. Running Skills. In the development of running skills, individuals may require instruction to improve their proficiency. Some important skills to consider are-

Figure 3-15 is available in the illustration section at the end of the book.

(1) Action of the Arms. Arm action is important. Check to see that arms are held loosely and that the action is relaxed. The faster the run, the more rapid the arm action.

(2) Breathing. Allow the individual to breathe through the mouth as the body demands a large supply of oxygen. Oxygen can be inhaled in greater quantities through the mouth.

3602. TYPES OF RUNNING

a. Double Time

(1) Description and Objective. Double timing is marching at the rate of 180 steps per minute, each step being 36 inches in length. It takes practice to double time with precision in formation. The Marines should keep in step, placing their feet flat on the ground. This, however, should not be a stamping motion, but should be done with as slight a jolt as possible. Double timing is like a jog, the difference being that in a jog the feet are lifted well off the ground and the running motion is bouncy. In double timing, the feet skim the ground and there is no bounce to the ran. Double timing is a vehicle for teaching proper running form and for the development of the cardiovascular system.

(2) Area and Equipment. This type of running can be completed over a variety of surfaces. Usually a Marine uses a field or road. There is no equipment requirement.

(3)Progression. There is no set standard for alternating quick time and double time in the early conditioning of Marines. A general rule is to begin with enough quick-time marching to ensure a thorough warming up, then double time about 100 paces. Change again to quick time until the individuals have made a reasonable recovery from the running, then double time another 100 paces. The amount of double time can be increased and the quick time decreased from week to week, until the individuals are double timing about 1800 yards. This type of training should be given at least twice a week, but by no means is it adequate as the sole means of conditioning.

(4) Leadership

(a) The instructor should be to one side of the column or group and toward the rear so there is a full view of all Marines. Inexperienced instructors have a tendency to supervise from a position too far forward.

(b) Select an individual who can maintain the proper cadence to act as the guide during double-time running.

(c) There are several ways for the instructor and group to count cadence while double timing. If not contrary to local policy, learn several methods and use them for variety.

b. Wind Sprints

(1) Description and Objective. This type of running involves a series of 30- or 40-yard dashes, usually conducted in successive waves of squads. Each squad is in line and the squad leader on the right flank. Wind sprints assist indeveloping speed and cardiovascular endurance. Any flat and level area may be used which will permit the squad to form a line and run the required distance.

(2) Progression. One or two 30-yard sprints will be adequate at the beginning. As time passes, sprints can be lengthened and up to six or seven sprints may be used.

(3) Leadership

(a) At the command READY (given by squad leader), each runner assumes the sprinter's starting position. At the command GO, the squad sprints approximately 30 yards, takes 10 yards to stop, and lines up immediately with the squad leader who repeats, READY, GO, and again the squad sprints. At the conclusion of the third sprint, the squad waits until all the squads of the platoon have made three sprints. Then they all line up and the squad leaders conduct three more wind sprints in the opposite direction.

(b) Valuable time is gained by having each squad ready to go when the preceding squad has moved off its second sprint mark.

c. Cross-Country Running

(1) Description and Objective. Cross-country running is a distance run conducted on a course laid out along roads, across fields, over hills, through woods, and on any irregular ground. The cross-country ran may be utilized as a conditioner or as a competitive event; the objective is to cover the distance in the shortest possible time. The course should be 2 to 2.5 miles in length and be laid out to avoid heavy vehicular traffic. The course should be marked by directional arrows until the runners know the course. These runs build leg muscles, increase lung capacity, and develop endurance. Any local area of varied terrain is suitable.

(2) Progression. In the mass training of a large group, leaders should be stationed at the head and the rear of the column and should make every effort to keep the individuals together .After determining the abilities of the individuals in cross-country running, it is advisable to divide the unit into three groups. The poorest conditioned group starts first, and the best conditioned group, last. The starting time of the groups should be staggered so that all of them finish about the same time. In preliminary training, the running is similar to ordinary road work in that it begins with rather slow jogging, alternating with walking. The speed and distance of the run is gradually increased. As the condition of the individual improves, occasional sprints may be introduced. At first the distance run is from 1/2 to 1 mile. It is gradually increased to 2 or 2.5 miles. Well-conditioned personnel can run 2 to 2.5 miles within a 15-minute period.

(3) Leadership. Marines should not be required to take part in distance running until they have been through a progressively scheduled training program which requires a considerable amount of running. Cross-country runs should be scheduled occasionally to provide variety in the program. Cross-country running has the advantage of allowing mass participation. Interest can be stimulated by putting the runs on a competitive basis. As a single activity, short cross-country runs can be scheduled once a week, gradually increasing the distance as the physical conditioning improves.

(4)Use With Other Programs. Cross-country running can be combined with other activities such as conditioning exercises.

d. Fartlek Training

(l)Description and Function. Fartlek training is various running exercises conducted along an intense combat-like course. The running is conducted to overwork the lungs, allowing only partial recovery which is followed by another intense period of overload. This sequence is repeated for the duration of the workout. Fartlek training is a useful combat training method and general cardiovascular conditioner. The entire unit must be in uniformly good to excellent condition to qualify for this training.

(2) Progression. The following is an example of one fartlek session:

(a) Warm up by stretching 3-5 minutes.

(b) Warm up running easily 5-10 minutes.

(c) Run at a fast, steady pace for 3/4 to 1 3/4 miles (dependent on terrain).

(d) Walk/jog at a moderate pace for 5 minutes (recovery).

(e) Run easily sprinting 15 to 20 meters occasionally.

(f) Run full speed uphill for 175 to 200 meters.

(g) Warm down by running easily for 1/2 to 1 mile.

(h) Warm down by stretching 3 to 5 minutes.

(3)Leadership. To add variety, try incorporating combat movements, terrain association, a series of exercises or carrying table of equipment weapons (hit and roll). This will take some initiative in setting the course, but it allows for variety. In this manner, the fartlek course can be used as a training session or part of one. It works extremely well with highly motivated, competitive Marines.

3603.ORIENTEERING

Orienteering is land navigation over a prescribed course as a timed, competitive event. It is an excellent way to integrate land navigation training with physical conditioning and requires Marines to be skilled in both areas. Commanders adjust the training as needed to change the degree of difficulty or to emphasize either navigation or conditioning. Taken to the extreme, an orienteering exercise is conducted in full combat gear over rugged terrain against aggressors. Orienteering requires the following physical skills: endurance, speed, strength, and all-around conditioning.

Section VII. Circuit Training

3701. GENERAL

a. Description and Objective. A strength circuit consists of a series of stations where individuals in small groups exercise vigorously for a short period of time and then move (on signal) to the next station where a different form of exercise is conducted. This rotation continues until all groups move through all stations. Strength circuits contain no set or specific types of exercise stations within the circuit. The objective of circuit training is to develop strength. There are three general types of circuits.

(1) Fixed Circuit. This is a circuit in which apparatus of an immovable type (fixed into the ground) is used.

(2) Movable Circuit. This circuit consists of individual exercise apparatus which is portable and can be moved to and from the training area.

(3) Simplified Circuit. This circuit requires no equipment or apparatus.

b. Formation. The exercises are done at will, but rapid, steady, and continuous work is required of all. Each Marine's nervous and muscular system reacts differently to timed vigorous exercises. His performance should be measured on how many movements per exercise he can complete as an individual. For example, one Marine may be able to complete 5 movements, while another may be able to complete 20, and yet each is receiving the maximum benefit. All three circuits contained in this chapter are designed for platoon-sized groups. Expansion beyond this capacity requires a large amount of equipment, as each Marine in the fixed and movable types of circuits must have an item of equipment available for exercise at each station. A group larger than a platoon could be exercised through use of the simplified type of circuit; however, the group would be unwieldy and control could be a problem.

c. Place in the Program. All circuits illustrated can be completed in a 15-minute period. This feature allows the exercise of a platoonor smaller group on the circuit for a single 15-minute period, or the scheduling of the circuit as a 15-minute period within a longer period. A circuit can thus be utilized within a rotating activity system of scheduling. Choice of a circuit by the unit depends upon area, facilities, and other local factors; however, there is a circuit for every need.

3702. FIXED STRENGTH CIRCUIT

a. Description and Objective. The strength circuit is an arrangement of various types of exercise apparatus which are fixed in position. (See fig. 3-16.)Seven basic exercises are used and each exercise requires an apparatus. All apparatus of one type are positioned together to constitute a station. Each station will accommodate 10 Marines. The circuit is designed to be accomplished in 15 minutes when conducted on a time-rotation basis, as normally executed for unit training. For individual use, a Marine may complete the circuit by executing a specific number of repetitions for each exercise. In this case, the time required for completion of the circuit would vary slightly depending on the number of repetitions the Marine completed. The objective of this circuit is to provide a series of exercises which will improve and maintain the strength of the body's major muscle groups.

Figure 3-16 is available in the illustration section at the end of the book.

b. Starting Level. Marines must be thoroughly warmed up prior to participating in the circuit system. If Marines have not engaged in vigorous exercise immediately prior to starting the circuit, then an instructor should execute the following warm up exercises. These exercises should be conducted in the normal formation for set drills. Seven repetitions of each exercise will normally provide sufficient warm up. The exercises are-

(1) High jumper, Conditioning Drill 1.
(2) Bend and reach, Conditioning Drill 1.
(3) Squat bender, Conditioning Drill 1.

c. Progression. The instructor can adjust the circuit system through several methods to accommodate participants of varying physical ability. One method of adjusting an exercise is to change the method of executing the leg lift or to select a heavier weight for the barbell curls. When it becomes apparent during a unit program that the overall fitness of the group has Improved, then the exercise may be made more strenuous in two additional ways. First, the instructor can have one or more of the supplementary stations added. Second, the time spent exercising at each station can be increased in 5-second increments to a maximum of 60 seconds. Also, as Marines learn the circuit, the 45-second period for movement and instruction between stations can be eliminated, thus allowing only enough time to change stations.

d. Starting Positions. The Marine starts on any station, exercises steadily for a certain period (45 seconds initially), then moves on command to the next type of apparatus where he again exercises steadily for an equal period. The Marine continues until the required exercise is completed at each station. If it is desired to expand the number of stations in the circuit to accommodate more participants at one time, the instructor can provide four additional supplementary exercises, requiring no equipment. This will increase the amount of time required to complete the circuit.

e. Leadership. Close control of all Marines is necessary to ensure that a minimum amount of time is spent in moving them to their initial stations and in moving them between stations. One instructor can control the activity on the strength circuit. A stop watch or wrist watch with a second hand is required. When Marines arrive at the strength circuit, the instructor will form them for exercise and conduct the warm up drill. The group is then reassembled and formed into a number of files equal to the number of stations being used in the circuit. Each file is then directed to a station. As soon as all participants have reached an exercise position at a station, the instructor gives the command READY, GO. After 45 seconds of exercise, the instructor gives the command STOP, CHANGE OVER. The instructor allows Marines 45 seconds for moving to the next station and for preparing for the next exercise before the command EXERCISE is again given. In lieu of verbal commands, a whistle may be used to stop and start the exercises. For large groups, a megaphone or loudspeaker is useful.

f. Individual Conditioning Program. For the Marine working alone on the strength circuit, it would be impractical to time the exercise periods. For individual exercise, the Marine should select a number of repetitions of each exercise to complete, then rotate to the next stations after completing these repetitions. The number of repetitions selected should be at or near the maximum that the Marine is capable of doing without halting for rest.

g. Fixed Circuit Stations

(1) Primary Stations (See fig. 3-16.)

(a) Pullups/Chinups. A horizontal bar placed 8 feet above the ground is necessary for each Marine at this station. The Marine will also need a space on the bar that is 45 inches wide. On the command**EXERCISE**,the Marine grasps the bar with both palms facing either forward or to the rear, arms fully extended, and feet free of the ground and executes the exercise as described in MCO 6100.3_.The exercise is repeated as many times as possible until the command **STOP** is given. Then the Marine moves to the next exercise. If a Marine has done his maximum number of pullups/chinups prior to the command **STOP**, he will remain in the "dead" handing position until the command **STOP** and move to the next station.

(b) Twist Grip. The apparatus is a horizontal bar, free to turn, held between uprights placed 30 inches apart. The bar is 52 inches above the ground. A weight of 20 pounds is attached to the center of the bar by a light rope long enough to permit the weight to rest on the ground. The Marine stands at arm's length from the bar and grasps it with his hands on either side of the rope, palms down, thumbs under the bar. On the command **EXERCISE,** the hands are rotated so that the backs of the hands are rotated away from the body, thus winding the rope on the bar. The elbows are kept straight to ensure that the exercise is performed by the hand and forearm. When the weight is drawn up to the bar, the bar is then rotated in the opposite direction to lower the weight to the ground. This exercise is continued until the command **STOP** is given. The Marine then moves to the next station.

(c) Pulley Weights. The apparatus is a T frame with a system of pulleys that suspends a weight of about 90 pounds. The weight is attached to a light steel cable which has a drawbar attached to the other end. The Marine grasps the drawbar and sits down directly under the bar, legs extended to the front and arms extended overhead. The exercise is executed by pulling the drawbar down behind the head, then extending the arms slowly again until they are fully extended overhead. The exercise is repeated as many times as possible until the command STOP is given. The Marine then will move to the next exercise. Upon completion of the exercise, the weight is lowered slowly to the ground.

(d) Barbell Curls. A barbell is necessary for each Marine at this station. The barbell is constructed of 1 1/4-inch pipe 5 feet long, and two concrete-filled No. 10 cans. Each barbell should weigh about 40 pounds. Variance in the weight of the barbells, up to about55 pounds, will allow appropriate overload to be applied to Marines who are above average in strength or weight. The Marine grasps the bar with the palms forward and assumes a standing position with the barbell held in front of the hips, hands approximately shoulder's width apart. On the command **EXERCISE,** the Marine flexes the elbows and draws the barbell up until it touches the upper chest. The elbows remain at the sides. Breath is inhaled with the upward movement and exhaled as the barbell is lowered to the starting position. The exercise is repeated as many times as possible until the command **STOP** is given. The Marine moves to the next apparatus.

(e) Step-Up. The apparatus is a platform or ledge 18 inches high and of such size. At accommodate 10 Marines. The Marine faces the platform and on the command **EXERCISE,** steps up onto the platform, bringing his trailing foot up beside the leading foot. He then steps back down to the original position, stepping down first with the same foot he initially used in stepping up. After 10 repetitions of the exercise, he changes the order of moving the feet to use the opposite leg for stepping up. Repeat this exercise until the command **STOP** is given. The Marine then moves to the next station.

(f) Leg Lift. The apparatus is a horizontal bar constructed as described in (a). To prevent the body from swaying, a horizontal back support is added 40 inches below the horizontal bar. The arms are kept fully extended. On the command **EXERCISE,** the Marine jumps up, grasps the bar with the palms forward and the back support behind him. The exercise is executed by raising the legs to a horizontal position then slowly lowering them to the vertical position. The Marine does not flex his knees. He does not swing the legs to the rear of a vertical position to gain momentum for raising them in the next repetition of the exercise. The movement is repeated until the command **STOP** is given. The Marine then moves to the next exercise. If unable to raise his legs to a horizontal position without flexing his knees, the Marine flexes his knees and draws them up to the chest, then lowers his legs to the vertical position.

(g) Rope Climb. The rope climb is 20 to 30 feet high with five ropes suspended from a horizontal bar which forms the uppermost part of the framework. To prevent the horizontal bar from sagging and to provide safety, only five ropes are attached to it. There are two frameworks per station. The ropes are 6 feet apart. Any method may be used to climb the rope, and the Marines climb as high as possible. Marines who are proficient should climb therope several times during the time allotted. Inexperienced Marines should be cautioned to take care during descent to avoid rope burns on their hands.

(2) Supplementary Stations. The following exercises are designed to expand the basic circuit by being inserted in specific places within the system. For each supplementary station used, there should be adequate room for 10 Marines to exercise.

(a) Bent Knee Situp or Bottoms-Up. These calisthenics are designed to strengthen the abdominal muscles. These exercises will be inserted between the pullup/chinup and twist grip stations. The primary stomach exercise is the situp. In case of inclement weather or other conditions that make ground contact undesirable, the bottoms up exercise is used. In situps on the command **EXERCISE,** the Marine lies on his back (supine position) with knees flexed and both feet flat on the ground and executes the exercise (minus an assistant) as described in MCO 6100.3_.The Marine then returns to the starting position, repeating the exercise until the command **STOP** is given. The Marine then moves to the next station. In bottoms up on the commandEXERCISE,the Marine assumes the front leaning rest and executes the bottoms-up exercise as described in Conditioning Drill 3. He continues this exercise at a moderate cadence until the command **STOP** is given. The Marine then moves to the next station.

(b) Pushup. This exercise is designed to strengthen the arm- and shoulder-girdle muscles. It should be included between the twist grip and pulley weight stations. Upon the command **EXERCISE,** the Marine executes the pushup as described in Conditioning Drill 1. He continues this exercise at a moderate cadence until the command **STOP** is given. The Marine then moves to the next station.

(c) Knee Bender. This exercise is designed to build leg muscles and is included between the pulley weight and barbell curl stations. On the command **EXERCISE,** the Marine executes the knee bender as described in Conditioning Drill 3.He continues this exercise at a moderate cadence until the command **STOP** is given. The Marine then moves to the next station.

(d) Trunk Twister. This exercise strengthens the major muscles of the trunk and is included between the step-up and pullup/chinup stations. On the command EXERCISE, the Marine executes the trunk twister as described in Conditioning Drill 1. He continues this exercise at a moderate cadence until the command **STOP** is given. The Marine then moves to the next station.

3703. MOVEABLE STRENGTH CIRCUIT

a. Description and Objective. The exercises in this circuit are progressive and the course is planned to gain and hold the interest of the participating groups. The circuit consists of a series of stations, with each station designed to develop a particular group of muscles. Along with muscular development, correct posture and deep rhythmic breathing should be stressed on this circuit at all times.(See figs. 3-17 and 3-18.)

Figure 3-17 is available in the illustration section at the end of the book.

b. Equipment. The equipment is set up in files. Six files of 8 stations will accommodate a platoon

c. Formation. The platoon marches to the area where the equipment is positioned and forms a file within each lane of stations, covering down on a piece of equipment. Movements are made on the double, the important factor being that no time is wasted in getting to work.

d. Progression. Initially 40 to 45 seconds per station is adequate. As individuals become stronger the time should be increased in 5-second increments until a minute to a minute and a half is reached.

e. Leadership

(1) The leader stands in front of the barbell station and controls the rotation from this position. The leader supervises the entire group, with the assistance of several instructors who move about in the platoon correcting and encouraging the Marines.

(2) The leader starts each group but does not count cadence nor lead them through the exercises. Each individual exercises rapidly but individually.

(3) As each Marine finishes his repetitions with the barbell, he places the barbells on the ground. The leader calls, READY, followed by the command, FALL OUT ONE. All will double-time to the station directly in front of them, while the individuals on the barbell stations will do an about face to the rear station in their lane.

f. Movable Circuit Stations. The Marine can obtain best results on the movable circuit if the exercises on the various stations are given in the following manner:

(1) Station 1—The Barbell. (See fig. 3-17, 1.)The exercises at this station stress proper posture; deep, rhythmic breathing; and development of the muscles of the arms, shoulders, and upper body. An instructor teaches the proper methods of lifting before the exercises begin. The methods are to lift with the legs, to keep the back straight, and to merely grip with the hands. Two recommended exercises are given below. The instructor will use only one exercise per period .He may use either exercise. At successive periods, the other exercise is used.

(a) Exercise 1--Two-Hands Military Press. (See fig. 3-18, 1.)Grasp the barbell with both hands, knuckles up at shoulder width, and lift to the chest. Steadily press to arm's length overhead; lower to the chest resisting weight all the way. Inhale as the weight is pressed up and exhale as the weight is brought down.

(b) Exercise 2--Two-Hands Regular Curl. (See fig. 3-18, 1.) Lift the weight to the waist, with the palms of the hands out, heels together, stomach in, chest lifted and arched, shoulders back, elbows in close to the sides; inhale deeply and curl the weight to the shoulders, using the arms only, at the same time keeping the elbows close to the sides; exhale Emphasize posture and the use of the arms only. This is a very valuable exercise for the development of the biceps and the grip and should be repeated from 8 to 16 times, depending on the ability of the Marine.

(2) Station 2--The Jump Rope. (See fig. 3-18, 2.) This exercise develops strength and agility in the legs and stamina of the whole body. It makes the Marine agile on his feet and increases his footwork efficiency and timing. Each Marine should progress until able to jump rope at least 3 minutes at top speed.

(3) Station 3--The Twist Grip. (See fig. 3-18, 3.) The twist grip is an excellent exercise for the hands and forearms, and adds greatly to the Marine's ability in hand-to-hand combat. The handle is gripped and twisted, winding the rope until the weighted can is level with the height of the hands, which are held horizontal. The weight is lowered in the same manner; the Marine resists the weight all the way, occasionally stopping the twisting motion and alternately removing first one hand, then the other, from the handle. A variation of the above exercise is to wind the handle with the palms up and the arms bent and with the elbows held close in to the sides. Each Marine should maintain a good posture and keep the stomach muscles taut throughout this exercise.

(4) Station 4--The Incline Plane. (See fig. 3-18, 4,)The use of the incline plane is a very strenuous exercise and well designed for the development of the abdomen. Although 6 to 10 repetitions are sufficient for the beginners, more can be added as ability increases. Marines with hernias or recent operations will be excused from participation at this station.

(5) Station 5--The War Club. (See fig. 3-18, 5.) The war club is a simple and effective means of exercising the principal muscle groups of the body, especially those of the trunk, back, and shoulders. To gain the maximum benefit from this exercise, the Marine must keep both feet flat on the ground at all times. Throughout the exercise period, the weight is swung from arm's length as follows:

(a) As in chopping wood, first on one side then on the other.

(b) As a batter warming up with a number of bats.

(c) In large circles, first with one hand and then with the other.

(6) Station 6--The Bicycle Ride. (See fig. 3-18, 6.) The bicycle ride is well suited to exercising many of the muscle groups of the body, particularly those of the abdomen. Vary the speed of the exercise, but keep everyone "riding" the entire period. A variation exercise may be performed by placing the legs together, raising them slowly to a height about 2 feet from the ground, and then lowering them slowly to the ground.

(7) Station 7--The Step-up. (See fig. 3-18, 7.) The step-up exercises the legs. The step-up is performed by initially stepping up with the left foot, followed by the right, then stepping down with the left foot followed by the right. Continue for 20 seconds, then change to the right foot as the lead foot for 20 seconds.

(8) Station 8--The Isometric Pull. (See fig. 3-18, 8.) Two trainees work at this station with a cable pull and perform the following exercises:

(a) Initially start with one Marine in the supine position and one Marine sitting. The sitting Marine lowers the upper body to the ground and pulls the partner up to the sitting position. The partner then performs this same action and this is continued for 20 seconds at a rapid rate.

(b) During the last 20 seconds, the same action takes place but, in this case, the Marine in the supine position resists the pull of the partner for approximately 5 seconds before allowing to be pulled up into the sitting position.

3704. CIRCUIT-INTERVAL TABLE

a. Description and Objective. The circuit-interval table is designed to develop strength and endurance within a short period of time, with no equipment requirement, through a rapid and vigorous routine of exercise. Fifteen minutes is an adequate period to execute all exercises and to secure a vigorous workout with the circuit-interval principle.

Figure 3-18 is available in the illustration section at the end of the book.

b. Formation and Starting Positions

(1) A leader forms platoons or smaller groups in an oval or circular formation with 3- to 5-yard intervals between Marines. The Marines face to the right and move forward at quick time and then double time. (See fig. 3-19, A.)After running several platoon circle laps, the leader calls out the name of an arm and shoulder exercise from the list below, orders quick time and commands, for example, PUSHUPS. On this command, all Marines immediately hit the ground and individually and rapidly begin doing pushups. No cadence is counted. (See fig. 3-19, B.) After 30 seconds of exercise, the leader commands. **QN YOUR FEET, FORWARD, MARCH** The platoon resumes the quick time cadence and the leader, when ready, gives the necessary commands for double time. The double time is continued for one or more laps and the leader calls out the name of the next exercise and the process is repeated. This continues, with running between each exercise, until every body part has been exercised.

(2) The instructor controls the running and quick time to observe the effects of the exercise upon the Marines. Cadence, step, and precision are not important to the objective and the instructor should not use them. What is important is speed and the instructor should stress this. After the exercise period is started, the Marines do not stop. This circuit method emphasizes stress and recovery, the recovery occurring during the quick time periods.

c. Activities. The leader can use the following exercises. These exercises can be repeated if necessary, during a second round. The leader can use other calisthenic exercises in the circuit-interval table.

(1) Arms and shoulders-push-ups.
(2) Stomach situps.
(3) Back-squat thrusts.
(4) Legs-bicycle (on back).

d. Progression. lhe progress is controlled by the leader, who must pace the running, quick-time movement, and exercise in such a way that Marines will receive a vigorous workout yet be able to participate throughout the 15-minute period. Marines who are in the initial stages of physical condition will not be able to double time or exercise as long as those who are better conditioned. The idea is to set a pace which can be increased during each workout, thus progressing gradually to a higher level of physical fitness.

e. Leadership. The platoon leader, platoon sergeant, or section leader can lead the group. The leader must execute the exercise with the unit to feel the effects and thereby adjust the pace.

f. Place in the Program. This activity may be scheduled whenever a short period of time is available. The only requirement is that enough space, indoors or out, be available to form the circle.

Figure 3-19 is available in the illustration section at the end of the book.

Section VIII. Basic Physical Skills and Obstacle Courses

3801. GENERAL

The purpose of this section is to list the basic military skills and the methods for their development. Many of these skills are best developed by obstacle courses but other drills are also discussed. The section explains types of obstacle courses, details of construction, and methods of negotiating the various obstacles.

3802. BASIC PHYSICAL SKILLS

a. Objective. The objective is to develop proficiency in the various military physical skills which are essential to personal safety and effective combat operations. In travel by foot over rugged terrain and in the execution of combat duties, Marines must be trained to perform basic skills, such as running, jumping, climbing, and carrying. During training, Marines will develop agility and coordination in these skills .Fast and skillful execution of these skills may mean the difference between success and failure on the battlefield.

b. Place in the Program. These skills are practiced throughout the entire program in many different activities. Many of these skills are best practiced on obstacle courses.(See par. 3803.)

c. Description. These basic skills are the minimum skills required by all Marines. The basic skills are as follows:
(1) Running. Running is used to strengthen the legs and develop cardiovascular endurance. Marines should be exposed to running in various situations: on roads, over rough ground, up and down hills, cross-country, and running over low obstacles.
(2) Jumping. In broad jumping, the take-off foot is planted firmly and the spring comes from the extension of this leg as the other leg reaches for the far side of a ditch or similar obstacle. (See fig. 3-20.) The arms are forcibly raised forward and upward to assist in propelling the body up and forward. Landing may be on one or both feet depending upon the length of the jump. In vertical jumping downward from a height, the jumper should aim his feet at the desired landing spot and he should also jump with the knees slightly bent and feet together, with the trunk inclined slightly forward. As the feet touch the ground, the shock is absorbed by bending the knees into a full squatting position. If the height is too great or the ground too hard to absorb the shock, then the jumper should forward roll or side roll thus eliminating some of the momentum.

Figure 3-20 is available in the illustration section at the end of the book.

(3) Dodging In combat situations, it is often times necessary to change directions quickly. To execute this movement while running, a lead foot is firmly planted, left foot if the direction is to the right and right foot if the direction is to the left. The opposite foot is moved toward the new direction. The knees are slightly flexed during the movement and the center of gravity is low and balanced. (See fig. 3-21.) At the time of the change of direction, the head and trunk are turned quickly in the new direction.

Figure 3-21 is available in the illustration section at the end of the book.

(4) Climbing and Surmounting. All Marines should know how to climb and surmount various types of obstacles. (See fig. 3-22.)

Figure 3-22 is available in the illustration section at the end of the book.

(a) Vertical Climbing, as in Climbing a Rope or Pole. The hands grasp the rope or pole overhead with the palms toward the face. Gripping the object, the body is pulled upward with the arms and shoulders, assisted by the feet which grip the object and assist by pushing downward. If shoulder girdle strength and body coordination are not adequate to permit alternating the hands, the arms act together in pulling upward.

(b) Climbing as in Surmounting a Wall. Ingoing over a wall, the body should be kept as close to the top as possible, since in combat operations it is important to offer as small a target as possible to the enemy. If an individual climbs a wall while carrying a rifle, both hands should be freed by slinging the rifle over the back. There are two methods commonly used for surmounting a wall of moderate height, but only one for dropping from it.The methods are as follows:

1 Run, Jump, and Vault. Approach the wall at a run, jump forward and upward at the wall and place one foot against it as high up as possible. Use the foot in contact with the wall to help push the body upward while grasping the top of the wall with the hands. Pull the body up with the arms, assisted by pressure of the foot against the wall and swing the legs over, propelling the body weight over the wall.

2 Hook and Swing. Approach the wall at a run and jump forward and upward. Hook one elbow over the wall, locking the arm in place by pulling up until the top of the wall is underneath the armpit. Grasp the top of the wall with the other hand. Draw the leg which is closer to the wall up as far toward the top as possible. Then swing the other leg over the top of the wall.The body is then carried over with a rolling motion. A variation of this leg action can be used by Marines who are unable to draw up the leg as described. While hanging with both legs fully extended, start a swinging motion with the legs together. When the legs have enough momentum, swing the outside leg over the top of the wall with a vigorous kick, then follow with the body.

3 **Dropping.** All drops from a height are executed in the same manner regardless of the method used to gain the top. One hand is placed against the far side of the wall while the other hand grasps the top. From this position, the body is rolled over the wall and vaulted away from it with the legs swinging clear. As the body passes over the wall and drops, it should at all times face the wall. This will keep the rifle and other equipment clear. Break the fall by retaining a grasp on the top of the wall as long as possible.

(c) Climbing Ladders and Cargo Nets. Rope ladders, stationary vertical ladders, and cargo nets employ the same general technique. The important element is to grasp the side supports firmly in the hands about shoulder height and place the feet on a rung which would cause the body to be fully extended. In movement upward, one hand is moved upward and a new grasp is secured and, at the same time, the opposite leg moves up a rung. As the knee straightens, the body is elevated. This process is repeated using the opposite arm and leg. Alternation continues in this manner until the climber reaches the objective.

(5) Traversing Horizontal Objects. The traversing of horizontal objects puts heavy stress on the arms and shoulder girdle area as the feet are usually suspended in the air with all of the body weight on the arms and shoulders.(See fig. 3-23.)

(a) Traversing Horizontal Ropes or Pipes. The hands grasp the horizontal support overhead with the palms facing. To propel the body forward, one hand is released and moved forward to secure a new grasp. At the same time, the opposite side of the body is swung forward (some people are able to "walk" in the air, keeping the body to the front and moving the legs in time with the arms as in walking on the ground). The other hand is then released and moved forward; this alternation is continued until the objective is reached.

Figure 3-23 is available in the illustration section at the end of the book.

(b) Traversing Horizontal Ladders. In this situation, the movement is the same as used in traversing a rope or pipe. The hands, however, are placed on the rung with the palms away from the face. Other than this difference the technique is the same.

(6) Crawling. Crawling in combat situations is an often used skill. Crawling may be high or low.(See fig. 3-24.)

Figure 3-24 is available in the illustration section at the end of the book.

(a) High Crawl.

In the high crawl, the Marine moves on hands and knees, moving one hand and the opposite knee and then continuing to move the hands in alternation with the opposite knee following the companion hand.

(b) Low Crawl. The Marine is in the prone position usually with the forearms and palms of the hand on the ground. He propels forward by bending the knee of one leg and pushing with the inside edge of the shoe. At the same time, the opposite arm moves forward and pulls to the rear. The body remains low and movement is continued by bending the opposite knee and pushing, and at the same time sliding the opposite arm forward and pulling. Alternation of hands and legs continues until the objective is reached.

(7) Throwing. Throwing may be executed from the kneeling or standing positions. The object to be thrown is held in the hand, and the throwing arm is bent at the elbow; the hand is then moved to the rear until the hand is behind the ear. The body is turned so that the lead foot and balance arm on the side toward the target are pointing at the target. (See fig. 3-25.) The balance arm is used to sight over and align the throwing hand and the target. When properly aligned, the elbow is moved rapidly forward until it is at a point just in front of the body where the arm is straightened and the wrist snapped. This whip motion propels the object to the target. Underhand throws secure momentum by the thrower bending his knees and swinging the throwing arm to the rear. As the knees are straightened, the arm is forcefully swung forward from the shoulder and the object released.

Figure 3-25 is available in the illustration section at the end of the book.

(8) Vaulting. Vaulting is employed to overcome low barriers or fences. (See fig. 3-26.)The object to be surmounted is approached at an angle. The hand on the side next to the obstacle is placed on the top of the obstacle and, with a straight arm, the body weight is pushed upward. At the same time, the leg on the side next to the obstacle is thrown upward and over the top followed by the other leg (side approach). In landing, the weight comes down on the leading leg first followed by regaining the balance on both legs. The free arm serves as a balance. A direct (front) approach can be used at which time both legs go over the object together.

(9) Man Carrying. There are three basic individual means of carrying personnel in combat situations and one of these methods may be used in carrying objects.

Figure 3-26 is available in the illustration section at the end of the book.

(a) Fireman's Carry."A" stands sideways in front of "B", "A" bends his knees and leans forward, placing one arm through "B's" crotch, grasps the wrist of "B's" arm, which is hanging over the shoulder, and then "A" runs forward.

(b) Saddle-Back Carry. With his back toward "B", "A" stands in front of "B". "B" mounts "A's" hips and clasps his arms in front of "A's" chest. "A" grasps "B's" thighs.

(c) Single-Shoulder Carry. "A" stands facing "B"."A" assumes a semi-squatting position. "B" leans forward until "B" lies across "A's" right shoulder. "A" clasps his arms around "B's" legs and straightens up, lifting "B" from the ground. "A" then runs forward. This method may also be used to carry heavy objects.

(10) Balancing. Balancing the body while walking or running on a narrow object when crossing obstacles is a skill which requires practice and confidence. Balance is required in negotiating a log placed across a stream, in crossing a narrow beam or rail, and in similar situations.(See fig. 3-27.) To perform this skill, place the feet on the object to be crossed, hold the arms to the sides at shoulder level, and fix the eyes on the object approximately 5 yards in front of the feet. Generally, it is not a good practice to look down at the feet. Walk the beam by placing first one foot and then the other in the center of the beam, thereby moving forward, using the arms to aid in maintaining balance.

Figure 3-27 is available in the illustration section at the end of the book.

(11) Falling. Injury can be avoided if Marines are taught to fall properly. They should know how to use the body momentum to their advantage during a fall rather than to try resisting that force. (See fig. 3-28.) If enough force is present, such as occurs during a fall while running or in jumping downward from a height, individuals can extend their hands to catch the weight. At the same time, duck the head and roll forward onto the feet. The key to falling without injury from the standing position is relaxation and rolling on the outside of the leg, hip, and buttocks to take the brunt of the fall.

Figure 3-28 is available in the illustration section at the end of the book.

3803. OBSTACLE COURSES

a. General. Obstacle courses are a valuable part of physical readiness training. The challenge presented by the obstacles assists in developing and testing the basic physical skills. In many combat situations, success will depend upon the Marine's ability to perform one or more of these skills, often while carrying field equipment and when fatigued. In this section, two types of obstacle courses will be discussed-Conditioning Obstacle Course and Confidence Obstacle Course.

b. Course Safety Precautions. Commanders and course instructors should take certain precautions to prevent injury on obstacle courses. A few of the precautions are:
(1) Inspect the course for faulty construction of obstacles, protruding nails, rotten logs, condition of the landing pits, and other hazards to safety.
(2) Conduct warm up exercises before the unit runs the course.
(3) Explain and demonstrate the correct techniques for negotiating all the obstacles before allowing the Marines to try them.
(4) Give Marines at least two weeks of conditioning exercises before scheduling the obstacle and/or confidence courses.
(5) Ensure that negotiation of the higher and more difficult obstacles are under the supervision of an instructor.
(6) Do not permit individuals who have neither practiced the basic skills nor run the conditioning obstacle course to participate in the confidence obstacle course.
(7) Weather conditions may cause footing or handhold surfaces to be slippery. If such is the case, postpone training on the course.

3804.CONDITIONING OBSTACLE COURSES

a. Description and Objective. The Conditioning Obstacle Course is commonly known as the Obstacle Course. This course consists of fairly low obstacles which are designed to be negotiated quickly. The obstacles serve to test various basic skills, and running the course is a test of the Marine's physical condition. After receiving instruction and an opportunity to practice the skills, Marines run the course against time.

b. Area and Equipment
(1) Complete standardization of obstacle courses should not be attempted since topographical conditions always vary. Commanders should use ingenuity in constructing a course, making good use of streams, hills, trees, rocks, and other natural obstacles. Since the course is eventually run at high speed, it should not be dangerous.

(2) The course should be wide enough for six or eight men to run simultaneously, encouraging competition. The lanes for the first several obstacles should be wider and the obstacles themselves easier than those that follow. This avoids congestion until the contestants scatter out over the course. The last two or three obstacles should not be too difficult and should not involve high climbing. This prevents injuries and falls resulting from fatigue.
(3) The total distance of the course should range from 300 to 450 yards and include from 15 to 25 obstacles. Normally the obstacles should be 20 to 30 yards apart and arranged so that those which exercise the same groups of muscles are separated.
(4) The obstacles should be substantially built. Peeled logs, 6 to 8 inches in diameter, are ideal for many of the obstacles. Sharp points and corners should be eliminated. Landing pits for jumps or vaults should be filled with sand or sawdust to prevent injuries.
(5) The course should be constructed and marked so that it is not possible to sidestep or detour obstacles. However, it is desirable to provide alternate obstacles of varying degrees of difficulty.
(6) The course should be in the shape of a horseshoe or figure eight so that the finish is close to the start and signs should be placed to indicate the course route.

c. Leadership. Before Marines run an obstacle course, they should be instructed in the proper technique of negotiating each obstacle. In each case this technique should be explained and demonstrated in detail, with emphasis on avoiding injury. Every individual should be given an opportunity to practice on each obstacle until he becomes reasonably proficient at negotiating it. Before the course is run against time, it is advisable for individuals to make several runs at a slower pace. During such practice or trial runs, the instructor should observe the performances and make appropriate corrections. Marines should never be permitted to run the course for time until they have practiced on all obstacles. The best method of timing the runners is to have the timer stand at the finish and call out the minutes and seconds as each individual finishes. If several watches are available, each wave may be timed separately. If only one watch is available, the different waves should be started at regular intervals, such as every 30 seconds. If an individual fails to negotiate an obstacle, a previously determined penalty should be exacted.

d. Types of Obstacles
(1) Jumping-Type Obstacles. These obstacles may be ditches which are cleared with one leap, trenches which the individuals can jump into, heights which require jumping downward, or hurdles which an individual can jump over. (See fig. 3-29.)

Figure 3-29 is available in the illustration section at the end of the book.

(2) Dodging-Type Obstacles. Obstacles of this type are usually mazes consisting of posts set in the ground at irregular intervals. (See fig. 3-30.) The intervals between posts should be rather narrow so that the Marines must pick their way carefully through and around them. Lane guides may be constructed to guide the Marines to dodge and change direction. Obstacles may be put into a maze pattern to cause the Marines to change direction.

Figure 3-30 is available in the illustration section at the end of the book.

(3) Vertical Climbing and Surmounting Type Obstacles. These obstacles may be climbing ropes, either plain or knotted and 1-5 inches in diameter; cargo nets or walls 7 or 8 feet high; or vertical poles 6 to 8 inches in diameter and 38 feet high.(Seefig. 3-31.)

Figure 3-31 is available in the illustration section at the end of the book.

(4) Horizontal Traversing-Type Obstacles. Horizontal obstacles may be pipes, beams, ladders, or ropes. (See fig. 3-32.)

(5) Crawling Type Obstacles. Obstacles which require crawling may be constructed of large pipe sections, low rails, and wire. (See fig. 3-33.)

Figure 3-32 is available in the illustration section at the end of the book.

(6) Vaulting-Type Obstacles. Obstacles of 3 to 3.5 feet in height such as low walls or fences may be used as a vaulting obstacle.(See fig. 3-34.)

(7)Balancing-Type Obstacles. Beams, logs, and planks may be used as balancing-type obstacles.(See fig. 3-35.)These items may be used to span water obstacles and dry ditches, or raised off the ground somewhat to simulate natural depressions.

Figure 3-33 is available in the illustration section at the end of the book.

3805.CONFIDENCE OBSTACLE COURSES

a. Description and Objective. The Confidence Obstacle Course is commonly known as the Confidence Course. This course is composed of higher and more difficult obstacles than those used in the conditioning course. The confidence obstacle course is designed to give Marines confidence in their mental and physical capacities and to cultivate their spirit of daring. They are encouraged but not compelled to negotiate this course and the course is not run against time. The negotiation of a confidence course, however, is strenuous enough to be an excellent physical conditioner. Marines should NEVER attempt to take the obstacles at high speed and should not compete for speed. The obstacles vary from fairly easy to extremely difficult. Some are of considerable height to accustom Marines to climbing such heights without fear. Considerable emphasis is placed on obstacles that train and test an individual's balance.

Figure 3-34 is available in the illustration section at the end of the book.

b. Area and Equipment

(1) The confidence course accommodates four platoons, one platoon at each group of six obstacles. The course should be made up of about 24 obstacles, numbered and marked as follows:! to 6, white numbers on red background; 7 to 12, black numbers on a white background;

13 to 18, white numbers on a blue background; and 19 to 24, white numbers on a black background.

(2)A few simple pieces of equipment should be provided for individuals who do not have the strength, courage, or ability to negotiate the obstacles.

Figure 3-35 is available in the illustration section at the end of the book.

c. Formation. The obstacles should be divided into groups of six, and each group is designated by a different color. Each platoon starts at a different color. Individuals are separated into groups of 8 to 12 at each obstacle. At the starting signal from the company commander, they proceed numerically through. Anyone may skip an obstacle who is afraid to try. Individuals proceed from obstacle to obstacle until time is called, then assemble as ordered.

d. Leadership. If the Marines are new to the confidence course, an instructor will demonstrate or will give a brief orientation at each obstacle, including an explanation and demonstration of a method of negotiating it. Marines are encouraged to try the various obstacles, but they are not compelled to do so. No compulsion is to be used. The manner of negotiating any obstacle is left to the discretion of the Marine. However, the instructor assists anyone who experiences difficulty. Instructors must supervise closely at all times to prevent injuries, as some of the obstacles are quite high. Also, some of the obstacles should not be used when slippery or wet. The example of instructors and especially selected demonstrators will serve to inspire the individuals to greater effort.

e. Negotiating the Obstacles. Although personnel need not conform to any one method of negotiating the obstacles, there should be some uniformity in the approach to them. A general method of negotiating the obstacles is indicated below.

(1) Red Group This group contains the first six obstacles. (See fig. 3-36.)

(a) The Belly Buster Individuals may vault, jump, or climb over. Warn them that the log is not stationary.

(b) Reverse Climb. Climb the reverse incline and go down the other side to the ground.

(c) The Weaver. Move from one end of the obstacle to the other by weaving the body under one bar and over the next.

(d) Hip-Hip. Step over each bar, either alternating legs or using same lead leg each time.

(e) Balancing Logs. Step up on log, retaining the balance, walk or run along log.

(f) Island Hopper. Jump from one log to another until the obstacle is negotiated.

(2)White Group. This group is composed of the second six obstacles. (See fig. 3-37.)

(a) Tough Nut. Step over each X in the lane.

(b) Slide for Life. Climb the tower, grasp the rope firmly, and swing the legs upward. Hold the rope with the legs to distribute the weight between them and the arms. Braking the slide with the feet and legs, proceed down the rope. Warn Marines that there is danger of getting rope burns on their hands. When the rope is slippery or wet, this can be a dangerous obstacle.

(c) Low Belly Over. .Mount the low log and jump onto the high log, both arms grasping over the top of the log, the stomach area in contact with it. Swing the legs over the log and lower the body to the ground.

(d) Belly Crawl Move forward under the wire, belly down, to the end of the obstacle.

(e) The Dirty Name Mount the low log and jump to or reach the higher logs in succession, then jump or drop to the ground. Warn the Marines about the height of the final log.

(f) The Tarzan Mount the lower log and walk the length of it and each successive, higher log until reaching the horizontal ladder. Grasp two rungs of the ladder and swing the body into the air. Negotiate the length of the ladder by releasing one hand at a time and swing forward, grasping a more distant rung.

(3) Blue Group. This group is formed by the third group of six obstacles. (See fig. 3-38.)

(a) High Step over. Step over each log, alternating the lead foot or using the same lead foot.

(b) Swinger. Climb onto the swinging log and over to the ground on the opposite side.

(c) Low Wire. Move under the wire on the back, using the hands to raise the wire to clear the body.

(d) Swing, Stop and Jump. Gain momentum with a short run, grasp the rope, and swing the body forward to the top of the wall. Release the rope while standing on the wall and jump to the ground.

(e) Six Vaults. Vault over the logs, using one or both hands.

(f) Easy Balancer. Walk up one inclined log and down the one on the other side to the ground.

(4) Black Group. The last group is formed by the final six obstacles. (See fig. 3- 39.)

(a) Incline Wall. Approach the underside of the wall, jump up and grasp the top and pull the body up and over. Slide or jump down the incline to the ground.

(b) Skyscraper. Jump or climb to the first floor, climb up the corner posts or assist each other to any desired floor. Descend to the ground in any desired manner.

(c) Jump and Land. Climb up the ladder to the platform and jump to the ground.

(d) Confidence Climb. Climb the inclined ladder to the vertical ladder. Go to the top of the vertical ladder, then down the other side to the ground.

(e) Belly Robber. Step on the lower log and assume the prone position on the horizontal logs. Crawl over the logs to the opposite end of the obstacle.

(f) The Tough One Climb the rope or pole on the higher end of the obstacle, then go down the ladder and across the log platform. Climb over or between the logs at the end and go down the rope or pole to the ground. Vault over the final log.

Figure 3-36 is available in the illustration section at the end of the book.

Figure 3-37 is available in the illustration section at the end of the book.

Figure 3-38 is available in the illustration section at the end of the book.

Figure 3-39 is available in the illustration section at the end of the book.

Section IX Individual Exercise Programs

3901.GENERAL

Marines at times will **be** stationed on independent duty and consequently will be responsible for their own physical fitness program. This section will assist them in understanding the need for exercise and will aid in the planning and execution of an individual exercise program. Exercise activities included are the Bench Conditioner, 6-12 Plan, Weight Training, and Isometric Contraction. Keeping physically fit is a problem that faces every Marine. Even though we are frequently engaged in training that requires some physical effort, in many cases, it is not enough to prepare us to meet the intense physical demands of combat. Attaining a satisfactory level of physical readiness is not an insurmountable objective. Available time appears to be the most difficult obstacle to the development of physical readiness. In most cases, regular physical training programs are centralized, requiring the individual to temporarily leave the work area. The problems involved in setting an hour aside two or three times each are numerous. However, most of us can devote a few minutes each day to physical fitness with little, if any, impact on our daily work schedule, especially if it does not require us to leave our work area.

a. Type of Program. There are many good physical fitness programs available to the individual or group. Regardless of the type or duration, to be effective, the program must contain exercises that are strenuous and are challenging to the individual. Space will not permit the inclusion of all available means of individual exercise. The programs selected for this chapter have met the requirement of minimum space and minimum time.

b. Need to Augment Program. These programs are quite strenuous and will develop a satisfactory level of physical readiness. However, if the Marine desires additional development of endurance, it is recommended that he supplement these programs with a 15- to 30-minute period of wind sprints and double timing on an alternating daily basis.

c. Progressive Training. If Marines are performing duties which require little or no physical activity, they must plan a physical conditioning program that assures a moderate beginning, moderate but steady progression, and sufficient warmup before starting the vigorous exercise. To avoid organic of bodily harm, a Marine activity without adequate warmup. He should conduct conditioning programs on a daily basis over an extended period of time, never on an unduly accelerated or crash basis.

3902.THE BENCH CONDITIONER

a. Description and Objective. The bench conditioning program uses a modified bench to employ both isotonic (moving) and isometric (stationary) exercises are as the nucleus of the program. The exercises are designed to develop strength and endurance in all the major muscle groups of the body. The principles of progression, overload, and balance are employed when the exercises are performed properly.

b. Area and Equipment. The conditioning apparatus can be constructed in any unit motor pool with welding equipment found in most salvage yards.(See figs. 3-40, 3-41.)Additionally, there are available any number of commercially produced apparatus available. The important thing is that in utilizing this equipment the following exercise routine be adhered to.

c. Starting Level and Progression.. The program consists of two tables, each with 10 exercises. The Marine controls progression by required repetitions or, in some cases, by application of maximum effort. Each table can be completed within 15 minutes.

d. Starting Position. To start the program, the Marine begins with the Table 1 and executes each exercise for the required number of repetitions as indicated. The Marine controls the starting level and progression. When the Marine executes the maximum repetitions for Table I within a 15-minute exercise period, he progresses to Table II. To maintain this level of development, the Marine should also execute the maximum repetitions for Table II within a 15-minute exercise period. The Marine should keep substitution of exercises to a minimum. However, it he completes a full 15 minutes of strenuous exercise and exercises all muscles, then there should be no appreciable difference in the overall development.

e. Bench Conditioning, Table I

(1) Exercise 1: Side-Straddle Hop. This is a two-count warmup exercise done at moderate cadence. The starting position is the position of attention. On count ONE jump slightly into the air, swinging the arms out to the sides and up to a vertical position, hands touching.(See fig. 3-40, A.)At the same time, spread the feet wider than shoulder-width apart. On count TWO, using a slight flexing of the knees and ankles, jump slightly into the air and return to the starting position by swinging the arms back down to the sides. Twenty repetitions of this exercise is the standard dosage throughout the program.

(2) Exercise 2: Hand Walk. Remove the lower horizontal bar. Adjust the upper horizontal bar so that it is high enough to permit a "dead hanging" position, with the feet off the ground.(See fig. 3-40, B.j From the "dead hanging" position, release one hand and drop the arm to the side of the body. Then raise that arm and regrasp the horizontal bar. Release the bar with the other hand and drop that arm to the side. Repeat this as many times as possible.

(3) Exercise 3: Situps. Lie down with the fingers interlocked and placed behind the head. Hook the toes under the foot braces. Raise the trunk and upper body to an upright sitting position, twisting it to the left and then forward and downward until the right elbow touches the left knee.(See fig. 3-40, C.)Lower the body to the starting position .Sit up again but twist the body to the opposite direction as before, touching the left elbow to the right knee. Again lower the body to the starting position. The starting dosage is 20 situps. The Marine should continue the progression until he has attained 40 situps.

Figure 3-40 is available in the illustration section at the end of the book.

(4) Exercise 4: Double Step-upStarting at one end of the bench, step up onto the bench, and walk across it. Step down from the other end; turn around and repeat the process to return to the starting point.(See fig. 3-40, D.)Each return to the starting point constitutes a repetition. The starting dosage is 20 repetitions. Maximum dosage is 35 repetitions. This exercise should be done at a rapid cadence.

(5) Exercise 5: Isometric Bar Lift. Adjust the lower bar so that it is slightly higher than the beltline. Placing the feet on the footplates at the base of the frame, grasp the lower bar so that the hands are spread shoulder-width apart. Assume a crouched position and lift with maximum effort using the arms, back and legs.(See fig. 3-40, E.)Starting dosage is 4 repetitions of a stress time of 5 seconds followed by a 5-second rest prior to the next repetition. The Marine obtains progression by lengthening stress periods to 6 and later 7 seconds. Do not increase the number of repetitions.

(6) Exercise 6: Knee Lift. Adjust the upper bar to the same height used in Exercise 2.Adjust the lower bar so that is stops rearward movement of the hips when the "dead hanging" position is assumed.(See fig. 3-40, F.) Keeping the arms extended, flex the legs and raise the knees as high as possible. Hold this position for 5 seconds, then return to the starting position. After 2 seconds in the starting position, raise the knees again. Each return to the starting position constitutes one repetition. The dosage is five repetitions. Progression is the same as in Exercise 5.

(7) Exercise 7: Isometric Pull. Adjust the lower horizontal bar to a position where it is slightly higher than the beltline. Grasp the handles and pull outward.(See fig. 3-40, G.)Apply maximum effort and hold for approximately 5 seconds. Relax for 5 seconds between repetitions; perform four repetitions. Moving the body closer to or farther away from the bar will change the stress from the upper arms to the forearms. Progression is the same as in Exercise 5.

(8) Exercise 8: Isometric Compression. Maintain the position as in exercise 7.(See fig. 3-40, H.)Grasping the handles in the same manner, press in **with** maximum effort and hold for approximately 5 seconds. Relax for 5 seconds between repetitions. Progression is the same as in Exercise 5.

(9) Exercise 9: Isometric Press. Remove the lower horizontal bar. Adjust the upper horizontal bar until it is about 6 inches lower than the extended arms can reach. Stepping on the footplates at the bottom of the frame, grasp the bar with both hands and push up.(See fig. 3-40,1.)Keep both the legs and arms slightly flexed and the back straight. Apply maximum effort for 5 seconds then relax for 5 seconds. Complete four repetitions. Progression is the same as in Exercise 5.

(10) Exercise 10: Pushups. Grasping the foot braces with both hands, assume the front leaning rest position.(See fig. 3-40, J.)Keeping the back and legs straight, lower the body until the chest is lower than the **hands,** then return to the starting position. The Marine should complete the maximum possible number of repetitions.

f. Bench Conditioning, Table II. There is no limit on the maximum number of repetitions attainable in Exercises 3, 4, and 6 of Table II. The only limit imposed is that the entire program of 10 exercises outlinted in either table should not exceed 15 minutes.

(1) Exercise 1: Side-Straddle Hop. This is a two-count warm up exercise done at a moderate cadence. The starting position is the position of attention. On count ONE, jump slightly into the air, swinging the arms out to the sides and up to a vertical position, hands touching.(See fig. 3-41, A.)At the same time, spread the feet wider than shoulder-width apart. On count TWO, using a slight flexing of the knees and ankles, jump slightly into the air and return to the starting position by swinging the arms back down to the sides. Twenty repetitions of this exercise is the standard dosage throughout the program.

(2) Exercise 2: Pull-up. Adjust the horizontal bar so that is is high enough to permit a "dead hanging" position with the feet off the ground. Grasp the bar with both hands, palms facing forward. By flexing the arms, raise the body to a position where the chin is higher than the bar.(See fig. 3-41, B.)Then lower the body to the "dead hanging" position. Repeat as many times as possible.

(3) Exercise 3: Bench situps. Sit on the bench and hook the feet under the foot braces. With the fingers interlocked behind the head, lean back until the head touches the floor.(See fig. 3-41, C.)Return to the starting position. The starting dosage is 15 situps.

(4) Exercise 4: Step-Up. Face the bench and step up on it with one foot, bringing the trailing foot up next to the leading foot. Step back down again, leading with the same foot used first in stepping up.(See fig. 3-41, D.)Perlorm half of the total repetitions, then change the sequence of moving the feet to use the other leg in stepping up, and repeat the same amount of exercise. The starting level is a total of 40 step-up. **This exercise should be done at a rapid cadence.**

(5) Exercise 5: Isometric Bar Lift\djust the lower bar so that it is slightly higher than the beltline. Placing the feet on the footplates at the base of the frame, grasp the lower bar so that the hands are spread shoulder-width apart. Assume a crouched position and lift with maximum effort using the arms, back, and legs.(See fig. 3-41, E.)Starting dosage is four repetitions of a stress time of 8 seconds followed by a 5-second rest prior to the next repetition. Progression is obtained by lengthening stress periods to 10 seconds. Do not increase the number of repetitions.

Figure 3-41 is available in the illustration section at the end of the book.

(6)Exercise 6: Leg Lift. Adjust the bars and assume the starting position as shown in figure 3-41, F. Keeping arms and legs extended, raise the legs to a horizontal position and hold in that position for 2 seconds. Then lower the legs slowly to the starting position. Five repetitions is the starting level.

(7) Exercise 7: Isometric Pull. Adjust the lower horizontal bar so that it is slightly higher than the beltline. Grasp the handles and pull outward.(See fig. 3-41, G.)Apply maximum effort and hold for approximately 8 seconds. Relax for 5 seconds between repetitions; perform four repetitions. Moving the body closer to or farther away from the bar will change the stress from the upper arms to the forearms. The Marine obtains progression by lengthening the stress period to 10 seconds.

(8) Exercise 8: Isometric Compression. (See fig. 3-41, H.)Maintain the position as in exercise 7.Grasping the handles in the same manner, press in with maximum effort and hold for approximately 8 seconds. Relax for 5 seconds between repetitions; perform four repetitions. The Marine obtains progression by lengthening the stress period to 10 seconds.

(9) Exercise 9: Isometric Pres. Remove the lower horizontal bar. Adjust the upper horizontal bar until it is about 6 inches lower than the extended arms can reach. Stepping on the footplates at the bottom of the frame, grasp the bar with both hands and push up.(See fig. 3-41,1.)Keep both the legs and arms slightly flexed and the back straight. Apply maximum effort for 8 seconds, then relax for 5 seconds. Complete four repetitions. The Marine obtains progression by lengthening the stress period to 10 seconds.

(10) Exercise 10: Inclined Pushup. Assume the front leaning rest position with the feet on the bench.(See fig. 3-41, J.)Keeping the back and legs straight, lower the body until the nose touches the ground. By extending the arms, raise the body to the starting position. Repeat as many times as possible.

3903. THE 6-12 PLAN

a. Description and objective. The 6-12 plan of physical fitness has been developed to assist in regulating quantity and progression and to provide a convenient set of exercises. This is a basic program and will take 18 weeks to complete if you follow the moderate progression as to the time prescribed for each level of achievement. The time can be shortened as explained below. This plan consists of six basic exercises a day which can be completed in 12 minutes. There are six tables of six exercises each, thus allowing you to progress from table to table. The plan is progressive, fits any age group, contains balance and variety, and applies the principle of overload in a safe and gradual manner. Begin at Table I, Progression Guide, with the number of repetitions as indicated by age.

b. Formation. If just starting an exercise program, do not rush through the first table. Remember, individuals should remain at each level for about a week before moving upward. The time allotment stated for each exercise at the bottom of the tables is a guide; some people may take more and some less time on the individual exercises. At the end of a one-week period, when the individual can comfortable perform the six exercises in 12 minutes, move on to the next level. To a certain degree, the individual must be the judge of his ability to progress form level to level and table to table. If starting with a certain degree of fitness, some of the beginning tables may present little challenge.

c. Starting Level **and Progression.** There are three levels of achievement for each age group, indicated as A, B, and C. Start at the C level for the appropriate age group. At the end of a one-week period, or when the individual can do all exercises at that level within 12 minutes, progress to the B level. At the end of the second week, or when the individual can accomplish that level within 12 minutes, progress to the A level. At the conclusion of the third week or when the individual can achieve the A level within the time limitation, move on to table II.(See Tables I through VI, Progression Guide.)

d. Maintenance Level. Attempt to work through all six tables. If this proves to be too difficult, then maintain exercise at the—

(1) A-level on Table IV, Progression Guide it in the 45 to 49, 50 to 59, or over 60 age group.

(2) A-level on Table V, Progression Guide if in the 17 to 29, 30 to 39, or 40 to 44 age group.

e. Precautions. To achieve the maximum benefit, perform each exercise exactly as specified. Read the descriptions and study the illustrations. Do not slight the movements. Use a sensible approach and follow these four points as they apply before starting or during your exercise program.

(1) If you have the slightest doubt about your ability to participate in this exercise program, consult a physician.

(2) Stop immediately if you notice unusual breathlessness or chest pain while taking part in these exercises. If these conditions persist, consult a physician.

The following Tables of Progression Guides and Exercises are the progression of the 6-12 Plan program.

Table I. Progression Guide is available in the illustration section at the end of the book.

Table I. Exercises.is available in the illustration section at the end of the book.

Table II. Progression Guide.is available in the illustration section at the end of the book.

Table II. Exercises.is available in the illustration section at the end of the book.

Table III. Progression Guide.is available in the illustration section at the end of the book.

Table III. Exercises.is available in the illustration section at the end of the book.

Table IV. Progression Guide.is available in the illustration section at the end of the book.

Table IV. Exercises.is available in the illustration section at the end of the book.

Table V. Progression Guide.is available in the illustration section at the end of the book.

Table V. Exercises.is available in the illustration section at the end of the book.

Table VI. Progression Guide.is available in the illustration section at the end of the book.

Table VI. Exercises.is available in the illustration section at the end of the book.

(3) Unless you have exercised regularly and know yourself to be in good physical condition, start at table I with the C level appropriate to your age.

(4) If you are out of shape, admit that fact to yourself. Hide your pride; after all, you are in the privacy of your own quarters. Set your goal for the longer, steadier pull toward fitness. Resist the urge to pass over the lower numbered tables to find a table that will test your fitness. You are not trying to test, but rather to develop.

3904.WEIGHT TRAINING

Weight or barbell training should not be confused with the more common types of weight lifting used as a competitive sport. Weight lifting is designed to develop specific muscle groups so that the individual is capable of lifting a large amount during a single lift. In contrast, weight training is the systematic development of all the major muscle groups by the use of calisthenics reinforced with weight to provide resistance.

3905.WEIGHT LIFTING

a. Description and Objective. The weight lifting program is progressive and applies the principle of overload in a safe, gradual manner. The exercises of the table can be completed in 15 minutes. The objective of the exercises is to develop strength and muscular endurance, and muscle tone of the five major muscle groups:legs, arms, back, trunk, and shoulder girdle.(See fig. 3-42.)

b. Warmup. A warm up exercise is important to prepare the body for the more vigorous exercises that are to follow. Ten repetitions of the high jumper exercise are excellent for a warm up period. Also, an exercise which will require fast body movement is needed.To provide such exercise, 3 to 5 minutes of rope skipping is recommended to increase the individual's development potential.

c. Formation. Marines should take care in the completion of these weight lifting exercises. They should ensure that their back is straight during the lifting phase of all exercises. When exercises require assuming the standing position with the weight, they should always grasp the weight while in a squatting position and then rise to a standing position.

d. Progression. Each exercise has a starting number of repetitions and specified pounds of weight. After each 4th or 5th day of exercise, the Marine should increase the repetitions by one until he has reached the maximum of 10 repetitions. At this time, the Marine should increase the weight by 5 pounds. Then he should repeat the process again with the initial number or repetitions.

e. Weight Training

(1) Exercise 1:Squat(See fig. 3-42, A.)Starting level—six repetitions, 50 pounds (commonly called the flatfoot deep knee bend). Place the bar upon the shoulders. Stand with feet about 18 inches apart. Keeping the feet flat, lower the body into the low squat position. Stand erect and repeat. Exhale while lowering into the squat position and inhale while standing. This constitutes one repetition.

(2) Exercise 2:Waist Bender(.See fig. 3-42, B.)Starting level-6 repetitions, 40 pounds. Assume the standing position with the bar across the shoulders and the feet a shoulder-width apart .Bend forward at the waist until the upper body is parallel to the ground; return to the starting position. Each return to the upright position constitutes one repetition.

(3) Exercise 3:Curl (See fig. 3-42, C.)Starting level—6 repetitions, 40 pounds. Grasp the barbell with the palms facing to the rear and assume the standing position with the feet a should-width apart. With the barbell held in front of the hips, flex the elbows and lift the weight until the bar touches the upper chest. Lower the barbell back to the hip level position. Inhale deeply with the upward movement and exhale on the downward movement. Each time that the bar touches the chest will constitute one repetition.

(4) Exercise 4:Side Bender(.See fig. 3-42, D.)Starting level—6 repetitions per side, 40 pounds. Assume the standing position, with the bar across the shoulders, with feet a shoulder-width apart. Bend to the left as far as possible and return to the starting position. Repeat six times and then execute the same procedure to the right for six repetitions.

(5) Exercise 5:Standing Press(See fig. 3-42, E.)Starting level—6 repetitions, 45 pounds. Grasp the bar with the palms facing forward and assume the starting position. Curl the weight to the upper chest position. Inhale deeply and press the bar upward to an overhead position. Exhale while lowering the bar to the chest position. Each time that the bar is pressed upward constitutes one repetition.

(6) Exercise 6:Upward Row. .see fig. 3-42, F.)Starting level—6 repetitions, 40 pounds. Grasp the bar, hands close together, palms to the rear, and assume the standing position. Starting with the bar held in front of the hips, flexing the elbows and the shoulder girdle muscles, lift the bar straight up to an overhead position. Inhale deeply while lifting the bar. Exhale while lowering the bar to the hip position. Each time that the bar returns to the hips constitutes one repetition.

(7) Exercise 7:Shoulder Curl. See fig. 3-42, G.)Starting level—6 repetitions, 25 pounds. Grasp the bar, palms down, and assume the standing position. Keeping the elbows locked, curl the bar, pivoting the arms at the shoulders until the bar is in an overhead position and as far to the rear as possible. Return the bar in the same manner to the hip position. Each time that the bar returns to the hip position constitutes one repetition.

Figure 3-42 is available in the illustration section at the end of the book.

3906.ISOMETRIC EXERCISING

a. Description and Objective. Isometric exercising is the application of maximum effort during an exercise period. The isometric principle is to apply force gradually over a 5- to 10- second period until the maximum effort is applied. Relaxation follows for approximately 5 seconds and then force is applied again. The Marine continues this process at the prescribed level for each exercise for a period of 15 minutes or less. The objective of isometric exercises is to create muscle growth. Isometric exercises are the fastest means of creating muscle growth. They are founded on the fact that a muscle will grow only so fast regardless of the type or duration of the activity. The principle of overload—that the muscles develop commensurate with demand—reinforces that fact. However, isometrics will not develop cardiovascular or muscular endurance. Consequently, Marines who choose to use isometrics in their exercise routine must also include running or some other type of aerobic activity.

b. Area and Equipment isometric exercises may be designed to be performed with or without equipment.

c. Use With Other Programs. The isometric exercises presented in this section will provide a variety from which to choose. In addition to the exercises contained in this section, the application of isometric force is used in some of the exercises in section II.

d.Door Frame Exercises. The following exercises are designed for use with a standard door frame found in all offices or barracks.(See fig. 3-43.)

(1) Exercise l:Arm Press. Stand in the doorway with the legs straight, knees locked. Using the arm muscles, press hard upward against the top of the door frame. Repeat for three repetitions applying gradual effort to maximum contraction.

(2) Exercise 2:Leg Press. Stand in the doorway with the hands on the top of the door frame, elbows locked .With knees bent, press hard with the leg muscles .Repeat for three repetitions beginning with a gradual effort and increasing to maximum contraction. A low platform may be necessary to reach the top of the door frame and still maintain a bent knee position.

(3) Exercise 3:Side Press. Extend both arms to the side of the doorway. Palms are shoulder high, facing outward. With both arms, press hard against the sides of the door frame. Repeat for three repetitions. Begin gradually and increase to maximum contraction.

(4) Exercise 4:Lateral Raise. Extend both arms to the sides of the doorway, arms down, palms facing inward. With the back of the hands, press hard against the sides of the door frame .Repeat for three repetitions. Begin with a gradual effort and increase to maximum contraction.

Figure 3-43 is available in the illustration section at the end of the book.

(5) Exercise 5:Neck Press?lace the forehead against the door frame, hands clasped behind the back Using the neck muscles, press hard against the door frame. Repeat for three repetitions, then reverse position so that the back of the head is resting on the door frame. Again do three repetitions .Begin gradually with both exercises and increase to maximum contraction.

(6) Exercise 6:Door Pull Stand facing the edge of the open door and grasp the doorknobs. Pull outward with both arms (if doorknobs are not available grasp the edge of the door).Applying outward pressure, move the body toward and away from the door .Repeat the three repetitions. Begin with gradual effort and increase to maximum contraction.

Chapter 4
COMBAT WATER SURVIVAL
Section I. Marine Corps Water Survival Program

4101.MILITARY SWIMMING

Military swimming emphasizes strokes that result in staying power rather than speed. Marines need to swim easily, quietly, and with adequate vision. Strokes should be used that allow the Marine to carry basic combat equipment and to tow or push a wounded buddy while keeping the face out of the water to allow breathing. The best strokes to use are the side stroke and the breast stroke. Marines must be proficient in floating, drown-proofing, and in using their basic combat equipment to make flotation devices. They must be drilled on using the Kapok and Mae West buoyancy compensators and on emergency egress from helicopters, landing craft, and amphibious assault vehicles.

4102.PROGRAM DEVELOPMENT

Commanders of the Landing Force Training Commands (Atlantic/Pacific) are tasked with developing and maintaining the individual water survival and swimming training program. They are responsible for training all Marine Corps water safety and survival instructors and integrating this training in to their combat readiness training program. All Marines are deployable and as soldiers of the sea must constantly practice the skills necessary to achieve mission accomplishment. Commanders must be imaginative in program development, integrating basic skills development which enhances the total mission accomplishment. An example is teaching waterproofing of the Marine's backpack so that it can be used as a flotation device in streams or rivers. Once learned, this skill would be practiced first in a pool and then in a stream or a river. Routine inspections during deployments would ensure that backpacks could be used as flotation devices.

4103.QUAL STDS AND TRAINING GUIDELINES

The inherent nature of Marine Corps operations and training requires that Marines achieve an ability to survive in water. Water survival and swimming training is designed to reduce fear of water, instill self-confidence, and develop a Marine's ability to survive in water. As a minimum, every Marine should be qualified as swimmer, third-class (S3).This qualification should be met during recruit training or Officer Candidate School. Once qualified, a Marine need not be requalified unless requalification at a specific level (S3 or higher) is required by unit mission, military occupational specialty or duty assignment, or for other reasons determined by the commander. The following classifications and standards apply to Marine Corps water survival and swimming training. Abbreviations will be used for service, medical, and training record entries citing MCO 1510.29A, Individual Water Survival and Swimming Training, as authority.

a. Qualification standards and test procedures are described in paragraph 4104.
b. Swimmer classification and abbreviations are:

(1) Unqualified - UQ.

(2) Swimmer, third-class (S3).Minimum aquatic skill level for all Marines.

(3) Swimmer, second-class (S2)-

(a) Minimum standard for all naval aviators, naval flight officers, and aircrewmen with written waiver from the commanding officer.

(b) Requirements for military occupational specialty 1803 (assault amphibian vehicle officer), 1833 (assault amphibian vehicle crewman), 1302 (engineer officer), 1371 (combat engineer), 1381 (shore party specialist), and 1379 (engineer operations chief).

(4) Swimmer, first-class (S1)-

(a) Naval aviator, naval flight officer, and aircrewman qualification requirement.

(b) Requirements for military occupational specialty 0321 (reconnaissance man), 8652 (reconnaissance man, parachute jump qualified), 8653 (reconnaissance man, self-contained underwater breathing apparatus qualified), 8654 (reconnaissance man, parachute/self-contained underwater breathing apparatus qualified), 9952 (self-contained underwater breathing apparatus Marine (officer/enlisted)), and 9953 (parachutist/self-contained underwater breathing apparatus Marine (officer/enlisted)).

(5) Water Survival Qualified (WSQ).Ultimate water survival/swimming goal for Marines.

c.Instructor and assistant instructor classifications are-

(1) **American Red Cross Advanced Lifesaver.** Individuals who are currently certified as American Red Cross advanced lifesavers may assist qualified instructors during Marine Corps water survival/swimming training and testing. Certification is valid for 3 years.

(2) American Red Cross Water Safety Instruction. Individuals who are currently certified as American Red Cross water safety instructors may conduct swimming training and qualify individuals as 3d, 2d, and 1st class swimmers. Certification is valid for 2 years.

(3) United States Marine Corps Water Safety and Survival Instruction. Those personnel certified as water safety and survival instructors by one of the landing force training commands may conduct Marine Corps water survival and swimming training and American Red Cross swimming and water safety training. These instructors may also participate as assistant instructors during the conduct of landing force training command-sponsored water safety and survival instructor courses. Certification is valid for2 years.

(4) American Red Cross Water Safety Instructor Traine. Individuals qualified who have been certified as instructor trainers by the American Red Cross may train American Red Cross water safety and basic swimming instructors. Certification is valid for 2 years.

(5) United States Marine Corps Water Safety and Survival Instructor Traine. Individuals qualified who are certified as instructor trainers through special training courses conducted jointly by the American Red Cross and the Marine Corps, under the direction of the Commandant of the Marine Corps, may train water safety and survival instructors and American Red Cross water safety and basic swimming instructors. Certification is valid for 2 years.

d. Training Guidelines Instructors engaged in training Marines as swimmers will use the current edition of the American Red Cross swimming and water safety courses instructors' manuals as the authoritative textbooks for swimming procedures, strokes, breaks, and holds.

4104.WATER SURV/SWIM QUAL STD AND TEST PROCS

a. The following qualification standards and test procedures are applicable to Marine Corps water survival and swimming training. The utility uniform without cover, boots/oxfords or socks will be worn for all swimming tests through swimmer, first class. The utility uniform with boots/oxfords will be worn as prescribed for the water survival qualification.

(1) Swimmer, Third-Class (S3) To qualify as a swimmer, third-class, a Marine must enter the water feet first from a minimum height of 5 feet and remain afloat for 5 minutes. During this time the Marine must swim 50 yards using any stroke or combination of strokes. This test should be taken by all Marines as early as possible in their initial training. Those unable to pass this test will be classified as UQ and should be given instruction in fundamental swimming skills. Those who barely meet requirements are swimmers who need help and should receive additional instruction in fundamental swimming skills, whenever possible.

(2) Swimmer, Second-Class (S2.) To qualify, a Marine must enter the water feet first from a minimum height of 10 feet and remain afloat for 10 minutes. The Marine must swim 100 yards and use the three basic survival swimming strokes (side stroke, backstroke, and breast stroke) for a minimum distance of 25 yards each.

(3) Swimmer, First-Class (S1)A prerequisite to qualification as swimmer, first-class, is successful completion of the test for swimmer, second-class. To qualify as a swimmer, first-class, a Marine must be able to do each of the following:

(a) Approach a person of approximately the same size while in the water, demonstrate a release, get that person in a carry position, and tow the "distressed swimmer" 25 yards.

(b) Enter water feet first and immediately swim underwater for 25 yards. Swimmer is to break the surface twice for breathing during this distance at intervals of approximately 25 feet.

(c) Remove trousers in water, inflate for support, and remain motionless for a minimum of 1 minute.

(d) Swim 220 yards using any survival swimming stroke or any combination of survival strokes.

(4) Water Survival Qualification (WSQ)A prerequisite to water survival qualification is successful completion of the test for swimmer, first-class. To be water survival-qualified, a Mannemust:

(a) Enter water by jumping from a height of 10 feet.

(b) Stay afloat in full utility uniform (less cover) for 1 hour. Boots/oxfords will be removed after 5 minutes in the water but will be retained by the swimmer.

(c) Traverse 75 yards in deep water in full utility uniform (less cover) with boots/oxfords.

b. All water survival and swimming training will include instruction in the following:

(1) Employment of safety lines or other comparable expedients to assist in swift stream/river crossings.

(2) Adverse physiological effects caused by cold water (hypothermia) and the precautionary measures to be taken prior to exposure to such an environment.

(3) Employment of a standard and expedient flotation devices in a water survival situation.

(4) Removal of individual field equipment after unscheduled water entry. Entry will be made from a minimum height of 10 feet with a complete set of field equipment, properly packed as an expedient flotation device and waterproofed. The ALICE pack will then be removed, the rifle laid across the top of the pack, the helmet remaining or put back on the individual's head, and the individual will traverse 25 yards to the side of the pool or water's edge.

c. The additional water survival qualification (flight physiology training) and requalification required of personnel assigned to flight status will be conducted in accordance with OPNAVINST 3710.7_,Promulgation of NATOPS General Flight and Operating Instructions. Swimming qualification (first-class, SI) is a prerequisite to water survival (flight physiology) training and is a one-time requirement.

Section II. Water Sports and Competitive Activities

4201.GENERAL

The commander, when developing the combat readiness program, needs to look carefully at the mission. If the mission requires Marines who must perform mission-essential water-related activities of long duration, he may want to look to water sports/competitive activities as a method of developing endurance. Endurance and confidence can only be developed by extended time in the water. While defined training can accomplish this goal, the goal can be enhanced through training which is integrated with the physical conditioning program. This approach will provide variety and enthusiasm. Two activities which will accomplish this goal are water relays and water polo. The important consideration is that the unit achieve total participation by all members of the unit.

4202.WATER RELAYS

Development of water relays will generally conform to the guidelines in chapter 5, section IV. When contemplating relays, the commander should review chapter 5 and adapt the principles to the size of the unit and the facility to be used.

4203.WATER POLO

The official rules provided in this paragraph will need to be adapted to the unit's unique situation, its size, and the size of the facility. Commanders should make every effort to have all members of the unit participate.

a. NCAA Rules This style of play is based on two popular sports: swimming and basketball. Playing is done on the surface of the water by teams of seven players each. The size of the playing area is 25 yards by 14 yards, which is the size of the typical high school or college indoor pool. The deeper the water, the better. If the entire playing area is deep-6 feet or more—this is ideal, although most indoor pools have a shallow end. At each end of the pool is a goal. In deep water, the goal is 10 feet across and 3 feet high. In shallow water, where the goalie can stand on and jump from the bottom, the goal is 10 feet across and 5 feet high. One of the seven players on each team is the goalie. He is the only player on the team who can stand on or jump from the bottom (if it is shallow enough to permit this), catch and pass the ball with both hands at the same time, or hit the ball with a clenched fist. However, the goalie may not swim across or pass the ball across the middle of the pool.

b.Equipment. The goalie on the visiting team must wear a white cap with the number 1 clearly marked on it; his teammates, who are designated as guards and forwards, must wear white caps numbered 2 through 7, with the substitutes wearing white caps numbered 8 through 21.The goalie on the home team must wear a dark-colored cap with the number 22 clearly marked on it; his teammates designated as guards and forwards, must wear similar dark-colored caps numbered 29 through 42.The guards and forwards on each team may swim freely up and down the pool as they see fit, interchanging positions as often as they wish. They may not stand on or jump from the bottom, touch the ball with more than one hand at a time when catching or passing or shooting it, or enter inside the opposing team's 2-yard line unless preceded by the ball. The ball is similar to a soccer ball, except that it is yellow and covered with a waterproofing substance that makes it easier to handle with one hand.

c. Playing Time. In intercollegiate competition the game consists of four 7-minule quarters, the teams changing ends after every period of play. In high school competition the game consists of four-,five-, or six-minute quarters, depending on the ages of the participants. Here again the teams change ends after every period. There is a very brief interval between each quarter. At the start of each period the teams line up at their respective ends, and with ablast of the whistle, the head referee throws the ball into the middle of the pool. The fastest swimmers on each team then race to gain possession.

d. Officials. The head referee walks along one side of the pool watching for infractions. He is aided by the assistant referee, who patrols the opposite side of the pool. The referees are both equipped with whistles and two official's flags. These flags should be 12 inches square and should correspond in color to the caps worn by the respective teams. Whenever a referee sees an infraction, he blows his whistle and signals with his flag indicating that the player nearest to the ball with the appropriate colored flag will be given possession of the ball. Whenever the whistle is blown, all players should react by looking immediately to see which color flag is being waived which will indicate which team should be moving into the defense. Other officials are a timer and a scorer. As in all sports, it is important that the officials be skilled and competent.

e. Technical Fouls. The following are some of the common technical fouls: starting before the referee blows his whistle to open play; holding onto or pushing off from the side of the pool during play; taking hold of the ball underwater when tackled by an opposing player; swimming inside an opponent's 2-yard line unless preceded by the ball; touching the ball with both hands at the same time (goalie excepted); standing, walking on or jumping from the bottom when taking an active part in the game. When a technical foul occurs, thereferee blows the whistle and with his flag awards possession. The player on the team awarded possession nearest the point of infraction then has 5 seconds to put the ball into play by passing to a teammate or by dropping the ball into the water and swimming after it.

f. Personal Fouls. The following are some of the more common personal fouls: committing any of the technical fouls mentioned above for the purpose of scoring or preventing a goal; holding, ducking, pulling, pushing off from, swimming over, or impeding the arm or leg movement of any opponent who is not touching the ball; and splashing water in the face of an opponent. When a personal foul occurs, the referee will follow the same procedure as with a technical foul. While doing this, the referee will also call out clearly and loudly the number of the player who was guilty of the infraction and a personal foul will be marked against that player by the official scorer. When a player accumulates five personal fouls, he fouls out of the game and must be replaced by a substitute.

g. Penalty Shot. A penalty shot can be awarded by either of the officials or the scorer when:

o An offensive player inside the opponents' 4-yard line not touching the ball is held, ducked, pulled-back, kicked, or struck.

o A team has accumulated a total of ten personal fouls. When the former occurs, the head or assistant referee should immediately blow the wh stle and by holding a two-flagged stick in a vertical position above his head, signal that a penalty shot has been awarded. When the latter occurs, the scorer should use a buzzer located at the scorer's tabic to signal that one team has accumulated ten personal fouls, thereby entitling the other team to a penalty shot. A penalty shot is taken from the 4-yard line in front of the goal. All players except the defending goalie must leave the 4-yard line until the shot is taken. No player can be within 1 yard of the shooter. After ascertaining that the shooting player is on the 4-yard line and the goalie is onThegoal line, the referee will ask the shooter to lift up the ball. When he docs, the referee will give a sharp blast from the whistle. At the whistle, the shooter must shoot without delay and without any faking at the goal. The goalie may try to block the shot. If he does, or the shot is otherwise missed, it is immediately in play and action continues.

h. When a Goal is Scored. When a goal is scored, either from a shot taken by a player out in the field or by penalty shot, the defending goalie must then pass the ball to the nearest referee. The referee will then pass the ball back to the goalie, who puts it into play by a pass to one of his teammates as soon as possible.

Section III. The Battle Swimming Test

4301.GENERAL

The battle fitness test is the evaluation contained in the Guidance for Basic Warrior Training (BWT) as part of the Basic Warrior Training Concept Plan.

4302.THE BATTLE SWIMMING TEST

The battle swim test consist of the following:
a. Be able to climb a three meter high board, while wearing a field uniform, carrying a M-16, web gear, and four canteens. Jump off blindfolded, recover and swim 100 meters with the M-16.
b. Repeat the same process, without the blindfold, and upon entering the water a second time, recover and retrieve all submerged gear.
c. Remain afloat and tread water for five minutes.
d. Demonstrate a knowledge of how to cross a stream using field expedient techniques and a single strand rope bridge.
e. Successfully complete helicopter egress training.

Chapter 5
COMPETITIVE CONDITIONING ACTIVITIES
Section I. Organization of Competitive Activities

5101.LEADERSHIP OF COMPETITIVE ACTIVITIES

a. Place in the Program. Competitive conditioning activities consist of dual combatives, relays, team contests, and team sports in which individuals or teams compete against an opponent to win. Competition is one of the best ways of maintaining interest in the physical training program during the sustaining stage. Organized competition provides enjoyable, vigorous physical activity that has proved to be one of the best supplements to conditioning drill activities. The benefits of competition are the development of aggressiveness, unit pride and identity, teamwork, and the will to win. These activities help develop cardiovascular and muscular endurance, strength, and coordination. Scheduling competitive activities in an orderly and progressive manner is desirable. The progression should be from relays to dual combatives, to team contests, and finally to team athletics. These activities should be conducted as part of the program after a basic level of conditioning has been developed, usually during the slow-improvement stage of conditioning. Muscles and joints should be strengthened by preconditioning to withstand the strain placed upon them by sudden stops and turns, body contact, bearing of weight, and falls. Competitive activities, however, should not be allowed to dominate the physical training program.

b. Area and Equipment. Some of the competitive activities included in this chapter require specific types of areas and equipment; others do not. The area requirement can usually be satisfied on available training fields. When items of equipment are required or specific courts or field layouts are to be marked off, such information is included.

c. Leadership. The principal factor for success as a leader of competitive activity is an energetic, dynamic, enthusiastic approach. The leader's attitude is reflected by the group, so the commander must carry on the activity in a snappy and vigorous manner. Confidence on the part of the commander will create an impression of decisiveness and certainty. Confidence grows out of experience and a thorough knowledge of the activity. Mastery of subject is the first step in developing confidence, assurance, and poise.

(1) The following suggestions are offered for leaders of competitive combatives, contests, and sports:

(a) Get the activity underway quickly by selecting and teaching only the essentials.

(b) Use rules to add to the enjoyment of the activity and do not allow them to interfere with the spirit of competition.

(c) Stop the activity before interest begins to lag.

(d) In team contests, clearly distinguish sides by attempting to maintain unit identity.

(e) Always insist on fair play, enforcing the rules impartially.

(2) The following procedure is recommended for presenting a competitive activity.

(a) Name the activity.

(b) Briefly explain the objective of the activity and give only the pertinent rules.
(c) Have a demonstration at slow speed and answer questions.
(d) Organize groups into teams and appoint captains.
(e) Arrange teams in the proper starting positions.
(f) Conduct the activity.

d. Competitive Units. Units for competition should be the same organizations in which Marines train: battalions, companies, platoons. In most situations, the unit is the squad.

e. Provisions for Instruction. One of the most effective methods of maintaining interest and participation in competitive activities is to provide instruction in activities with which most Marines are unacquainted. Such instruction can be conducted during regular physical training periods. Careful planning is required to keep all individuals continuously engaged in vigorous activity. The materials in the following sections should be used as a guide for instructional purposes.

f. Officiating and Control. Every effort should be made to provide good officials for all competitive activities. Poor officials quickly cause dissatisfaction among participants in team activities and create a situation which can turn an organized activity into a brawl. Each company should have or develop several competent and qualified officials available for games on company and platoon levels.

5102.MILITARY FIELD MEETS

When units reach the latter part of the slow-improvement and sustaining stage of training, interest in the program may lag. A change in course content can arouse the desire to participate. An event which does not require a high degree of skill yet demands strenuous activity is ideal for the military field meet. This is a series of team contests conducted on a station-to-station basis during a given period of time. Team contests carried on simultaneously provide essential training, vigorous exercise, and stimulating competition in an atmosphere that is enjoyable for all. Because of the healthy rivalry that a field meet arouses, it is an excellent form of inter-unit competition.

a. Objective and Advantages. The objective of the military field meet is to provide activity for everyone in the participating units. Activities are chosen that will develop aggressiveness, teamwork, a will to win, and competitive spirit, and that will stimulate interest and build esprit de corps. The military field meet can **be** included as part of the physical training program or as part of the off-duty recreational program. It is a form of contest that can be conducted in nearly all circumstances because it can be easily modified, requires a minimum amount of equipment, and can be readily organized.

b. Level of Competition. The military field meet is flexible. It can be adjusted to large or small groups. A company-size unit is the most desirable, but it may also be administered within a larger unit. If it is conducted within a company, the participating units will be squads. If it is held within a larger unit, platoons compose the teams. A larger-size unit requires more extensive organization and administration.

c. Selection of Events. In organizing a military field meet, select events that are simple and easy to administer. All rules and regulations should be clearly understood by everyone. No event should require previous practice. In selecting the events, consider the interest and capabilities of the Marines and available equipment and facilities. Select events which will require that all members of the units participate. Events should not be dominated by the athletes; rather, MOS-related events which have significance in producing skills and combat readiness should be used.

d. Equipment and Facilities. The site must be large enough to permit events to be grouped about a central control point. Use available facilities such as parade decks, Softball fields, or tracks. A public address system is desirable at the control point for the initial orientation of teams and for subsequent announcements of time lapses, cumulative scores, and final standings. A tally board is necessary so that all units can constantly monitor how the events are progressing.

e. Administrators. Efficient administration of the military field meet depends on the referees, judges, and scorers. For these positions, choose individuals within the units who have had athletic or officiating experience. Prior to the day of the event, all administrators should be briefed and assigned a specific task in order that they may become familiar with the rules and organization of the contests they will conduct. The following officials should be available:

(1) A primary instructor or supervisor who is in charge of the control center and who is responsible for the successful operation of the athletic carnival. The supervisor must have an assistant to act as a timer and scorer.

(2) One assistant instructor in charge of each event. This person should be-

(a) Familiar with the rules of the event.

(b) Effective as a leader to ensure proper supervision and control over the participating teams.

(c) Enthusiastic to provide proper motivation.

(d) Self-confident of the ability to judge infractions of the rules. The assistant instructor must be fair in judgment and penalize without hesitation when infractions occur.

(3) Runners between event stations and the control point. These individuals collect and deliver scores.

f. Team Organization for Competition. The size of the teams is determined by the level on which the field meet is organized. Maintaining the integrity of the unit promotes esprit de corps, but this does not preclude grouping two squads into one team.

g. Conduct of Events

(1) The assistant instructor at each station takes charge of the group and gives a brief explanation of the major rules of the event. The assistant instructor speaks clearly and distinctly from a position to be seen and heard by all. A short demonstration is desirable if it will help clarify the event.

(2) The assistant instructor should make certain that teams can be clearly distinguished by the use of, for example, T-shirts and fatigue jackets, or caps and no cap, or colored jerseys. The assistant instructor should teach a whistle response (teams stop play immediately upon hearing whistle), get the event started as quickly as possible, and make any necessary corrections as the contest progresses. The rale of good officiating is to use a minimum of calls, yet maintain control of the contest. Penalize when necessary, but refrain from disqualifying contestants or teams.

(3) Keep the activity moving as rapidly as possible. When the central control point sounds the whistle to stop the play, all competition ends immediately. The assistant instructor then assembles the group, forwards the team scores to the central point, and, upon a signal from the central control point, rotates the teams to their next station. It is essential that rotation and orientation of teams be quick and orderly to reduce time spent between contests.

(4) Upon completion of the final event, assistant instructors move their teams to the central control point for the announcement of winning teams, presentation of awards (if any), and final critique.

h. Scoring System. The system for determining the winner of the military field meet should be simple and efficient. At the completion of competition, the scorer totals the points that each team has scored in all contests. The scorer then subtracts the number of points scored against a team (penalty points for nonparticipation,etc.) from the total. The resulting scores are placed in a column with the highest score at the top and the lowest at the bottom. The team with the highest total is the winner. This type of scoring system encourages a team to ensure total unit participation.

Section II.Combatives

5201.DESCRIPTTON AND OBJECTIVE

Combatives are strenuous, short competitive contests in which two individuals attempt to overcome each other in a bout of skill and strength. These contests help to develop Marines' resourcefulness, confidence, strength, agility, coordination, and will to win. Any level ground area can be used. However, extremely hard ground should be avoided as some of the combatives require ground contact. A whistle is needed to control the bouts during the competition since voice commands may go unheeded. The extended rectangular formation is used for dual combatives. Combatives are conducted on an informal basis. Marines are allowed to remain at ease between activities and are allowed to brush themselves off after being on the ground. To get the most out of combatives, individuals must be urged to overcome their opponents as quickly as possible as would be required in combat.

a. Benefits for Marine Training. The possibility of close contact with an enemy in combat faces Marines at all times. They must be trained to react aggressively and violently. Combatives may be used as an introduction to such hand-to-hand contact and should be followed by hand-to-hand combat training. Marines enjoy competition, and this type of activity is a welcome change from formal conditioning activities.

b. Instructor Responsibility

(1) The instructor tells the individuals that all combatives begin and end on the whistle signal. The instructor demonstrates each activity before having it performed, explaining it in simple terms.

(2) After stopping one activity, the instructor gets everyone in place for the next.

(3) The instructor must closely supervise combatives to insure that contestants do not use unfair or unsportsmanlike tactics. To avoid unnecessary injury, instructors must see that the bouts are closely controlled and opponents equally paired. Adjustments should be made in apparent cases of mismatched abilities.

5202.COMBATIVES TABLES

There are three tables of combatives, each of which can be completed in 15 minutes. These tables become progressively more difficult from lower to higher numbers.

a.Combatives Table I

(1) Open-Hand Slap Boxing(See fig. 5-1, A. individuals assume a boxer's stance, palms open, fingers extended and joined. Each contestant tries to slap the opponent about the head and upper body with the open hand. This is a good warm- up activity.

(2) Wrist Tug-of-War(See fig. 5-1, B.)Two individuals sit on the ground with the soles of their feet in contact. Each grasps the opponent's wrists so that the hands are directly over the feet. At the whistle, each individual tries to pull the opponent from sitting to standing position.

(3) Arm-Lock Wrestling.(See fig. 5-1, C.)Two individuals sit back-to-back with legs spread and arms locked at the elbows. Each right arm is inside the opponent's left. Each tries to force the opponent's left arm or shoulder to the ground. The individual who first wins three bouts is the winner.

(4) Bulling(See fig. 5-1, D.)Two individuals assume the Westmoreland wrestling hold, each grasping the opponent's neck with the right hand and the opponent's right elbow with the left hand. Each tries to force the other to move one foot by pushing, pulling, or otherwise manipulating. The individual who first wins two boutsis the winner.

(5) Indian Wrestling(See fig. 5-1, E.)Two individuals lie on the ground, side by side, with their heads in opposite directions. They link right elbows. On the instructor's signal or by mutual agreement, each individual raises the right leg approximately straight and far enough to engage the opponent's heel. To start the contest, each person usually raises the leg three times rhythmically and on the third time engages the opponent's heel. Each tries to roll the other over backward. The right leg is used for three bouts, then the left leg for three bouts.

Figure 5-1 is available in the illustration section at the end of the book.

b. Combatives Table II

(1) Wrist Bending.(See fig. 5-2, A.) Opponents pair off, face each other, raise their arms forward, and, with palms forward, interlock their fingers. At the starting signal, each individual attempts to bend the opponent's wrist. The arms are kept up and forward and are not allowed to swing around and down to the sides. The individual who first wins two bouts is the winner.

(2) Back-to-Back Push.(See fig. 5-2, B.)Two individuals stand back-to-back with elbows locked. Each right arm is inside the opponent's left arm. At the starting signal, each pushes backward, trying to move the opponent. Opponents are not allowed to lift and carry each other; only pushing is permitted. The one who pushes the opponent the farthest wins the bout. The individual who first wins two bouts is the winner.

(3) Hop and Pull Hand.(See fig. 5-2, C.) The individuals are matched in pairs. Each grasps the opponent's right hand and, hopping on the right foot, attempts to pull the opponent off balance. Contestants automatically lose if they touch their free hand or their lifted foot to the ground. For successive bouts, they alternate hands and feet.

(4) Westmoreland Wrestling. (See fig. 5-2, D.)Each contestant grasps the back of the opponent's neck with the right hand and the opponent's right elbow with the left hand. In this position, each attempts to pull, push, or force the opponent to touch the ground with any part of the body other than the feet. The individual who first wins two bouts is the winner.

(5) Crab Fight.(See fig. 5-2, E.)Two individuals sit on the ground facing in opposite directions with their hands on the ground behind them. At the whistle, they raise their hips and push with their shoulders and bodies, each trying to make the other's hips touch the ground. The individual who first wins two bouts is the winner.

Figure 5-2 is available in the illustration section at the end of the book.

c. Combatives Table III

(1) Hand Wrestling.(See fig. 5-3, A.) Two individuals stand facing each other. Their right feet are forward and braced side by side. They grasp right hands for the first bout, left hands for the second bout. Each pulls, pushes, makes sideward movements, and otherwise maneuvers to force the opponent to move one or both feet from the original position. The contestant who first wins two bouts is the winner.

(2) Back-to-Back Tug.(See fig, 5-3, B.) Two individuals stand back-to-back with both arms linked at the elbows. Each has the right arm inside the opponent's left arm. At the starting signal, each attempts to pull the opponent. Lifting and carrying are permitted. The contestants must maintain their original direction and keep their arms linked. After a predetermined time, the player pulled or carried the farthest is the loser.

(3) Wrestling to Lift off Feet.(See fig. 5-3, C.) Contestants face each other. Each places the right arm under the left arm of the opponent and around the body. The left arm is over the opponent's right shoulder. Each tries to lift the other off the ground. The individual who first wins two bouts is the winner.

(4) Arm Pull Between Legs.(See fig. 5-3, D.) Two individuals are paired off, back-to-back. Each bends forward and, extending the right arm between the legs, grasps the opponent's right wrist. At the starting signal, each person attempts to pull the opponent After a predetermined time, the player who has pulled the opponent the farthest is the winner of the bout. The one who first wins two bouts is the winner. Repeat with the left hand and then both hands.

(5) Rooster Fight.(See fig.5-3, E.) Each contestant grasps the left foot with the right hand from behind, and the right arm with left hand. Each hops on the right foot, and, by shoulder-butting the opponent or by feinting and sudden evasions, forces the opponent to let go of the foot or arm. The contestant who first wins two bouts is winner.

Figure 5-3 is available in the illustration section at the end of the book.

Section III .Relays

5301. DESCRIPTION AND OBJECTIVE

Relays are races in which each member of a team runs one leg of the race. The team effort decides the winner. Relays provide stimulating competition and contribute to the conditioning of personnel. They also develop aggressiveness, team spirit, and the will to win. Relays should be dispersed throughout the program for short periods of time to provide a change of activity.

a. Team Organization. Relays are conducted most efficiently in platoon-size groups. Teams of equal size must be organized. Competitive spirit is encouraged and teams are organized better by basing teams on units such as squads, crews, or sections. Team captains should be designated. Extra Marines may be used as officials. The number of individuals on a relay team should be limited to squad size. If larger teams are used, the runners will spend too much time awaiting their turns and too little time actually participating. Two to six teams are ideal for relay competition. It is difficult to keep track of winners when more teams compete.

b. Administration of Relays

(1) The time spent on any one relay should be relatively short. If one team achieves a substantial lead in a long relay, the competitive spirit and enthusiastic participation of the other teams may decrease. Several short relays are generally better than one long relay.

(2) To maintain competitive spirit throughout a number of relays, determine the teams that win, place, and show in each relay and their total points for all relays. This can be done by awarding points to all teams on the basis of position at the finish of each relay. The team with the greatest number of points is the winner of the entire set of relays.

(3) Difficulties commonly encountered in conducting relays may be avoided by the following procedures:

(a) The last player in a relay race should be conspicuously identified. For example, the last runner can tie a handkerchief around the head or arm, take off the shirt, put on a hat or take it off, or use some other means.

(b) Another way to keep track of the progress of the race is to have each player sit or squat as soon as each is finished.

(c) Judges at the starting line can keep the runners from starting too soon.

(d) To prevent contestants from turning before they run the full distance, they should be required to run around a peg, pole, or assistant instructor.

(e) Batons, handkerchiefs, tent pegs, or other objects should be passed from one runner to the next when relays are run on a circular track.

(4) Before a unit's first participation in a relay, inform participants of the rules and scoring system. Violation of the rules should not result in disqualification. Instead, impose point penalties. A point penalty is imposed by subtracting a point from the team total at the conclusion of the relay.

(5) Careful administration will prevent most violations. For efficient conduct of relays, follow this procedure:

(a) Announce the name of the relay.

(b) Form the individuals in relay position.

(c) Briefly explain the relay and the rules for running it.

(d) Demonstrate.

(e) Have a definite finish line, and ensure that all know where it is.

(f) Answer questions.

(g) Conduct the relay.

(h) Determine winner and award points.

5302. RELAY TABLES

a. Events. The following relays are grouped into a table of activities. Each relay table can be completed in 15 minutes. Thus, relays can be used as a sole activity or as a part of a longer period. Each table is planned for a platoon-size group (30-60Marines). Adequate warmup for participants is provided by conducting several repetitions of exercises 1 and 2 of a conditioning drill. The recommended relay tables require an area 40 by 60 meters in size. Each table provides a variety of activity. The tables are progressive in the overload applied and should be scheduled in numerical order although not necessarily on successive days.

b. Relay Table I

(1) **60-Meter Lane Relay** (See fig. 5-4, A.) Each team is assembled in single file behind the starting line. On signal, the first individual of each team runs to the turn-around line 30 meters away, then runs back and touches the next teammate waiting at the starting line. The winning team is the first team to get its last member across the finish line. If an individual starts before being touched by the preceding runner, the team may be penalized.

(2) Wheelbarrow Race.(See fig. 5-4, B.) The players of each team pair off and line up in single file behind the starting line. The first individual walks on the hands while the partner grasps the ankles. They advance to the distance line (25 meters) behind which they exchange positions and return to the starting line. The rear individual must always hold the partner's ankles. After the first pair returns across the starting line, the next pair starts.

Figure 5-4 is available in the illustration section at the end of the book.

(3) Squad Front Relay .(See fig. 5-4, C.) The teams form in a line along the starting line with a 10-foot interval between teams. The members of each team lock elbows so that they are linked together. At the starting signal, the teams run to the distance line (20meters) where the left flank individual acts as a pivot. The team swings around on the pivot and returns to the base line. If a team breaks its links, it must reform before continuing. The first team to completely cross the base line intact is the winner.

(4) Crab-Walk Race.(See fig. 5-4, D.) The players of each team line up in single file. The first person of each team assumes the crab-walk position with feet forward on the starting line, hands on the ground behind it. At the starting signal, contestants move forward to the distance line (10 meters). They touch the line with their feet and then return to the starting line in the reverse position with the head and hands leading. The second person may not start until the first individual touches the finish line.

c.Relay Table II

(1) 100-Meter Lane Relay.(See fig. 5-5, A.)This relay is conducted exactly as the first relay of table 1 except that the start and turn-around line are 50 meters apart. This relay provides progression in sprinting.

(2) Frog-Jump Relay. (See fig. 5-5, B.) Each team lines up in single file. The first individual assumes a squatting position on the starting line. At the starting signal, the contestant progresses to the distance line (15 meters) and back by leaping forward, catching the weight on the hands, and bringing up the legs to the squat position for the next leap forward.

(3) Simple Relay. (See fig. 5-5, C.)Each team lines up in single file. Place a marker on the distance line (20 meters) in front of each team. Each team member, behind the first one, grasps the belt of the individual ahead. At the starting signal, each team runs as a unit to the marker, circles it, and returns to the starting line. The first team to completely cross the starting line intact is thewinner.

(4) Fireman's Carry Relay.(See fig. 5-5, D.)The players of each team line up in pairs, one behind the other. One individual in each pair carries the partner to the distance line (30 meters), using the fireman's carry. At the distance line, individuals exchange places and return to the starting line. As a variation, the person to be carried lies on the ground. The one carrying lifts the partner to the proper position. This relay may be performed with the other carries described in guerrilla exercises in chapter 3.

d.Relay Table III

(1) 200-Meter Circle Relay .(See fig. 5-6, A.)A course is laid out in a circular, rectangular,or oval pattern that is 200 meters around. Each team provides one runner on the starting line. On signal, the runner races around the 200-meter track and touches the next teammate, waiting at the starting line, who runs the same course. Each team member runs one lap of the course.

Figure 5-5 is available in the illustration section at the end of the book.

(2) Bear-and-Crab Race.(See fig. 5-6, B.) Each team lines up in single file. At the signal to start, the first contestant in each column assumes the bear-walk position (hands on ground in front), walks to the distance line (15 meters), and then runs back to the starting line. The returning contestant touches off the second individual and goes to the rear of the line. The second contestant assumes the crab-walk position (hands on ground behind) and crab-walks with feet leading to the distance line. This contestant also runs back to the starting line and touches off the next individual, who walks bear fashion. The rest of the members of each team alternate in this manner. The relay ends when the first person is back at the head of the line.

(3) Pilot Relay.(See fig. 5-6, C.) The players are grouped in threes, arms interlocked at the elbows, and outside players facing backwards. The middle person runs forward; the two outside individuals run backward. They run to the turning point (15meters), where they start back, this time with the middle person running backward and the two outside individuals running forward. The next set of three players starts when the first set crosses the starting line.

(4) Saddle-Back Relay.(See fig. 5-6, D.) Mark two parallel lines 15 meters apart. Each team selects a rider. The remaining members of each team count off. The even-numbered players from each team form in single files behind one line and the odd-numbered players from each team form in single files behind the other line directly across from their teammates. At the starting signal, the rider mounts the back of the first player of the team who carries the rider across the other line. There the rider changes mounts to the first player in the second line without touching the ground. This person carries the rider to the next player waiting in the first line. The relay continues until all of the mounts have carried the rider. A rider who falls off must mount again at the point of the fall. A rider who falls in changing mounts must get back on the original mount before making the change.

e. Relay Table IV

(1) 100-Meter Circle Relay (See fig. 5-7, A.)A course is laid out in either a circular, rectangular, or oval pattern that is 200 meters around. Each team is divided in half with each half positioned at starting lines on opposite sides of the track. Each runner races halfway around the track and touches a teammate who completes the lap. Each runner then waits in file at the halfway point until touched by the runner and then completes the second half of the lap. The first team to return all runners to their original starting line is the winner.

Figure 5-6 is available in the illustration section at the end of the book.

(2) In-and-Out Relay.(See fig. 5-7, B.) Each team lines up in a file with players 2 meters apart. At the starting signal, the first player runs back through the column in a zigzag fashion. He alternates, going to the right of one teammate and to the left of the next. Upon completing the run, the player lines up 2 meters behind the last one in the column. As soon as the first runner has passed the second person, the latter starts to run. This continues until all the players have realigned their original order. The team that finishes first is the winner. It may be desirable to have this relay continue until all individuals have run through their entire team two or three times in succession.

(3) Circle Race.(See fig. 5-7, **C.**) Each team forms a circle and holds hands with all individuals facing out except one who faces in and is the "driver."At the starting signal, the teams race to the distance line (20 meters) and back, keeping the circle intact. All the individuals in the circle must completely cross the distance line. The "driver" gives directions and orders. When the circle breaks, it must be reformed before it can continue. The first team completely over the starting line is the winner.

(4) Horse-and-Rider Relay.(See fig. 5-7, D.)Each team lines up in a single file. At the signal to start, the second player in each column leaps upon the back of the first who carries the rider across the distance line (30 meters). At the distance line, the rider dismounts and runs back to the starting line. There the rider picks up the third individual in the column, and carries this player to the distance line where the first player has remained. This continues until the last person is carried across the distance line.

Figure 5-7 is available in the illustration section at the end of the book.

Section IV. Team Contests and Athletics

5401. DESCRIPTION AND OBJECTIVE

Team contests are competitive activities in which Marines as a team compete with another team to win. They are guided by simple rules and organization. The function of team contests is to provide competition and an opportunity for body contact and to contribute to the development of physical readiness. In competing and working together as a team, individuals develop aggressiveness, the will to win, and teamwork. Team athletics deserve a prominent place in the physical training program because they contribute to increased combat efficiency. Because of the competitive nature of athletics and their natural appeal, individuals take part with enthusiasm. Athletic teams formed at intramural and higher levels are a strong unifying influence and provide one of the best means of developing esprit de corps.

a. Area and Equipment A level training field is sometimes the only area required. Many contests need no equipment. In contests requiring equipment, the need is for standard items such as logs, balls, nets, goals, and similar types of equipment. Specific requirements for area and equipment are listed with each contest description.

b. Progression. Team contests of a strenuous nature should be introduced after a basic period of conditioning has been completed and individuals are in the slow improvement stage of conditioning. Progression can take place from the less active to the more vigorous contests, and then from the non-contact to the contact or combative-type contests.

c. Necessity for Preconditioning Individuals must undergo conditioning prior to participation in athletics. Muscles, organs, joints, and ligaments not accustomed to stress and strain from sudden stops and starts, falls, body contact, rapid turns, prolonged running, and other rigors of athletic competition are subject to injury. Although athletics should not be introduced until players are physically prepared, there is still opportunity to engage in competition through lead-up contests. Individuals learn many of the skills required for athletics while participating in team contests.

d. Benefits. Athletics are beneficial primarily to sustain interest in the program and to maintain an achieved level of physical fitness. Athletics are a supplement, not a substitute for other types of conditioning activities which should continue. All of the desirable traits of physical fitness cannot be developed through athletics, yet their contribution is significant. For athletics to make a proper contribution to physical conditioning, the selected sports must bevigorous. The team contests recommended here are carefully selected for their simplicity, aggressiveness and applicability to squad- teaching teamwork. They are generally strenuous and involve the whole team throughout the entire duration of the contest.

5402. TEAM CONTESTS

a. Pushball

(1) **Players.** 10 to 50 players on a side. (See fig. 5-8.)

(2) Equipment A large pushball 5 to 6 feet in diameter.

(3) Area. A field 240 to 300 feet long, 120 to 150 feet wide. Mark a center line in the middle of the field parallel to the end lines. Mark a line 45 feet on either side of this center line and parallel to it, extending it across the width of the field. Mark another parallel line 15 feet from each end line, extending it across the width of the field.

(4) Game. Four 10-minute quarters are played. Give 2-minute rests between quarters and 5-minute rests between halves. The object of the game is to propel the ball over the opponent's goal line by pushing, rolling, passing, carrying, or any other way except kicking.

(a) The ball is placed on the center line with the opposing captains 3 feet from the ball. The rest of the players are 45 feet from the ball, on their half of the field. On the referee's starting whistle, the captains immediately play the ball with their respective teams coming to their assistance.

(b) At quarter time, the ball remains dead for 2 minutes at the spot where it was when the quarter ended. At half time, the teams exchange goals. The play is then started as it was in the beginning.

Figure 5-8 is available in the illustration section at the end of the book.

(c) Players may use any means of interfering with an opponent's progress except striking and clipping. Clipping is throwing the body across the back of an opponent's leg, while the opponent is running or standing. Legal use of force may be applied to all opponents whether or not they are playing the ball. For striking an opponent, the offender is removed from the game. The team penalty is half the distance to their goal. The penalty for clipping is the same.

(d) When any part of the ball goes out of bounds, it is a dead ball. The teams line up at right angles to the side lines and 3 feet apart at the point where the ball went out. The referee then tosses the ball between the teams.

(e) When, for any reason, the ball remains in one spot for more than 10 seconds, the referee declares the ball dead. The ball is then put into play as it is for an out-of-bounds situation.

(5) Scoring. A goal is scored when the ball, or any part of it, is propelled across the opponent's end line. A goal counts 5 points. The team scoring a goal has the privilege of trying for a point after the goal. To try for this extra point, the ball is placed on the opponent's 5-yard (or 15-foot) line. The teams line up on either side of this line separated by the width of the ball. Before the whistle blows, one player on each team may place hands on the ball. On the referee's signal, the ball is put into play for 1 minute. If any part of the ball is driven across the goal line in this 1-minute period, the offensive team scores 1 point. The defending team may not score during the opponent's try for the extra point.

b. Line Rush

(1) Players. Any number up to 50 on each side. (See fig. 5-9.)

(2) Equipment. None.

(3) Area. A field, 75 by 100 feet.

(4) Game. One team lines up behind one goal line and the other in midfield. On the starting signal, the team standing behind the goal line seeks to cross the goal line at the opposite end of the field within 30 seconds. The team in the center seeks to prevent it by catching and holding the runners. At the end of 30 seconds, the teams change.

(5) Scoring. Count the number of individuals who have crossed the far goal at the end of 30 seconds. After each team has had from three to five tries, the scores are added and the winner declared. A player scores 1 point when any part of the body is across the goal line.

Figure 5-9 is available in the illustration section at the end of the book.

c. Human Tug-of-War

(1) Players. 10 to 20 on a team.(See fig. 5-10.)

(2) Equipment. None.

(3) Area.40 to 60 feet.

(4) Game. Draw a line in the center of the area. Divide the players into two equal teams. Place them in single file on opposite sides of the center line facing each other. Each individual places the arms around the waist of the teammate in front. The two leaders of the opposing teams grasp each other around the waist. On signal, the teams try to pull each other over the center line within 30 seconds.

(5) Scoring. The team pulled across the center line loses. If neither team is pulled over the center line, but one team breaks its file, that team loses the match.

(6) Variation. Use a 3/4- or 1-inch rope. Space the leading players on each team 10 feet apart, each holding one end of the rope. The team pulled across the center line loses.

Figure 5-10 is available in the illustration section at the end of the book.

d. Master of the Ring

(1) Players. Any number. (See fig. 5-11.)

(2) Equipment. None.

(3) Area. A clearly marked circle large enough to contain all the players.

(4) Game. All the players stand inside the circle. At the signal, all players attempt to throw each other out of the circle. All tactics are fair except unnecessary roughness. When any part of the body touches across the line, the player is out and leaves the circle at once. Several officials are needed to spot the players who cross the line.

(5) Scoring The player who remains in the circle when all the others are out is the master of the ring.

(6) Variation 1. The players are divided into two equal teams. Each team is clearly marked. On signal, each team tries to throw the opponents out of the circle. The winning team is the team that eliminates all opponents from the circle.

(7) Variation 2.The players are divided into two equal teams. Each team sends only one individual into the circle. When one person has been forced out of the circle, the losing side sends in only one player. The team which eliminates all opponents is the winner.

(8) Variation 3.A pit, approximately 4-feet deep, is used rather than a circle on level ground. The contest may then be an individual or team activity.

Figure 5-11 is available in the illustration section at the end of the book.

e. Log Pivot Circle. Space teams far enough apart so each team can pivot in a circle without colliding with any other team. Each team holds a log in the bend of the arms in front of the chest. (See fig.5-12.) At the command **CIRCLE RIGHT, MOVE**,the left-flank individual holds the pivot while the log is carried around 360 degrees in a counter- clockwise motion, back to the original position. This movement may also be performed to the left by facing in the other direction and pivoting clockwise (**CIRCLE RIGHT, MOVE**).Other commands may be used such as: **CIRCLE HALF-RIGHT, CIRCLE HALF- LEFT** and so on. The first team to complete the prescribed movement is the winner.

Figure 5-12 is available in the illustration section at the end of the book.

f. Log-Rolling Race. Each team tries to roll its log a measured distance by pushing it with the hands and driving the body forward with the legs. (See fig. 5-13.)The first team to get the entire length of the log across the finish line wins.

Figure 5-13 is available in the illustration section at the end of the book.

g. Prone Pushing Contest.. Two teams lie prone, facing each other with a log between them. (See fig. 5-14.)Both teams place their hands against the log, keeping their arms straight. Then, by driving with the legs, each team attempts to push the other a measured distance to the rear.

Figure 5-14 is available in the illustration section at the end of the book.

h. Shuttle Relay Race. Each team in this race divides into two parts, A and B. Members of part A run 50 yards with a log held under their right arms. (See fig 5-15.)At the distance line, they give the log to their teammates in part B who bring it back to the starting line. The team pair finishing first is the winner.

Figure 5-15 is available in the illustration section at the end of the book.

5403.CROSS-COUNTRY AND DISTANCE RUNNING

a. Value of Running. Long-distance running gives some benefits that cannot be obtained in the same degree from any other sport. It builds powerful leg muscles, increases lung capacity, and develops endurance. For these reasons, cross-country and distance running should be included in the physical training program. These sports require only a few miles of open space which is generally available. Short cross-country runs and middle-distance runs can be used to supplement other activities, particularly team sports or the sports that develop precision or agility rather than endurance. Short cross-country runs can be scheduled once a week, gradually increasing the distance as the physical condition of the runners improves.

b. Cross-Country Runs. A cross-country run is a distance run held on a course laid out along roads, across fields, over hills, through woods, and on any irregular ground. A flat cinder or dirt track is not a suitable surface for cross-country running. Some runs are as long as 6 miles while others are as short as 3 or 4 miles. If cross-country running is being used to supplement other activities, the 3-mile course is long enough for most. Cross-country runs should be used only after individuals reach the sustaining stage of conditioning. This type of running should then be scheduled occasionally to provide variety. Cross-country running has the advantage of allowing mass participation. Interest can be stimulated by putting the runs on a competitive basis.

c. Practice Methods. Conditioning is essential to distance and cross-country running. Championship distance running depends on stamina, which can be developed only through constant training. An individual of only average ability can become an outstanding distance runner by steady and careful training. Hiking is an excellent method for getting into condition before the season opens. Long walks build up leg muscles. During the first month of the season, training should be gradual, starting with short distances, and increasing day by day. At first the legs will become stiff, but the stiffness gradually disappears if running is practiced for a while every day. To prevent strain, it is essential to limber up thoroughly each day before running.

(1) In training a large group, leaders should be stationed at the head and the rear of the column. They should make every effort to keep the runners together. After determining the abilities of the unit in cross-country running, it is advisable to divide them into three groups. The poorest-conditioned group is started first, the best-conditioned group, last. The starting time of the groups should be staggered so that all of them come in at about the same time.

(2) In preliminary training, running is similar to ordinary road work in that it begins with rather slow jogging, alternating with walking. The speed and distance of the run is gradually increased. As conditioning improves, occasional sprints may be introduced. Atfirst, the distance run is from one-half to 1 mile. It gradually increases to 2 or 3 miles. On completing runs, require runners to walk for 3 or 4 minutes before stopping to permit a gradual cooling off.

d. Facilities and Equipment. There should be at least one timer with a stopwatch (preferably three) for timing the runners. A course 3 or 5 miles long should be measured and marked by one of the three following methods:

(1) Fasten directional arrows to the top of tall posts and place them at every point where the course turns. Such signs should also be placed at every other point where there may be doubt as to the direction of travel.

(2) Place a lime line on the ground over the entire course.

(3) Place flags, clearly visible to the runners.

- **(a)** A red flag indicates a left turn.
- **(b)** A white flag indicates a right turn.
- **(c)** A blue flag indicates the course is straight ahead.

e. Rules/Scoring. A cross-country team consists of seven individuals, unless otherwise agreed. In dual meets, a maximum of 12 on one team may be entered, but 7 or fewer are scored. First place scores 1 point, second place 2, third place 3, and so on. All who finish the course are ranked and tallied in this manner. The team score is then determined by totaling the points scored by the first five of each team to finish. The team scoring the least number of points is the winner. If fewer than five (or the number determined prior to the race) finish, the places of all members of that team are disregarded. If two or more teams score the same number of points, the event is called a tie.

5404.TEAM ATHLETICS

Team athletics add variety to the combat readiness program. At no time should they be allowed to replace intramural activities or to provide practice time for intramural or base teams. If team athletics are integrated into the program, the goal should be total participation and competition. Therefore, the games selected can be played by the skilled or the unskilled. They require limited equipment, yet will teach competitiveness, aggressiveness, and teamwork while they improve conditioning.

a.Soccer. Soccer is one of the best athletic activities for developing endurance, agility, leg strength, and great skill fulness in using the legs. The most popular sport in the world, soccer is the national game of many European, Asian, and Central and South American countries. In recent years, it has become popular in the United States. A soccer ball is the only equipment needed for the game, and Marines can learn to play it easily. Players do not need much skill to participate, but the amount of skill they can develop is almost unlimited.

(1) Place in the Program. Soccer should be introduced during the latter part of the slow improvement stage of physical training. It can be used as a competitive activity in the sustaining stage. It is primarily a spring or fall sport. Any level field is suitable for competition. The boundaries for the soccer field are similar to the dimensions for a football field. (See fig. 5-16.) Goal posts are essential, but they are easily constructed and usually temporary so that they may be removed when not in use.

Figure 5-16 is available in the illustration section at the end of the book.

(2) Basic Skills

(a) Passing. Passing with the feet is the primary means of moving the ball. Short passes are easier to control. Emphasis should be continually placed on the skill of passing.

(b) Dribbling. The ball is dribbled by a series of kicks with the inside or outside of the foot. Do not kick with the toe. Keep the head over the ball when kicking and propel it only a short distance at a time. Keep it close to the feet. If a player lets a ball move far from the feet, an opposing player can easily take it away.

(c) Instep Kicking. The instep kick is the basic soccer kick. The toe does not come in contact with the ball. The toe is pointed downward, and the instep (the area under the shoe laces) is applied to the ball with a vigorous snap from the knee.

(d) Inside-of-the-Foot Kicking. The ball is kicked with the inside of the foot, and the leg is swung from the hip. The toe is turned outward, and the sole of the foot is parallel with the ground as the foot strikes the ball. This kick is used for short passes and for dribbling.

(e) Foot Trapping. The foot trap is the method of stopping the ball by trapping it between the ground and the foot. Place the sole of the foot on top of the ball at the instant it touches the ground, keeping the foot relaxed. This is an effective way to stop a fast-moving ball.

(f) Shin Trapping. .The shin trap is a method of stopping the ball with the shins. Stand just forward of the spot where the ball should strike the ground and allow it to strike the shins in flight or on the bounce. Use either one or both legs from the knee down so as to absorb the bounce.

(g) Body Trapping. .Intercept the ball with any part of the upper body except the arms and hands. Keep the body relaxed and inclined toward the ball. To keep the ball from bouncing, move backwards from it as it strikes the body. This drops the ball at the feet in position for dribbling or passing.

(h) Heading. Heading is a technique for changing the direction of the flight of a ball by butting it with the head. Tense the neck muscles and jump up to meet the ball. Butt the ball with the forehead at about the hairline to reverse its direction. Use the side of the head to deflect it to the side.

(3) Offensive and Defensive Positions. Forwards usually play on the offensive half of the field. Fullbacks usually play on the defensive half of the field. Halfbacks are the backbone of the team. They move forward on the offense and back on defense. The goalkeeper almost always remains within a few feet of the goal.

(4) Abridged Rules

(a) A soccer team is composed of 11 players. (See fig. 5-16.)

(b) The player propels the ball by kicking it with the feet or any part of the legs, by butting it with the head, and by hitting it with any portion of the body except the arms or hands.

(c) The goalkeeper is the only individual allowed to place hands on the ball, but only in the goalkeeper's area. The term "hands" includes the whole arm from the point of the shoulder down.

(d) A goal is made by causing the ball to cross completely the section of the goal line lying between the uprights and under the cross bar.

(e) Each goal scores 1 point for the team scoring the goal.

(f) The penalty for a foul committed anywhere on the playing field (except by the defensive team in its penalty area) is a free kick awarded to the opposing team.

(g) All opponents must be at least 10 yards from the ball when a free kick is taken.

(h) The penalty for a foul committed by the defensive team in its penalty area is a penalty kick.

(i) A penalty kick is a free kick at the goal from the spot 12 yards directly in front of the goal. The only players allowed within the penalty area at the time of the kick are the kicker and the defending goalkeeper.

(j) An official game consists of two 30-minute halves,

(k) Teams change goals at the end of every quarter.

(l) In the event of a tie, an extra quarter may played. This may be followed by a sudden death period. If the tie still exists, the game can be decided by penalty kick.

(m) After a team has propelled the ball across a side line and out of play, the ball is put back into play by a throw-in from the side line by a member of the opposing team. The ball is thrown in from the point where it crossed the side line as it went out of bounds. The ball must be thrown in with both hands using an over-the-head motion, keeping both feet on the ground.

(n) When the offensive team propels the ball across the defensive team's goal line but not in the goal, the defensive team is awarded a goal kick-a free kick taken from within its own goal area. The ball must come out of the penalty area to be in play.

(o) When the defensive team causes the ball to go behind its own goal line, excluding the portion between the goal posts, the offensive team is awarded a corner kick—a free kick taken by a member of the offensive team at the quarter circle, at the cornerflag post nearest to where the ball went behind the goal line. The flag post must not be removed.

(p) The game is started and, after a goal has been scored, is resumed by placing the ball in the center of the midfield line. Players must be on their side of the line until the ball is kicked. The ball must be kicked forward and must move at least 2 feet to be legal. The first kicker may not touch the ball twice in succession at the kick off. The opposing team must be 10 yards from the ball until it moves.

b.Speedball. Speedball offers vigorous and varied action with plenty of scoring opportunities. It is easy to learn and provides spontaneous fun. Little equipment is needed-a ball is all that is absolutely necessary. Speedball combines the kicking, trapping, and intercepting elements of soccer; the passing game of basketball; and the punting, drop kicking, and scoring pass of football.

(1) Place in the Program.Speedball, like soccer, should be introduced into the physical training program during the latter part of the toughening stage and used as a competitive activity in the sustaining stage. Speedball may be played any time weather permits but is primarily a spring or fall activity.

(2) Method of Play.The game is played as follows:

(a) Two teams of 11 each play the game under official rules, but any number of players may successfully constitute a team.An inflated leather ball, usually a soccer ball, is used. The playing field is a football field with a football goal post at each end. (See fig. 5-17.)

(b) The game starts with a soccer-type kickoff. The kicking team tries to retain possession of the ball and advance it toward the opposite goal by passing or kicking it. Running with the ball is not allowed, with the result that there is no tackling or interference. After the ball touches the ground, it cannot be picked up with the hands or caught on the bounce but must be played as in soccer until it is raised into the air directly from a kick. Then the hands may be used.

(c) When the ball goes out of bounds over the sidelines, it is given to a player of the opposing team and is put into play with a basketball throw- in. When it goes over the end line without a score, it is given to a player of the opposing team who may either pass or kick it onto the field.

(d) When two opposing players are contesting possession of a held ball, the official tosses the ball up between them as in basketball.

(e) Points are scored by kicking the ball under the crossbar of the goal post, drop kicking the ball over the crossbar, or completing a forward pass into the end zone for a touchdown.

Figure 5-17 is available in the illustration section at the end of the book.

(3) Basic Skills

(a) Shared Skills. Skills include kicking, passing, heading, and trapping (from soccer); punting, dropkicking and forward passing (from football); and passing, receiving, and pivoting (from basketball).

(b) Kickups. The kickup is a play in which a player lifts the ball into the air with the feet so as to legally play the ball with the hands. The kickup generally makes the transition from ground play to aerial play. The technique of making the play depends upon whether the ball is rolling or stationary. If the ball is rolling or bouncing toward the player, the foot is held on the ground with the toe down until the ball rolls onto the foot. Then the foot is raised, projecting the ball upward. If the ball is stationary, the player rolls it backward with one foot. Then the player places a foot where the ball will roll onto it. The player then raises the foot, projecting the ball as before. If a ball is rolling away, the player should stop it with a foot and play it as a stationary ball. The player can also raise the ball by standing over it with a foot on either side. The player presses the feet against the ball and jumps into the air, propelling the ball into the hands.

(4) Offensive and Defensive Play. The positions of players in speedball are much the same as in soccer. However, some of the positions have different names. Each team has 11 players. The forward line is composed of five players, the right end, right forward, center, left forward, and left end. The second line consists of right halfback, fullback, and left halfback. In the next line is the right guard and left guard. The player who defends the goal is the goal guard. The strategy employed in speedball during offensive play is very similar to that of soccer. There are two types of defensive formations in speedball: man-for-man and position defense. Man-for-man defense is recommended for beginning players.

(5) Abridged Rules

(a) Field. A speedball field is 100 yards long with additional 10-yard end zones on each end and 55 yards wide, the same as a regulation football field.

(b) Players. Eleven players are on a team. The goal guard has no special privileges.

(c) Time. Ten-minute quarters. Two minutes between quarters. Ten minutes between halves. Five minutes for extra overtime periods.(Begin the first overtime by a jump ball at center, same goals; change goals in the event of a second overtime period.)

(d) Winner of Toss. The winner of the toss has the choice of kicking, receiving, or defending a specific goal.

(e) Kickoff. A kick off starts the first and second half of a game and starts play after a score. The kick off is made from any place on the 50-yard line. The kicking team must be behind the ball when it is kicked. The receiving team must stay back of its restraining line (10 yards distance) until the ball is kicked. The ball must go toward the opposing team's goal before it can be played. Kick off out of bounds goes to the opponent at that spot. A kick off touched by the receiving team before going out of bounds, with no impetus added, still belongs to the receiving team. A kick off, in possession and control of the receiving team and then fumbled out of bounds, belongs to the kicking team at that spot.

(f) Starting Second and Fourth Quarters. The ball is given to the team that had possession at the end of the previous quarter, from out of bounds, as in basketball.

(g) Second Half. The first-half receiving team kicks off at the beginning of the second half.

(h) Fly and Ground Ball. The most characteristic playing rule of speedball is the difference between a fly ball (or aerial ball) and a ground ball. A player is not permitted to touch a ground ball with the hands and must play it as in soccer. A fly ball is one that has risen into the air directly from the foot of a player (for example, a punt, drop kick, place kick, or kickup).Such a ball may be caught with the hands provided the catch is made before the ball strikes the ground again.

(i) Kickup and Overhead Dribble. A kickup is a ball that is so kicked by a player that he can catch it himself. A bounce from the ground may not be touched with the hand because it has touched the ground since being kicked. This rule prohibits the ordinary basketball dribble, but one overhead dribble (throwing the ball into the air and advancing to catch it before it hits the ground) is permitted.

(j) Out of Bounds. If a team causes the ball to go out of bounds over the sidelines, a free throw-in (any style) is given to the opposing team. When the ball goes over the end line without scoring, it is given to the opponents who may pass or kick from out of bounds at that point.

(k) Tie Ball. In case two players are contesting the possession of a held ball, even in the end zone, a tie ball is declared. The ball is tossed up between them.

(6) Scoring

(a) Field Goal(3 points).A soccer-type kick, in which a ground ball is kicked under the crossbar and between the goal post from the field of play or end zone, is a field goal. (A punt going straight through is not a field goal for it is not a ground ball. The ball must hit the ground first.) A drop kick from the field of play that goes under a crossbar does not count as a field goal. A drop kick from the end zone that goes under the crossbar counts as a field goal;if it goes over the crossbar, it is ruled as a touch back, and the other team takes possession of the ball.

(b) Dropkick(2 points). A scoring drop kick must be made from the field of play and go over the crossbar and between the uprights. The ball must hit the ground before it is kicked (usually with the instep).

(c) End goal(1 point). This is a ground ball which receives its impetus (kicked or legally propelled by the body) from any player, offensive or defensive, in the end zone and which passes over the end line but not between the goal posts.

(d) Penalty Kick(1 point).A ball kicked from the penalty mark that goes between the goal posts and under the crossbar is a penalty kick. The penalty mark is placed directly in front of the goal at the center of the goal line.

(e) Touchdown(1 point). A touchdown is a forward pass from the field of play completed in the end zone. The player must be entirely in the end zone. If the player is on the goal line or has one foot in the field of play and the other in the end zone, the ball is declared out of bounds. If a forward pass is missed, the ball continues in play, but it must be returned to the field of play before another forward pass or drop kick may be made.

(7) Substitutions. Substitutions may be made any time when the ball is not in play. A player who is withdrawn may not return during that same period.

(8) Timeout. Three legal timeouts of 2 minutes each are permitted each team during the game.

(9) Fouls

(a) Personal Foul. Kicking, tripping, charging, pushing, holding, blocking, and unnecessary roughness of any kind, such as running into an opponent from behind, are personal fouls. Kicking at a fly ball and thereby kicking an opponent is a personal foul. Four personal fouls disqualify a player from the game.

(b) Technical Foul. Illegal substitution, more than three timeouts in a game, unsportsmanlike conduct, and unnecessarily delaying the game are technical fouls.

(c) Violation. Traveling with the ball, touching a ground ball with the hands or arms, double overhead dribble, violating tie ball, and kicking or kneeing a fly ball before catching it are violations.

(d) Penalties. The offended player shall attempt the kick.

(e) Officiation of Fouls

1 Fouls in the field of play allow no follow-up while fouls in the end zone always allow follow-up.

2 On penalty kicks, with no follow-up, only the kicker and goalie are involved.

3 On penalty kicks, with a follow-up, the kicking side is behind the ball and the defending side behind the end line or in the field of play. No one is allowed in the end zone or between the goal posts except the goal guard. The kicker must make an actual attempt at goal and cannot play the ball again until after another player plays it.

c. Volleyball. Volleyball is a popular sport. The game entails much physical activity, yet it is not strenuous. It is, therefore, a game for young and old alike, for beginners and for skilled players. It may be played indoors or outdoors on any type of terrain. As an informal activity, volleyball can be played by any number; as an organized activity, it provides, as few other sports do, a game for 12 to play in a limited area. While volleyball requires no great skill to play, it permits a high degree of proficiency. Players naturally get more enjoyment when they know the game and play it well. For this reason, instruction in the basic skills should be provided.

(1) Place in the Program Volleyball may be used occasionally as a competitive activity during the sustaining stage. Though it is a year- round sport, it should be included in the physical training program only when a more strenuous activity is impractical. Volleyball is an excellent off-duty activity.

(2) Instruction. Usually during the first three or four classes, a 10- to 15- minute period of instruction, followed by scrimmage, is enough to teach the basic skills, rules, and techniques of volleyball. More time, if available, can be given to teaching basic skills, but emphasize competitive play rather than formal instruction. Divide the unit into 6-person teams, organized on the basis of ability .Teams should be as nearly equal as possible.

(3) Basic Skills

(a) Passing the High Ball. The chest pass is the most effective method of playing the ball. To receive the ball, the knees are fixed, and the body is tilted forward. The elbows are raised sideward to a point in line with the shoulders. The wrists are extended in line with the forearm. The arms, wrists, and hands are rotated inward. To pass the ball, the hands are chest-high, thumbs pointing inward. The fingers are flexed, forming a cup, allowing them to contact the ball. On contact with the ball, the wrists are snapped while the fingers and elbows are pushed upward, sending the ball upward. A high ball is much easier to handle than a low one.

(b) Passing the Low Ball. A ball that is lower than the waist is one of the easiest to hit, but it is also a frequent cause of fouls (holding or carrying the ball). The best position for handling a low ball is to have the knees flexed and arms flexed at the elbows and rotated so the thumbs are pointing outward, the palms up. When the fingers contact the ball, the entire body reacts in a lifting motion. The arms and hands swing upward in a scooping action. It is important that the fingers, not the palms, contact the ball, and that the ball is batted, not thrown.

(c) Underhand Serve. Take a position behind the back line facing the net, left foot forward, holding the ball in the palm of the left hand. The left knee is flexed; the right knee is straight. Swing the right arm back and at the same time move the left hand (holding the ball) across the body in line with the right hip. Then swing the right arm forward hitting the ball off of the left hand with the palm of the right hand, raising the hips and arching the back in the same motion. Be certain to swing the right arm in a straight line, or the ball will be difficult to control. When the opposition is in formation, the server should try to place the ball in the right or left back area, and not near the net.

(d) Setup. A setup is a ball hit into the air near the net by one player, so that a teammate may hit or "spike" it sharply downward into the opponent's court. The chest pass is the best pass to use. The ball is sent approximately 10 feet into the air toward the spiker so it will descend from 4 to 20 inches from the net.

(e) Spiking. The spike is a leap into the air and a sharp downward hitting of the ball into the opponent's court. A spiker must be able to spring easily from the floor, judge the movement of the ball, and strike it with a downward movement of the arm. To jump from the floor, step off with one foot and jump with the other. Stand with the right or left side to the net, facing the setup player. Much depends upon the setup player to place the ball in the proper position. The spiker jumps into the air and strikes the ball above its center so as to drive it downward. A snapping movement of the arm and wrist will drive the ball forward and downward with power and control. Aim for a weak spot in the opponent's defense.

(f) Blocking. The block is a defense technique used to prevent a spiker from driving the ball across the net. It is an attempt by one or more defensive players at the net to block a hard-hit shot by using the force of the ball to send it immediately back into the opponent's court. In an effective block, forwards on the defensive team spring into the air at the time of the spike, placing both hands and arms in the expected path of the ball. An effective block tends to upset the offense and presents another element for the spiker to worry about. To be effective, the blocker must anticipate the path of the ball and time the block with the spike.

(4) Abridged Rules

(a) The volleyball court is 30 feet wide by 60 feet long. (See fig. 5-18.)

(b) The top of the net is 8 feet high.

(c) A volleyball team consists of 6 players.

(d) A match consists of the best two out of three games.

(e) The first team scoring 15 points wins the game, provided that they have 2 points more than their opponents.

(f) A deuce game is a game in which both teams score 14 points. The game is continued until one team obtains a 2-point advantage over the other.

(g) Only the serving team can score. If the serving team commits a fault, it loses the serve to the opposing team.

(h) The team members who receive the served ball rotate one position in a clockwise direction.

(i) The ball is put into play by serving from behind the back line.

(j) A served ball touching the net results in the loss of the serve. At any other time during play, a ball touching the net is still in play.

Figure 5-18 is available in the illustration section at the end of the book.

(k) The ball is out of play when it touches the ground or goes outside one of the boundary lines.

(l) All line balls are good.

(m) The players must hit or bat the ball; they may not throw, lift, or grasp it.

(n) A player may not touch the ball with any part of the body below the knees.

(o) A player may not play the ball twice in succession. In receiving a hard-driven spike, a defensive player may make several contacts with the ball even if they are not simultaneous. All such contacts, however, must constitute one continuous play, and all must be above the knees.

(p) The ball may be touched no more than three times on one side of the net before being returned across the net to the opposing team.

(q) A player must not touch or reach across the net.

(r) A player may touch the line under the net, but may not cross it.

Chapter 6
EVALUATION OF PERFORMANCE DURING TRAINING

6001.OBJECTIVES OF COMBAT READINESS

The objectives of the Marine Corps physical fitness program are-

a. To contribute to the health and well-being of every Marine through regular exercise and health education.

b. To develop Marines who are physically capable of performing their duties in garrison and in combat.

c. To develop Marines with a reserve level of physical fitness that will help them win in combat.

d. To provide a medium for developing the self-confidence of the individual Marine and thereby enhancing overall discipline, morale, esprit de corps, unit efficiency, and the desire to excel within the Marine Corps.

6002. THE PHYSICAL FITNESS TEST

The physical fitness test (PFT) has distracted attention from the proper goal of the physical fitness program. Simply stated, the goal of the physical fitness program is the success of Marines in combat. Unit commanders must not allow the PFT to become an end in itself. The PFT should be used within the commander's program of combat readiness training as a standard, an index of the physical fitness of individual Marines at a given time and place. The PFT is not difficult, and it must be kept in perspective. All athletes should be first class, and first class is attainable for all Marines. The conduct of the PFT is described in Marine Corps Order 6100.3_.

6003. OTHER METHODS OF EVALUATION

The commander should use other ways of evaluation to provide variety and imagination. Intangible objectives, such as confidence and aggressiveness, are hard to measure. The use of either inspection or observation is necessary to evaluate objectives of this nature. The commander has several other methods available when evaluating the physical condition of the command. The additional methods of inspection, observation, medical examination, and testing are summarized as follows:

a. Formal Inspections. Formal inspections, using inspection officers and standardized rating criteria, may help evaluate unit physical fitness.

b. Daily Observation. Routine observation of physical performance and appearance can serve as an indicator of an individual's or unit's physical readiness. However, mere observation is not a totally reliable or accurate means of evaluation.

c. **Physical Examination.** Medical examination may detect any individual disability or detrimental physical condition. It may guide in application of remedial, therapeutic, or limited exercise programs.

d. Other Physical Fitness Tests. Some other tests are described in detail in this manual for information purposes. Commanders are cautioned, however, not to place too great an emphasis on testing.

6004. RESPONSIBILITY

The commander is responsible for the physical fitness of the command and for the measurement and evaluation of its physical readiness. The goal of combat readiness training remains to prepare Marines and units for the physical demands of combat. The bulk of the commander's combat readiness training program should be such activities as obstacle and confidence courses, endurance courses, and progressive load-bearing marches which will prepare Marines for combat. Group activities which are competitive and combative in nature are advantageous and add imagination and enthusiasm to the program. Activities such as orienteering, which is not only physically demanding but also competitive and which teaches essential subject skills, are excellent and motivate all concerned. A commander's goal should be to provide an integrated training program that is so aggressive and demanding of Marines and leaders alike that a PFT administered at any time will reflect that each and every Marine is both physically and mentally prepared for the demands of combat.

6005. COMMANDER'S PHYSICAL FITNESS PROGRAM

The Commandant of the Marine Corps has directed that the physical fitness test shall be the universal measure of adequate individual physical fitness. The focus of a commander's physical fitness training program, however, should not be to prepare Marines to pass the physical fitness test, but to prepare them for the physical rigors of combat.

a. Administration. The physical fitness program requires each Marine to spend a minimum of 3 hours per week in physical fitness training, to be tested semiannually, and to obtain a minimum standard of third class. The attainment of a higher score is a laudable individual goal which should be encouraged but should not become a unit objective. Overemphasis could be detrimental to training required to develop the complete Marine. All commanders should exercise sound judgment as well as positive, aggressive leadership in striving toward the training objective.

b. Variety of Activities. The program must include the following elements:
(1) Physical Readiness Training. physical readiness training is a complete physical training program which develops and maintains the strength, endurance, and physical skills needed to sustain the individual during combat.
(2) Remedial Physical Conditioning. Remedial physical conditioning is a process by which physically substandard Marines are conditioned to meet prescribed standards.
(3) Competitive Conditioning Activities. Competitive conditioning activities consist of teams or individuals competing against an opponent to win. This involves a combination of sports and military skills designed to foster competitive attitudes and develop unit pride and esprit de corps.
(4) Occupational Conditioning. Marching long distances, dry net training, etc., should not be ignored as a means to fulfill the time requirement.

c.Time. In order to develop the desired level of physical fitness, unit commanders are advised that a minimum program of 3 hours per week is mandatory. The program must encompass at least three exercise periods weekly. Each period should include calisthenics, running, or other forms of vigorous activity .The minimum desirable length for each exercise period is 60 minutes. Shorter periods may be authorized by the unit commander if dictated by the local training situation. Unit commanders are encouraged to use the normal working day to satisfy this training requirement. However, commanders are authorized to conduct required physical fitness training during off-duty hours when the mission, workload, personnel status or other significant factors preclude fulfilling it during duty hours.

d.Action. Marine Corps Order 6100.3_ requires that commanders-

o Establish and maintain an effective physical fitness program which ensures that all Marines maintain an acceptable level of physical fitness.

o Ensure that all individuals are medically qualified prior to participating in the physical fitness program.

o Establish a minimum physical fitness program of at least 3 hours per week. Physical training may be authorized on an individual basis at the discretion of the unit commander.

o Ensure that all Marines participate in physical fitness conditioning activities commensurate with their medical qualifications and limitations.

o Conduct physical fitness testing for all Marines.

o Place all Marines who fail to pass the PFT on a daily command-supervised remedial physical conditioning program until they pass the test.

o Ensure that results of physical fitness testing are entered on section A of the fitness report. Comments are placed in section C in connection with attainment of superior physical performance or a medical excuse from the PFT.

6006. OTHER FORMAL TESTS

Commanders must realize that the best index of combat readiness is personal observation of performance on conditioning marches of 15 miles while under load. This manual presents three additional indicators of unit fitness. These indices are the physical readiness test, which was formerly the Marine Corps standard and is now routinely administered at Officer Candidate School; the airborne trainee physical fitness test which many Marines who attend U.S. Army formal schools are required to take; and the battle fitness test, the ultimate test of the combat readiness of Marines. Commanders may wish to include these tests or portions thereof within their combat readiness programs.

6007. PHYSICAL READINESS TEST

a. Purpose. The purpose of this test is to measure the Marine's ability to meet the minimum standard of physical readiness for duty in the field. This test is currently administered to all male officer candidates at Officer Candidate School, MCCDC, Quantico, VA. Female candidates do not take this test.

b. Test Area. Administration of this test requires an area suitable for crawling, jumping, and running 160 yards to include a trench that is 8 feet wide.

c. Equipment. The equipment required for this test includes—
o A platform, rail on a ledge, or a box 18 inches high,
o A 20-foot climbing rope,
o A stopwatch.

d. Conduct of Test. Adequate timers and other supervisory personnel must be provided by the company staff. All events should be conducted in a single session of one morning or afternoon. The events may be run in any sequence.

e. Uniform. The uniform for the physical readiness test is as follows: full utilities, cartridge belt, belt suspender straps, two canteens (full of water), rifle, and helmet. Depending on the heat condition, the PT shirt may be worn in place of the utility jacket. All equipment will remain secure throughout the test. Except for emergencies, canteens are not to be emptied.

f. Events

(1) Event 1: Climbing Uphill (Step-ups)This event simulates marching uphill at a rapid and steady pace. The Marine stands in front of the step. On the command GO, the Marine places either foot on the step and steps up. Hand pressure on the knee may be used if desired. The Marine then stands erect and steps backward and down, on one foot at a time. The same leg may be used for each repetition or the legs may be alternated. For 100 points, Marines will complete 60 repetitions in the maximum allotted time of 1 minute and 20 seconds (80 seconds).

*(2) Event 2: Rope Climb. T*his event resembles entering and leaving a hovering helicopter, using ropes in house-to-house fighting, and ascending and descending landing nets. The Marine assumes a sitting position on the ground at the bottom of the rope with hands at the highest point that can be reached. On the command GO, the Marine jumps up and climbs to the top. The Marine must touch the 20-foot mark with one hand, then descend without sliding or dropping. Marines must reach the 20-foot mark in the maximum allotted time of 30 seconds or less.

(3) Event 3: Evacuation. This event simulates reaching and evacuating a wounded Marine under fire. The Marine assumes the prone position at a distance of 50 yards from a casualty who is of approximately equal weight. On the command GO, the Marine springs up and covers the distance in a sprint. The Marine lifts the casualty to a fireman's carry and returns to the starting point. Assistance by the casualty is permitted. The tested individual must carry all equipment; the casualty has none. Marines must negotiate the course in a maximum time of 47 seconds or less.

(4) Event 4.'Advance by Fire and Maneuven. This event simulates advancing and assaulting during an attack. The Marine assumes the prone position at the starting point. On the command GO, the Marine then alternately crawls, runs in a zigzag manner, rolls, jumps, and assumes different firing positions for the next 130yards. Before reaching the objective, the Marine must "hit the deck," roll over, and assume the firing position three times. While assuming the firing position, the Marine must pull back the charging handle, take aim, and squeeze the trigger before rising from the ground. After rising from the ground for the third time,the Marine continues running and jumps to clear the 8-foot trench in one leap. The Marine then finishes the course in

the maximum allotted time of 2 minutes. The Marine is allowed multiple attempts only within the specified time limit.

(5) Event 5: Forced March. This event is a combat run of 3 miles without halts and against time. The Marine must double-time. Marines must complete the 3-mile course in a maximum allotted time of 30 minutes or less.

g. Event Standards

(1) To successfully pass the physical readiness test, the Marine must pass each event with a minimum of 80 points for each. Points are then assessed for each event in accordance with the scoring tables. (See fig. 6-1.)

(2) If a Marine fails one event, the highest possible score that the Marine may receive for the entire test is 79 percent. (See example 1 in fig. 6-2.) If two events are failed, the highest possible score for the entire test is 69 percent, regardless of the total points accumulated. (See example 2 in fig. 6-2.) If three events are failed, the highest possible score is 59 percent. Failure in four or more events will result in a score of zero for the entire test.

(3) To convert the total points for all five satisfactorily passed events to a final percentage score, total the accumulated points for all five events and divide by five. (See example 3 in fig. 6-2.)

Figure 6-1 is available in the illustration section at the end of the book.

Figure 6-2 is available in the illustration section at the end of the book.

6008. AIRBORNE TRAINEE PHYSICAL FITNESS TEST

a. Use and Composition of the Test.

(1) Use. The airborne trainee physical fitness test is a means of determining the physical ability of the applicant for acceptance to and retention in the airborne course of instruction.

(2) Test Events The test battery consists of five events as follows: chinups, knee bender, pushups, situps, and an endurance run.

b. Method of Scoring and Standards

(1) Scoring. The Marine will be scored by a trained scorer who is thoroughly familiar with the minimum standards for the test events. The Marine will be scored on a pass or fail basis. The performance on each event may be recorded on the reverse side of DA Form 705 (Physical Fitness Testing Record).The examinee records the personal information on the face of the card as specified on the first, second, and third lines. The face of the card is identified by writing diagonally across the lower half of the card "Airborne Trainee PPT—See reverse side."Using line 7 on the reverse side of the scorecard, the scorer enters the test title in the block entitled "(Other (Specify))" and completes the required information on the remainder of that line. The "Remarks" section can be used to record the test event titles and the applicant's performance on each test event.

(2) Standards. To successfully pass the test, the examinee must reach the standard in each test event. The standards follow:

(a) Chinups

(b) Knee bender 80(2-minute period)

(c) Pushups 45(2-minute period)

(d) Situps45(2-minute period)

(e) Endurance runs:2 miles in 15:54 minutes or less in athletic gear and 4 miles in 32 minutes in utilities and running shoes.

c. Uniform for Testing

(1) Examinees. The prescribed uniform for test participation is boots and the work uniform of the season. No headdress is worn. When climatic conditions permit, jackets or outer shirts may be removed.

(2) Officials. Scorers and other test officials should be uniformly and distinctively dressed for contrast with Marines being tested.

d. Test Administration Procedure

(1) Preparation for Test. The administration of the test to a large group makes it mandatory that the test effort be organized and efficiently operated. All testing is not completed with large test groups; on certain occasions, individuals and small groups are tested. Care must be exercised to administer the test uniformly and to standardize the conduct of all elements of the test. Regardless of the size of the test group, the following elements of sound test administration should be included:

(a) An orientation to include the purpose, method of administration, scoring of the test, preparation of the score card, and required standards.

(b) A correct demonstration of each event to ensure that there is no misunderstanding of the proper form and required standards.

(c) Completion of all five test events in one test period with all Marines taking the events in the same sequence.

(d) Adequate rest periods between the test events to allow for recovery before the next event.

(2) Method of Administration with a Large Group. With a minimum of 12 lanes per test event, 14 officials can administer the test battery to 150 or 200 Marines in 2 hours. The officials are designated as follows: one officer in charge, one demonstrator,and 12 scorers. If more or fewer Marines are to be tested, a greater or lesser number of officials will be required. The following procedure is recommended:

(a) Conduct an orientation and ensure the examinees have properly completed their score cards.

(b) Assign Marines to lanes and caution them to remain in the same lane order throughout the test.

(c) Explain and demonstrate the chinup event, administer it, and score it. Then proceed to the knee bender and pushup events, and administer them in the same manner.

(d) Grant a 5- to 10-minute rest period after the pushup event. Advise against excessive consumption of water during the break period.

(e) Explain and demonstrate the situp event, administer it, and score it as prescribed. Then move to the run area, explain the running event, and complete it.

(f) Retain the scorecards at the completion of the running event.

(3) Method of Administration with a Small Group. A similar procedure is followed for the testing of individuals and small groups. The informality usually associated with small groups must not conflict with sound test administration. With fewer examinees, fewer officials are required.

e. Description and Explanation of Test Events

(1) Test Event 1:Chinups

(a) Purpose. This event is devised to test arm and shoulder flexor strength.

(b)Equipment There is one horizontal bar per lane, made of plumber's pipe or a gymnasium horizontal bar 1 1/2 inches in outside diameter. The bar must be rigidly supported at a height of 8 feet above the ground. The upright supports must be 5 feet apart. There must be a movable stand at each bar for short Marines to stand on to reach the bar.

(c) Officials There is one scorer per lane.

(d) Organization. The examinees, holding their scorecards, stand in order behind the restraining line in their respective lanes. The scorers take each scorecard when the Marine is called forward for the test.

(e) Starting Position The bar is grasped with the palms turned toward the face, the thumbs underneath the bar. The body is fully extended in a "dead" hanging position with the arms straight and the feet above the ground.

(f) Movement. Pull the body directly upward until the chin is placed over the bar. Lower the body until the elbows are completely straight and the body is again in the "dead" hanging position. Repeat as many times as required.

(g) Instructions. Explain and demonstrate the fully extended "dead" hanging position with the proper grasp. Show that the chin is placed over the bar at the top of the movement and that the arms are fully extended, the elbows completely straight, at the bottom of the movement (the hanging position). Explain that the body must be kept from swinging and that it is permissible to raise the legs and flex the hips when pulling up, but that any kicking, bicycling, or jerking motion with the trunk or legs is not acceptable. Inform the examinees that no penalty is exacted for hanging on the bar to rest in the bottom position but that this is not to their advantage. Tell them that half-completed chinups are not counted, and that the scorer will repeat the number of the last correct chinup when incorrect execution is detected.

(h) Administration and Scoring. Caution the examinees to assume the "dead" hanging position and wait for the scorer's command to begin. The scorer is at the examinee's left with a clear view of the bar. If the examinee begins to swing widely, the scorer should stop the swinging by extending the left arm across the front of the examinee's body, being sure not to hinder the execution of the chinup. The scorer counts aloud the number of chinups correctly executed. When a chinup is not correctly executed, the scorer repeats the number of the last correct one. The scorer records the number of correct chinups on the scorecard and returns the card to the examinee.

(2) Test Event 2:Knee Bender

(a) Purpose. This event measures the strength and endurance of the leg muscles.

(b) Equipment. None.

(c) Officials There is one scorer per lane.

(d) Organization. Marines stand in numerical order behind the restraining line in their respective lanes. The scorer takes each scorecard when the Marine is called forward for the test.

(e) Starting Position. The feet are spread less than shoulder width apart, hands on hips, thumbs in the small of the back, elbows back.

(f) Movement. Do a knee bend and at the same time bend slightly forward at the waist and thrust the arms between the legs until the extended fingers touch the ground. The hands are about 6 inches apart. The bend is approximately a three-quarters bend. From this knee bend position, recover to the starting position by moving the body upward, straightening the knees, and returning the hands to the hips. Repeat as many times as required.

(g) Instructions. Explain and demonstrate the correct starting position. Be certain examinees understand the correct knee bend and that only the tips of the fingers touch the ground. Tell them the scorer will repeat the number of the last correct knee bender when incorrect execution is detected. Some of the common errors are failure to correctly bend the knees, failure to touch the ground, and failure to assume the completely erect position after the bend has been executed.

(h) Administration and Scoring. The scorer stands to one side so as to see that the knees are properly bent and the fingers touch the ground as prescribed. From this position, the scorer can view the examinee to see that a properly erect position is assumed after each knee bend. The scorer counts aloud the number of correctly executed knee bends. When a knee bend is done incorrectly, the scorer repeats the number of the last correct one. The scorer records the number of correct knee bends on the scorecard and returns it to the examinee.

(3) Test Event 3:Pushups

(a) Purpose. Pushups measure arm and shoulder extensor strength.

(b) Equipment. None.

(c) Official. There is one scorer per lane.

(d) Organization. Marines stand behind the restraining line in their respective lanes until the scorer calls on them to perform. The scorer takes each scorecard when the Marine comes forward.

(e) Starting Position. The front-leaning rest position is the starting position. The body is straight from head to heels, palms are flat on the ground directly underneath the shoulders, and elbows are straight and locked. The body weight is supported on the hands and toes throughout the event.

(f) Movement. Bending only the elbows, lower the body in one straight plane until the chest touches the scorer's hand. Straightening and locking the elbows, raise the body in one straight plane, returning to the original front-leaning rest position. Repeat as many times as required, keeping the body in a straight line from head to heels.

(g) Instructions. Explain and demonstrate that the arms are straight at the beginning and completion of the movement and that the chest must touch the scorer's hand, but the stomach and thighs must not touch the ground. Also explain that the whole body must be maintained in a straight line as it is lowered and raised; that is, there is to be no breaking at the hips or shoulders so that any body part is lowered or raised in advance of the other or as a separate segment .Dipping or rolling through the shoulders is illegal, as is lowering or raising the body with one arm or shoulder at any time. Resting is not permitted during repetitions. Instruct

Marines that the scorer will repeat the number of the last correct pushup when incorrect execution is detected.

(h) Administration and Scoring. It is recommended that Marines assume a prone position while placing their feet and hands in the proper positions. This permits them to rest while the scorer gets into position and, at the same time, provides a feel of the body in a straight plane from head to heels. The scorer lies on the right hip and side to the right of the examinee. This gives the scorer a clear view of the examinee's body to see any errors. The palm of the scorer's right hand rests flat on the ground underneath the lowest part of the examinee's chest. By keeping the right forearm flat on the ground at an angle in front of the examinee's right arm, the scorer's position will not prevent the examinee from lowering the body completely. The scorer's left hand is free to test the straightening of the elbow at the completion of the movements and to point out body segments being lowered or raised separately. When in position and ready, the scorer has the examinee assume the starting position and begin doing pushups. The scorer counts aloud the repetitions done correctly and repeats the number of the last correct pushup if one is done incorrectly. There is no penalty if the contour of the examinee's body causes the hips to protrude slightly out of line, provided that the whole body is raised and lowered simultaneously. The scorer enters the number of repetitions on the scorecard and returns it to the examinee.

(4) Test Event 4: Situps

(a) Purpose. This event primarily measures abdominal strength.

(b) Equipment. None.

(c) Official. There is one scorer per lane.

(d) Organization Marines stand behind the restraining line in their respective lanes until the scorer calls on them to perform. The scorer takes each scorecard when the Marine comes forward.

(e) Starting Position. The examinee lies flat on the back with knees flexed, both feet flat on the ground. The correct angle of the thighs to the ground is 45 degrees. If the heels are too near the buttocks, the applicant will not be able to sit up. Legs are spread shoulder width apart. The examinee interlaces fingers and places them behind the head in contact with the ground. The feet are not held by another person.

(f) Movement. Bend forward at the waist and raise the upper body until the head is directly over the knees. Heels are not to leave the ground. Elbows remain in the same plane to the head and body throughout the event. The upper body is slowly lowered to the starting position until the head touches the ground. Repetitions are done at a slow cadence with no rest periods.

(g) Instructions. Explain and demonstrate the correct starting position and the proper execution of the situps to be sure that Marines understand the movement. Warn them that their knees must remain flexed during each situp, the heels cannot leave the ground at any time, and they may not roll up on one side and push up with one elbow. Tell them they must do the repetitions at a slow cadence with no rest periods. Instruct Marines that the scorer will repeat the number of the last correct situp when incorrect execution is detected.

(h) Administration and Scoring. When the performer is in position and ready, the scorer has the performer assume the starting position and begin doing situps. The scorer counts aloud the correct executions. When a situp is improperly done, the scorer repeats the number of the last correct one. No situp is credited if the hands are unclasped from behind the head, if the back is used to bounce up from the ground (which means the shoulders would not touch the ground), or if one shoulder or elbow is used to push up. The examinee is not penalized if the heels slide forward slightly so long as the knees remain flexed and the heels maintain contact with the ground. The scorer enters the number of repetitions on the scorecard and returns it to the examinee.

(5) Test Event 5:Endurance Run

(a) Purpose. This event measures cardiovascular endurance.

(b) Equipment One stopwatch or watch with a sweep second hand.

(c) Area. A large training field on which a quarter-mile track has been staked out or a level road over flat terrain may be used as a running surface. A 1-mile route is designated with wooden stakes marking the start point, finish point, and one-quarter mile intervals.

(d) Officials For large groups, a scorer times the event and controls the conduct of the run, and a guide runs with the group and sets the pace.

(e) Organization The run is conducted with groups of Marines in a column formation. Company-sized units may run at the same time with platoons serving as running groups. The scorer issues the command to assume double-time.

(f) Starting Position. Marines are assembled in proper column formation (column of twos, threes, or fours, as appropriate to the size of the group) with short men to the front. When all is ready, the column is moved forward a short distance before the running period is started.

(g) Movement. At the command DOUBLE-TIME, MARCH, the examinees retain their places in the column formation and execute the command. Length of steps is about 40 inches. The scorer has the group execute the run. The formation is maintained during therun.

(h) Instructions. Marines are instructed to maintain formation while running and are informed that the guide will set the proper pace. They will be instructed in the command used to control the column in the execution of the test. The scorer should announce the 4-minute, 2-minute, 1-minute, and 30-second remaining time intervals.

(i) Administration and Scoring. The event may be administered as previously prescribed to a large group, to several Marines, or to an individual. An individual examinee usually does not require a guide or pacer. If the event is administered on a training field, the scorer may stand in the center as the runner(s) circles about the field. This method of administration relieves the scorer of running with each group to be tested. Scoring is based on successful completion of the run as prescribed. The scorer should announce the remaining times as prescribed in (h).

6009. BATTLE FITNESS TEST

The battle fitness test is a physical fitness evaluation which is currently under development as part of the basic warrior training concept plan.

Chapter 7
THE HUMAN BODY

7001.GENERAL

To implement an effective physical conditioning program, a leader must understand how the human body functions and how exercise affects the body .This chapter provides only an introductory discussion of anatomy and functioning.

a. Body Functioning During the Stages of Conditioning. For more on the stages of conditioning, see chapter 1.

(1) During the toughening stage, the waste products of muscle activity (lactic acids) collect more rapidly than the blood can remove them. This acid waste builds up in the muscle tissue and irritates the nerves in the muscle fiber, causing pain and stiffness. As the exercise program continues, more blood is carried through the muscle, removing the waste materials more rapidly and eventually causing the soreness to disappear.

(2) During the slow improvement stage, the blood circulation in the muscles increases, and the body as a whole becomes a more efficient machine. The improvement is rapid in the first few weeks, but as a higher level of skill and conditioning is reached, the improvement becomes less noticeable. The body reaches its maximum level of performance after 6 to 10 weeks and should then be maintained at this peak.

(3) During the sustaining stage, it is possible to maintain this state of conditioning through 15 to 20 minutes of exercise a day, but the exercise must be quite strenuous.

b. Diet. Regular exercise increases the appetite. If the desire for greater amounts of food is satisfied by a balanced diet, the body benefits.

(1) There are two main types of foods: body-building and energy- producing.

(a) Body-building foods consist of proteins, which build up tissue and repair wear and tear.

(b) Energy-producing foods include carbohydrates and fats. Carbohydrates provide a quick source of energy, while fats act as a reserve store of energy.

(2) In addition, food contains vitamins, mineral salts, and water. During hot weather and strenuous training periods, the body requires greater fluid intake.

(3) Diet should be supplemented with proper rest to allow the digestive system to digest the lood.

(4) Occasionally, especially during early stages of conditioning, strenuous exercise may cause vomiting. Although vomiting is not a frequent occurrence, it is not usually a cause for concern.

7002. SYSTEMS OF THE BODY

The systems of the body include the skeletal, muscular, circulatory, respiratory, endocrine, digestive, genitourinary, and nervous systems. Each has a different function, but all must work in cooperation with one another to insure a sound body. Of these systems, the first four are the most affected by exercise and are discussed in succeeding paragraphs.

7003. THE SKELETON

a. Bones. The skeleton (see fig. 7-1) is composed of about 206 bones, which-
o Provide structure for the body and support for the attachment of muscles,
o Protect vital organs, such as the brain, lungs, and heart,
o Manufacture red blood cells, which carry oxygen through the body.

Figure 7-1 is available in the illustration section at the end of the book.

b. Joints. A joint is a place of union between two or more bones. Joints can be-
o Immovable, such as in the face and head.
o Slightly movable, such as in the vertebrae or sternum.
o Freely movable, such as in the shoulder, hip, knee, ankle, and elbow. The bones in a freely movable joint are held in place by ligaments. Freely movable joints are of the greatest importance in physical training because exercise improves their mobility and stability.

c. Cartilage and Ligaments. The joints in the body, where bones connect, are supported by cartilage and ligaments.
o Cartilage is a tough, elastic, translucent tissue that acts as a shock absorber or buffer between bones. Examples are the discs between the vertebrae, the tissue attaching the ribs to the sternum, and buffers in the knee joints. Damaged cartilage does not heal.
o Ligaments are connective tissue that bind bones together. They contract and extend but are not elastic. Consequently, ligaments can be torn or strained. Damaged ligaments will heal to some extent.

7004. THE MUSCLES

a. General. Muscles are tissue, or an organ composed of tissue, which contract or extend to produce movement. Muscles are of three classifications: involuntary, voluntary, and cardiac. The voluntary and cardiac muscles are the most affected by physical training; physical training has little effect on involuntary muscles.(Seefigs. 7-2, 7-3, 7-4.)
o People have no control over the involuntary muscles, an example of which is the diaphragm,
o Voluntary muscles are the larger skeletal muscles which cause movement and which the individual can control.
o Cardiac muscle is found only in the heart and for all practical purposes is an involuntary muscle. Physical training has a significant effect on the fitness of cardiac muscles.

b. Muscle Structure. Muscle is composed of cells which are grouped into bundles called fibers. Groups of fibers form muscles. These bundles of fibers are held in place by sarcolemma, a thin, sheath-like material that surrounds the muscle bundles and secretes a fluid that lubricates the muscle tissue. The fused ends of the sarcolemma form the tendons which attach the muscles to the bones.

c. Attachment of Muscles. The arrangement of muscles on the skeleton provides the proper angle of pull to make movement possible. Voluntary muscles are usually attached to the skeleton in two places.

d.Action of Muscles. To produce motion and to do work a muscle usually shortens its fibers. The movement may be flexion such as bending the arm at the elbow, or extension, such as the straightening the arm. Some muscles raise the arms or legs, others depress the raised limb. Some muscles have the primary function of rotating the trunk from side to side. In addition, muscles act as stabilizers as well as prime movers.

Figure 7-2 is available in the illustration section at the end of the book.

Figure 7-3 is available in the illustration section at the end of the book.

Figure 7-4 is available in the illustration section at the end of the book.

7005. FUNCTIONING OF THE SKELETON AND MUSCLES

a. General. The skeleton and muscles function in close coordination. Muscles move the body, and bones support the muscles.

b. Effect of Exercise on Bones. Continuous exercise, particularly among younger people, usually strengthens the bones, allowing them to withstand greater stress and strain. Bones which are not used lose much of their strength, a factor which should be considered when individuals return to physical training after a prolonged period of inactivity. People in this category should refrain from activities which might result in bone breakage before the bone is exercised back to normal condition. The condition known as "march fracture" is usually a result of this condition.

c. Muscular Strength. When exercised vigorously to improve strength, muscle grows in size. In general, the larger the muscle, the stronger the muscle. Furthermore, trained muscle functions more efficiently than untrained muscle. Trained muscles contract more vigorously and with less effort. Regular and strenuous exercise also toughens muscle, making it firmer and able to stand more strain.

d. **Muscular Endurance.** Muscles naturally become fatigued during continuous, repeated work. Through specific exercises, such as chinups or situps, local muscle groups become exhausted long before an individual fatigues. Training to develop muscular endurance enables people to continue a relatively heavy load of exercise over a long period of time. Lengthened exercise periods improve muscular endurance.

e. Muscular Coordination. Improved speed and strength result in part from improved muscular coordination. An unskilled individual may use irrelevant muscles to perform a particular activity, increasing the amount of work without increasing the mechanical output. This increase in skill is highly desirable, but it should be offset by greater effort in duration or intensity to compensate for the loss in overload due to increased skill.

f. Muscular Fatigue. When the rate of work is increased, the energy required is proportionately much greater than the increase in rate. For example, if an individual doubles running speed, the amount of power demanded to do this is increased eight times.

g. Circulation in Muscles Regular, strenuous exercise causes the creation of new capillaries and the opening of inactive, latent capillaries, increasing blood circulation within the muscle by as much as 400 percent. This increases the supply of food materials and oxygen to the muscle, improving its endurance. This process takes about 8 to 12 weeks of regular conditioning in young adults and longer as age increases.

7006. THE CIRCULATORY AND RESPIRATORY SYSTEMS

a. The Circulatory System. The circulatory system transports blood to all parts of the body, removes waste products for disposal, and delivers protecting and repairing substances where needed. The circulatory system includes the heart and blood vessels.

(1) The Heart. The heart is a pump which forces blood through the blood vessels to the parts of the body. The heart is a little larger than a fist and is located in the left center of the chest between the lungs.

(2) Blood Vessels. The vessels carrying blood away from the heart are the arteries. They eventually divide into capillaries, the very small vessels through which diffusion and osmosis take place. The capillaries gradually increase in size, forming the veins which carry blood back to the heart.

b. The Respiratory System. The respiratory system performs the function of breathing, whereby oxygen is inhaled and carbon dioxide is exhaled. The respiratory system consists of the mouth, nose, trachea, lungs, and diaphragm.

o Trachea. The trachea, or windpipe, is a hollow, tube-like structure that carries air from the mouth to the lungs.

o Lungs. The lungs are elastic bags in the chest where the exchange of oxygen and carbon dioxide takes place.

o Diaphragm. The diaphragm is a thin, sheet-like muscle just below the lungs. During inhalation, the diaphragm flattens out and lowers, allowing the lungs to expand and fill with air. During exhalation, the diaphragm raises into a dome shape, helping to expel air from the lungs.

7007. CARDIOVASCULAR FUNCTIONING

a. General. Cardiovascular functioning is the combined functioning of the circulatory and respiratory systems. The chief organs of these systems, the heart and lungs, function together to provide oxygen to the body.(See fig. 7-5.)

Figure 7-5 is available in the illustration section at the end of the book.

b. Heart Action. The heart, the organ which propels blood through the blood vessels, is the chief organ of cardiovascular endurance. A person tires quickly if the heart lacks the capacity to circulate the blood through the body. The heart is a muscular organ and is developed by exercise just like any other muscle. The heart cannot be exercised alone; any exercise that puts a sufficient load on the heart, such as running or cycling, will also exercise other body parts. A conditioned heart tends to beat more slowly but pump more blood when at rest. This is known as an increase in "stroke volume,"a desirable condition because it enables the heart to pump more blood with a slower contraction rate.

c. Functions of the Lungs. Strenuous and regular exercise can improve the efficiency with which the lungs transmit oxygen to the blood by as much as 25 percent In a poorly conditioned person, some of the alveoli (air sacs) within the lungs are closed or collapsed. The forced breathing created by exercise over a period of weeks causes the air sacs to slowly expand, increasing the ability to absorb oxygen.

d. Relationship of Heart and Lungs. The heart pumps carbon dioxide-laden blood through the pulmonary artery to the lungs. In the lungs the carbon dioxide is exchanged for oxygen and the purified blood is returned to the heart by way of the pulmonary vein. The heart then pumps the blood through the aorta for circulation throughout the body. Blood moving into the muscles exchanges oxygen for carbon dioxide, after which time it is ready for the return trip to the heart.

e. Cardiovascular Functioning in High Altitudes Marines to be employed in high altitudes should be acclimated in a similar area for 10 to 14 days prior to employment. Persons not accustomed to the rarefied air of higher altitudes tire more quickly and may collapse after rapid physical exertion. Air is much less dense at high altitudes than at sea level. Persons accustomed to sea level inhale only about 80 percent of the oxygen they are used to no matter how hard they breathe. Further, people accustomed to sea level or moderate altitude simply do not have enough red corpuscles in the blood to fulfill their needs at high altitudes. The red blood cells will increase over time.

f. Symptoms of Cardiovascular Malfunctioning. The brain is the first organ to react to a lack of oxygen. Unconsciousness results when the brain is denied sufficient oxygen. "Blacking out" is actually a defense mechanism to keep the body alive.

APPENDIX A
INSTRUCTOR HINTS AND INSTRUCTOR TRAINING

1. General. This appendix informs instructors on administration of exercise periods, commands, the extended rectangular and circle formations, methods of instruction and conduct of exercise activities, explanation of basic positions associated with the various activities, and instructor training.

2. Aim of the Physical Fitness Program. The aim of the physical fitness program is to prepare Marines physically for war. This can be conveniently split into two phases.

a. Preparation Phase

(1) Development of all-around physical fitness.

(2) Development of purposeful physical skills.

(3) Development of mental alertness.

(4) Development of character and leadership.

b. Maintenance Phase. Once the preparation is completed, the program must progress to a maintenance program. This phase revolves around the need to maintain the trained Marine's physical fitness for a particular role. This will include—

(1) Maintenance of a high standard of all-around physical fitness.

(2) Application of purposeful physical skills to combat skills.

(3) Development of leadership and self-confidence.

(4) Development of mental toughness and endurance.

3. Physical and Mental Development

a. Marines need to be developed physically and mentally. The physical training (PT) instructor must ensure that the program accomplishes this.

b. The system of the body functions best in response to progressive exercise. This means that Marines should master the easier exercises before they attempt the more difficult ones.

c. The body and mind are linked inseparably and must be considered as one in physical fitness training. Physical exercises assist in developing the mind.

d. The confidence in oneself and in one's comrades attained by physical achievement helps to promote the indomitable will to win. The Marine must be taught to realize this. It will help during arduous training, which at all times must be imaginative, realistic, and filled with enthusiasm. All physical fitness training must be associated with developing the qualities required of a Marine.

4. Planning and Preparation of Programs The following factors must be considered when compiling a program (see also ch. 2)~

a. Aim or desired skill level.

b. Duration of the complete training program and number of physical training lessons.

c. Purpose of included exercises,drills, or practical applications.

d. Beginning/inventory fitness levels.

e. Medical facilities, medical personnel, and Marine Corps or local command regulations governing their proximity to the training.
f. Number of Marines to be trained.
g. Availability of instructors, assistants, and facilities.
h. Apparatus and equipment available.
i. Alternative training areas for varying weather conditions,
j. Stages of progression and tests of achievement.

5. Five Basic Principles of Program Design
a. Regularity. Regularity of exercise is far more important than the amount of exercise performed on a given day. A Marine should exercise a minimum of three times per week. Human muscle begins to atrophy after 3 days without exercise.
b. Balance. The program must be balanced in terms of total physical fitness. Imbalance is created when, for example, an individual merely runs. While the cardiovascular system is improved, general upper body strength may be lacking. Conversely, a person who just does weight training may develop strength, but not cardiovascular endurance.
c. Overload. If total body fitness is to be developed, then the muscular and cardiovascular systems must be made to perform more work than previously. This is done by increasing the frequency, intensity, and duration of the exercise. The body adapts to the amount of stress placed upon it and becomes stronger and more efficient.
d. Progression. To be successful, any type of training program must be progressive. This is an area where untrained persons conducting PT can cause injury to Marines resulting in failure of the PT program. Close supervision must be made of untrained instructors. The result of progressing too fast is that the class begins to reject PT because it is no longer enjoyable or motivating. The program should start at a slow pace, and the pace should increase as proficiency improves. Average persons will encounter three stages of progressiom the toughening stage, the slow improvement stage, and the sustaining stage. (See ch. 1.)
e. Variety. Some programs fail because the training becomes boring. Variety is essential and perhaps presents the greatest challenge to the commander and staff.

6. Grouping. Providing for different levels of physical fitness is particularly recommended in the early stages of conditioning. Older Marines and those in poorer physical condition should be expected to attain a group level of fitness, but they should be given more time to do it.
a. Homogeneous Grouping One simple method of providing for the difference in levels is to group them according to their condition. A two-group classification would divide Marines into highly conditioned and average groups. A finer classification could be obtained by dividing them into three groups—a highly conditioned, a moderately conditioned, and an unconditioned group.
b. Performance or Age Grouping. The segregation of Marines into different exercise groups should be based on physical fitness test scores or on the level of fitness they demonstrate. They may also be grouped at first according to age. A common classification by ages is under 30, 30 to 34, and 35 and above.

7. Preparatory Commands and Commands of Execution. The prepreparatory command describes and specifies what is required. The command of execution calls into action what has been prescribed. All preparatory commands are given with a rising inflection. The interval between commands is long enough to permit the average Marine to understand the first one before the second one is given.

8. Extended Rectangular Formation. The traditional formation for carrying on many physical training activities is the extended rectangular formation (fig. A-1). This formation is more compact than any other. It is the best type to use for large numbers of Marines because it is simple and easy to assume.

a. For the formation of oneplatoon, the base man paces off five paces from the stand, faces left and moves five paces, halts, and again faces left. With the base man positioned facing the stand, the platoon leader then commands. **FALL OUT AND FALL IN ON THE BASE MAN**. At this command, all Marines run to the designated area and reform. This procedure is preferred to marching the unit into position. If more control is desired, the unit may march at double time to the vicinity of the base man and then be directed to fall out and fall in on him. Much valuable time is wasted in the field by needless maneuvering of Marines at quick time in an effort to position the platoon or unit on the exact spot for the exercises.

b. A company-size unit assumes the extended rectangular formation from a column of threes or fours at normal intervals between squads. This extension can also be executed from a company mass without interval between platoons. In extending either a platoon or company-size unit, take your place at the head of the column and command:
(1) TAKE INTERVAL TO THE LEFT, MARCH. At this command, the Marines in the right flank file stand fast with arms extended sideward. All others turn to the left and run forward at double time. After taking a sufficient number of steps, all Marines face the front with both arms extended sideward. The distance between finger-tips is about 12 inches, and dress is right.

Figure A-1 is available in the illustration section at the end of the book.

(2) ARMS DOWNWARD, MOVE. At this command, the arms are lowered smartly to the sides.
(3) LEFT, FACE
(4) EXTEND TO THE LEFT, MARCH. At this command, those in the right flank file stand fast with arms extended sideward. All others turn to the left and run forward at double time. Spacing is the same as in (1) above, and dress is right.
(5) ARMS DOWNWARD, MOVE. Same as in (2) above.
(6) RIGHT, FACE.
(7) FROM FRONT TO REAR, COUNT OFF. At this command, the leading Marine in each column turns the head to the right rear, calls of **ONE** and faces the front. Successive Marines in each column call off in turn **TWO, THREE, FOUR, FIVE** in the same manner.
(8) EVEN NUMBERS TO THE LEFT, UNCOVER. At this command, each even-numbered Marine stride jumps sideward to the left, squarely in the center of the interval. In doing this, each one swings the left leg sideward, jumps from the right foot to the left foot, and smartly brings the right into position against the left.

c. To assemble the unit, you commandASSEMBLE TO THE RIGHT, MARCHAt this command, all return to their original position in the column at double time and reform on the base man.

d.It is recommended that the area for grounding equipment and arms be at the edge of, or well away from, the area to be used for exercising. To conserve time and ensure proper position of the unit, the base man or, if the unit is composed of several platoon-sizegroups, the various base men may precede the unit and establish their positions in relation to the instructor's stand.

9. Circle Formation.The circle formation is effective for conducting various exercise activities (fig. A-2). This formation has an advantage in that supervision of all Marines is facilitated and a moving formation is available which provides control. Guerrilla exercises, grass drills, and some forms of running are examples of activities which are more easily conducted in the circle formation than in the extended rectangular formation.

a. It is not advisable to have more than 60 Marines in a circle. When more must be accommodated, separate circles should be used.Concentric circles may be employed to reduce the size of the circle or to accommodate more persons. If concentric circles are formed, the different circles are made by designating squads for each circle.Each additional circle requires more Marines than the one inside it.For example, one squad of a platoon may form the inner circle and theremaimng three squads the outer circle.

b. When a platoon is to form a circle, the commands ai€IRCLEFORMATION, MARCH, FOLLOW MEUpon this command, the right flank squad of the column moves forward at double time with the leader of the platoon group gradually forming a circle in a counterclockwise direction. Each succeeding file falls in behind that on the right. After the rough outline of the circle is formed, the leader commandsPICK UP A **5**-YARD INTERVAL.

Figure A-2 is available in the illustration section at the end of the book.

c.The group may be halted and faced toward the center, or, if instruction is not necessary, the exercise activity may be executed without stopping the platoon.

lO.Leadership Techniques

a. Unless you experience all the exercises, you cannot appreciate how arduous they are, what movements are most strenuous and difficult, where the errors in performance are likely to occur, and what the proper cadence should be.

b. You must give everyone careful supervision and participate in the exercises to show that you can do them. When you participate, your assistant instructors should supervise because it is difficult for you to supervise and exercise at the same time.

c. Marines should never be kept too long in one position, especially a constrained one. They should never have to perform an exercise more times than they can do it without losing the proper form. You should insist upon proper form in the execution of all exercise activities.Even slight deviations from the proper form reduce the value of the exercise.

d. Avoid long explanations.As a rule, it should be necessary to give a full explanation of new exercises only. Explain the most essential features of an exercise first; add details later. Too many details at one time are more likely to confuse than to assist them.Minor corrections should be made to the entire class while the exercise is in progress (for exampHEADS **UP,** KNEES STRAIGHT). If necessary, follow this direction by the name of the person who is particularly at fault. If a Marine requires special attention, givethat person separate instruction by an assistant instructor to avoidwasting the time of the group.

e. The heavy demand on your voice can be lightened by training assistant instructors to assume some of the instruction. Using mass cadence is also an effective method of lessening the demand on your voice.

“Illustrations”

DAY	TIME	ACTIVITY
MONDAY	60 MIN	RUNNING ACTIVITY
TUESDAY	60 MIN	COMBATIVES
WEDNESDAY	60 MIN	CONDITIONING & LOG DRILLS
THURSDAY	60 MIN	CONDITIONING & RELAYS
FRIDAY	60 MIN	RUNNING & CONDITIONING
SATURDAY	60 MIN	RUNNING ACTIVITY

Figure 2-1. Sample A- -Single Activity.

DAY	TIME	ACTIVITY	REMARKS
MONDAY	20 MIN 20 MIN 20 MIN	RIFLE DRILL COMBATIVES RUNNING	PROGRESSIVE ACTIVITY SYSTEM
TUESDAY	30 MIN	GUERRILLA EXERCISES TABLE I	SINGLE ACTIVITY
WEDNESDAY	30 MIN	RIFLE DRILL	SINGLE ACTIVITY
THURSDAY	30 MIN	CIRCUIT INTERVAL TRAINING	SINGLE ACTIVITY
FRIDAY	20 MIN 20 MIN 20 MIN	STRENGTH CIRCUIT CONDITIONING DRILL CROSS-COUNTRY RUN	ROTATION ACTIVITY
SATURDAY	60 MIN	WARM UP RUN AND GRASS DRILL	SINGLE ACTIVITY

Figure 2-2. Sample B- -Various Time Periods, Activities, and Systems.

DAY	TIME	ACTIVITY
MONDAY	45 MIN	RUNNING – CROSS-COUNTRY
TUESDAY	75 MIN	TEAM ATHLETICS – TOUCH FOOTBALL
WEDNESDAY	45 MIN	RUNNING – CROSS-COUNTRY
THURSDAY	20 MIN 40 MIN	LOG DRILL DUAL COMBATIVES
FRIDAY	90 MIN	SPEED MARCH
SATURDAY	90 MIN	TEAM CONTESTS

Figure 2-3. Sample C--Sustaining Activities.

Route	Kilometers from SP	Recommended rate of march (kmph)	Remarks
SP: RJ 8th Div Rd—Superhighway	......	4	Bridge; hard surface; two guides.
RJ Jamestown—Hersey Rd	2.25	4	Traffic heavy; two guides; Jamestown Rd bears to right.
RJ Jamestown—Yankee Rd	7.10	4	Two guides
RJ Jamestown—Lightning Rd	10.50	4	Two guides
RJ Jamestown—Sunshine Rd	12.40	4	Light traffic; one guide Sunshine Rd; gravel, poor traction when wet.
RP; RJ Sunshine—Sedan Rd	14.80	4	Two guides

Figure 3-1.Example of Route Reconnaissance Report.

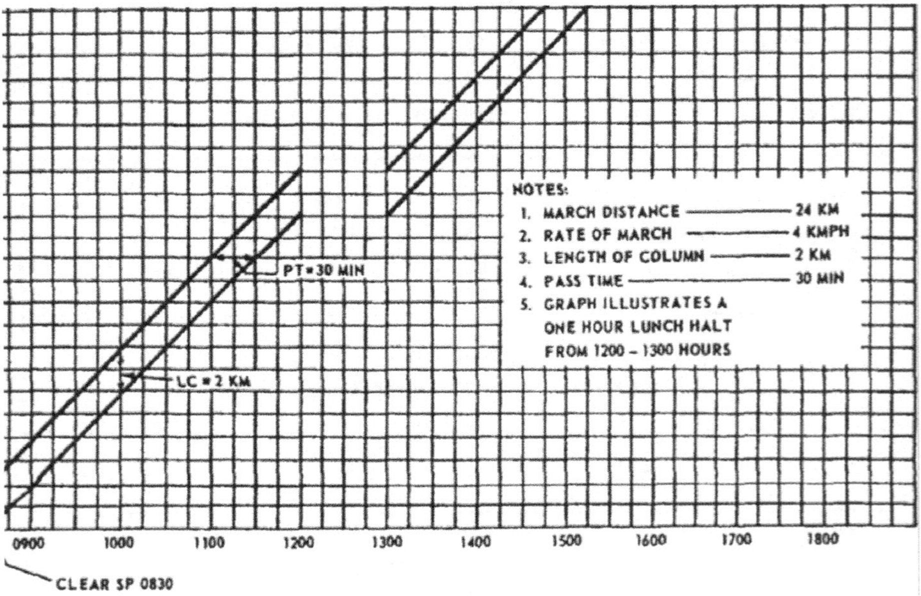

Figure 3-2.Road Movement Graph.

Figure 3-3.The Pace-Setter.

	SINGLE FILE	COLUMN OF TWO
2 m/MAN APART	2 4	1 2
5 m/MAN APART	5 4	2 7

GAP DISTANCES			
SERIAL:	(3-1) × 100 m	=	200 m
MARCH UNIT:	(12-1-2-) × 50 m	=	450 m
	TOTAL COLUMN GAP		650 m

Figure 3-4.Conditioning Drill One.

Figure 3-5.Conditioning Drill Two.

Figure 3-6.Conditioning Drill Three.

Figure 3-7. Rifle Drill.

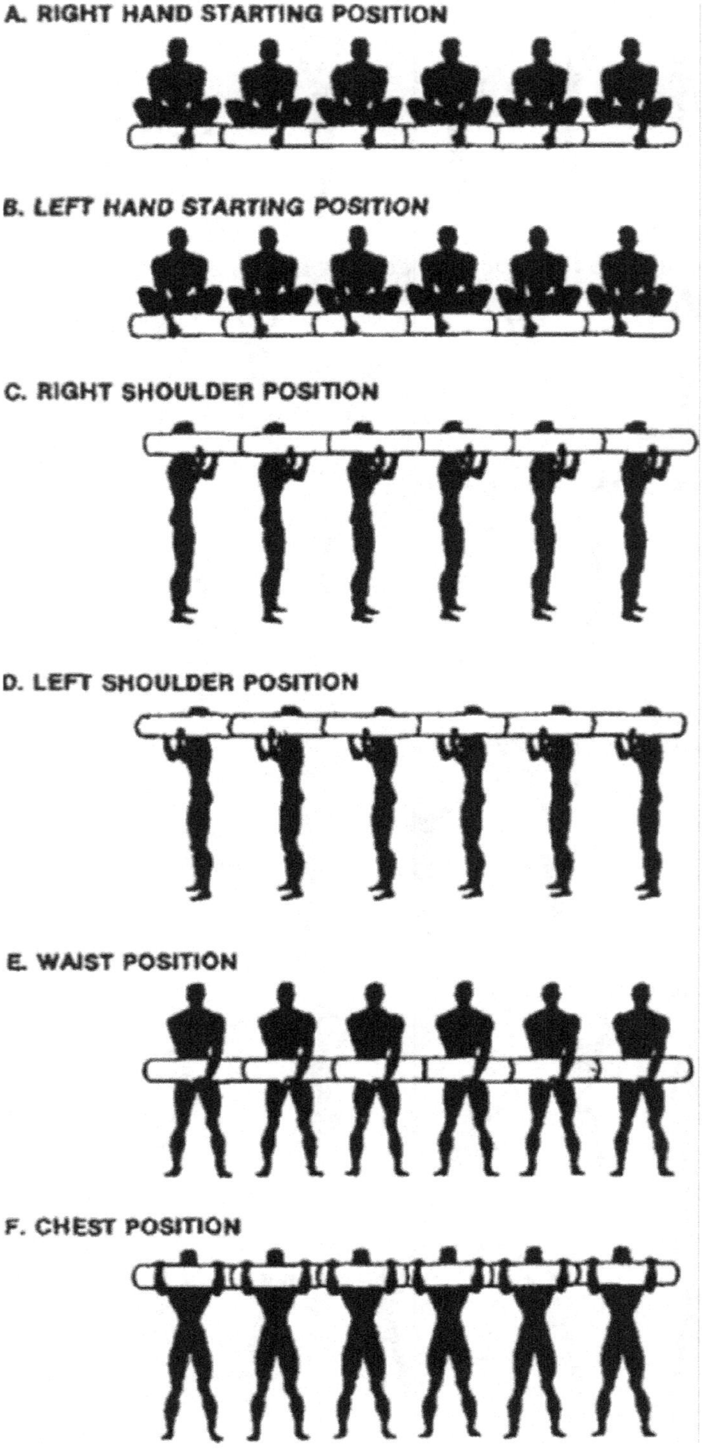

Figure 3-8. Starting Positions.

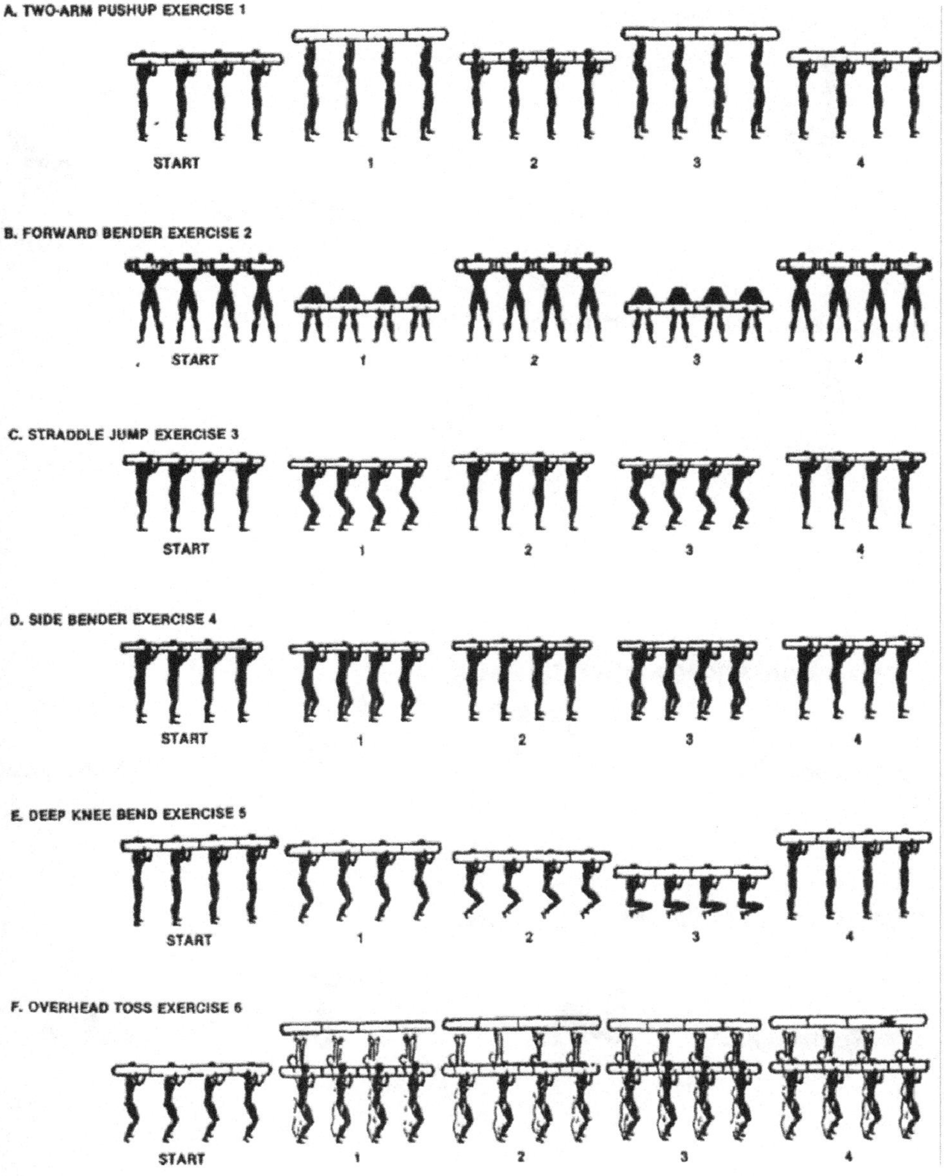

Figure 3-9. Log Drill

Figure 3-10.Basic Positions for Grass Drill.

Figure 3-11.Grass Drill One.

Figure 3-12.Grass Drill Two.

Figure 3-13.Guerrilla Table I.

Figure 3-14.Guerrilla Table II.

Figure 3-15.Proper Running Form.

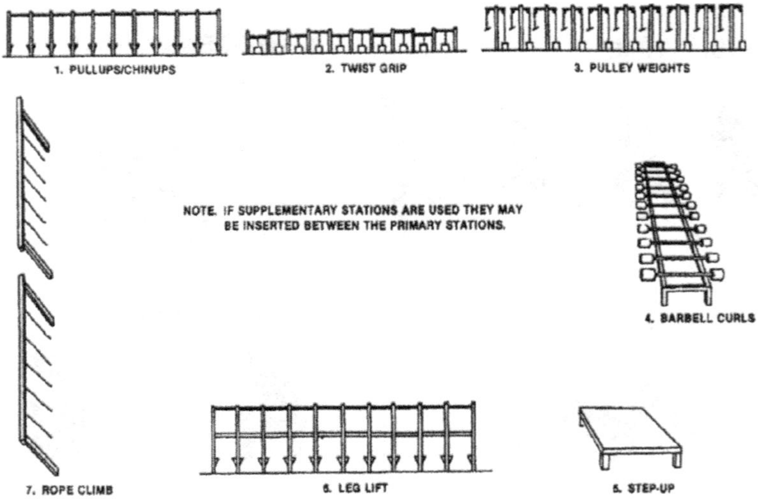

Figure 3-16. Fixed Strength Circuit.

STATION	ITEM	NO.	SPECIFICATION
1	BARBELL		1¼-INCH PIPE 5 FEET LONG WITH CONCRETE FILLED NO. 10 CANS
2	JUMP ROPE		¼- OR ⅜-INCH ROPE, 10 FEET LONG
3	TWIST GRIP		HANDLE 12 INCHES LONG, ROPE 4 FEET LONG, NO. 10 CAN CONCRETE FILLED.
4	INCLINE PLANE		¾-INCH PLYWOOD PLATFORM 2 FEET WIDE AND 6 FEET, 6 INCHES LONG, ELEVATED 10 INCHES AT ONE END. STRAP TO HOLD FEET DOWN.
5	WAR CLUB		HEAD IS 6 BY 12 INCHES, HANDLE IS 14 INCHES LONG BY 1¼ INCHES IN DIAMETER, ABOUT 20 POUNDS.
6	BICYCLE RIDE		PLYWOOD BOARD OR PLATFORM 2 BY 3 FEET WITH 2 BY 2 RUNNERS.
7	STEP – UP		A BOX OR STURDY PLATFORM 18 INCHES HIGH, 18 INCHES WIDE, 24 INCHES LONG.
8	ISOMETRIC PULL		TWO HANDLES 12 INCHES LONG WITH 4 FEET (BETWEEN HANDLES) OR LIGHT WIRE CABLE OR ¼-INCH ROPE

Figure 3-17.Movable Strength Circuit.

Figure. 3-18.Stations of Movable Strength Circuit.

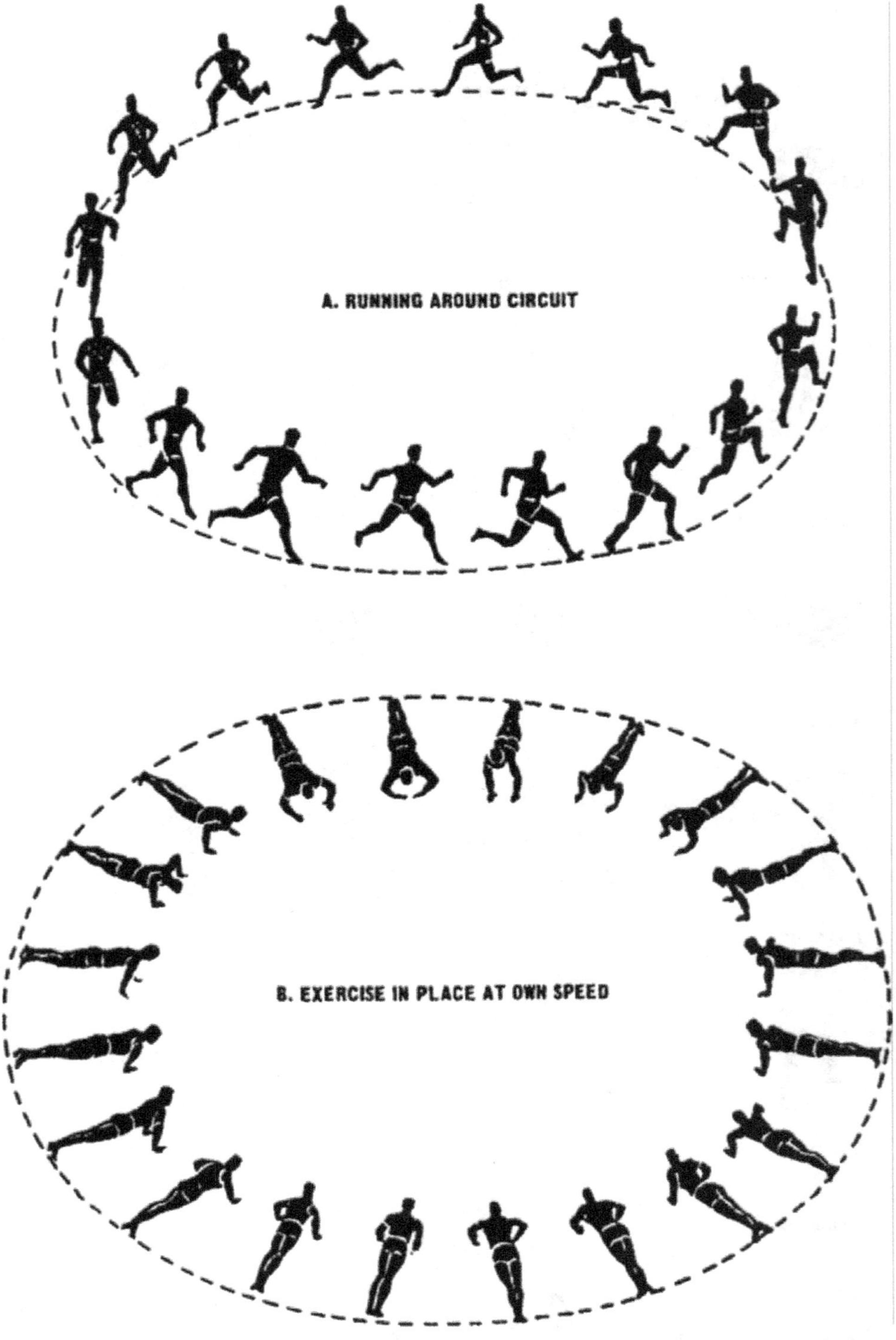

Figure 3-19. Circuit-IntervalTable.

Figure 3-20. Jumping.

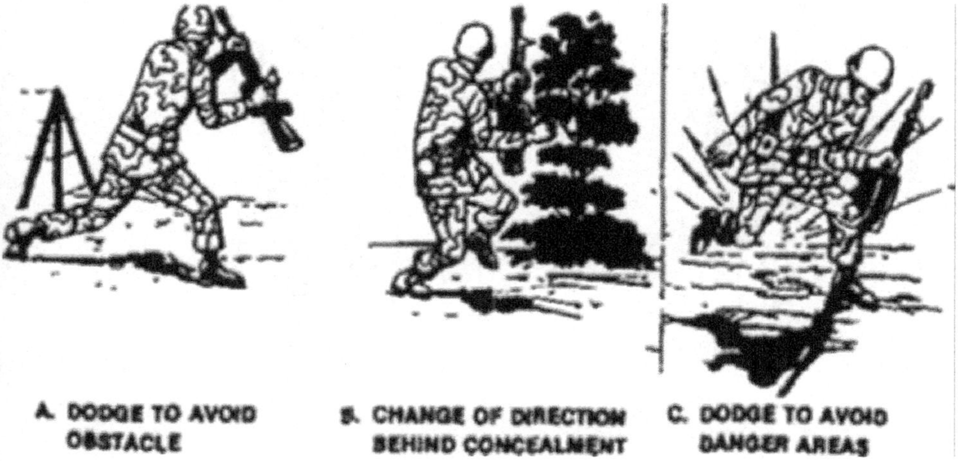

Figure 3-21.Dodging.

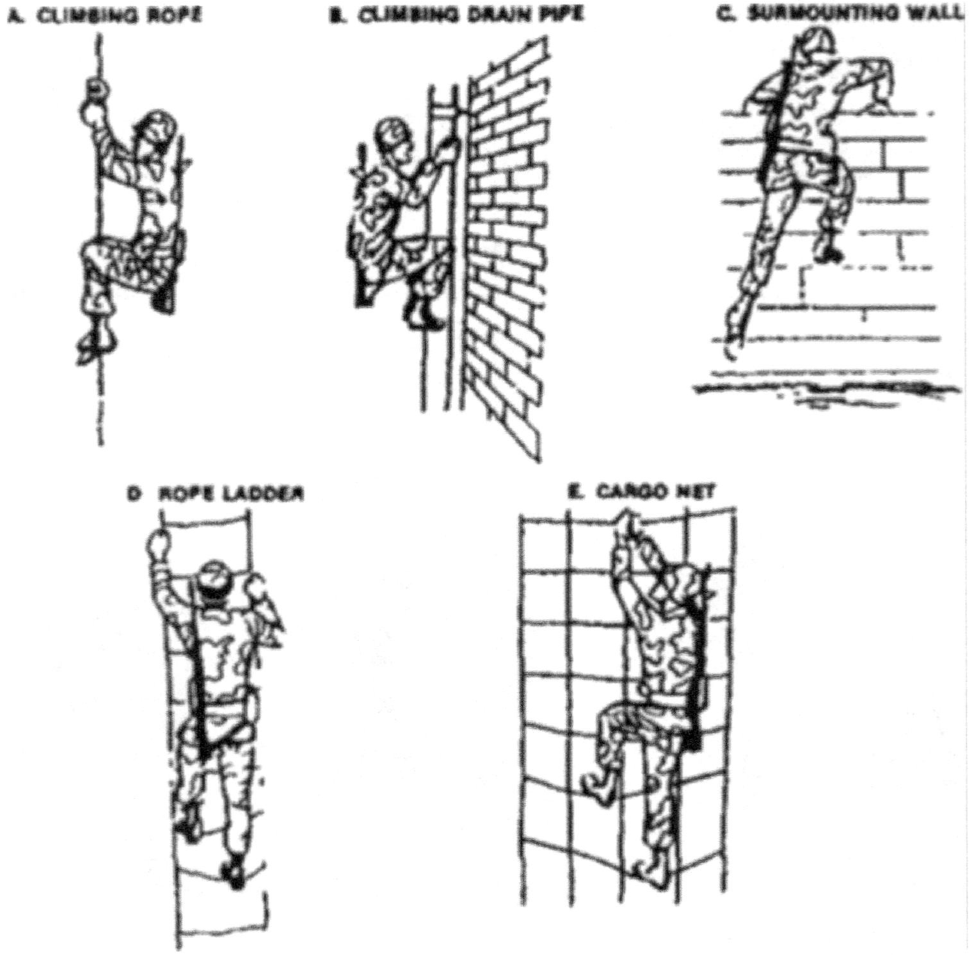

Figure 3-22.Climbing and Surmounting Vertical Objects.

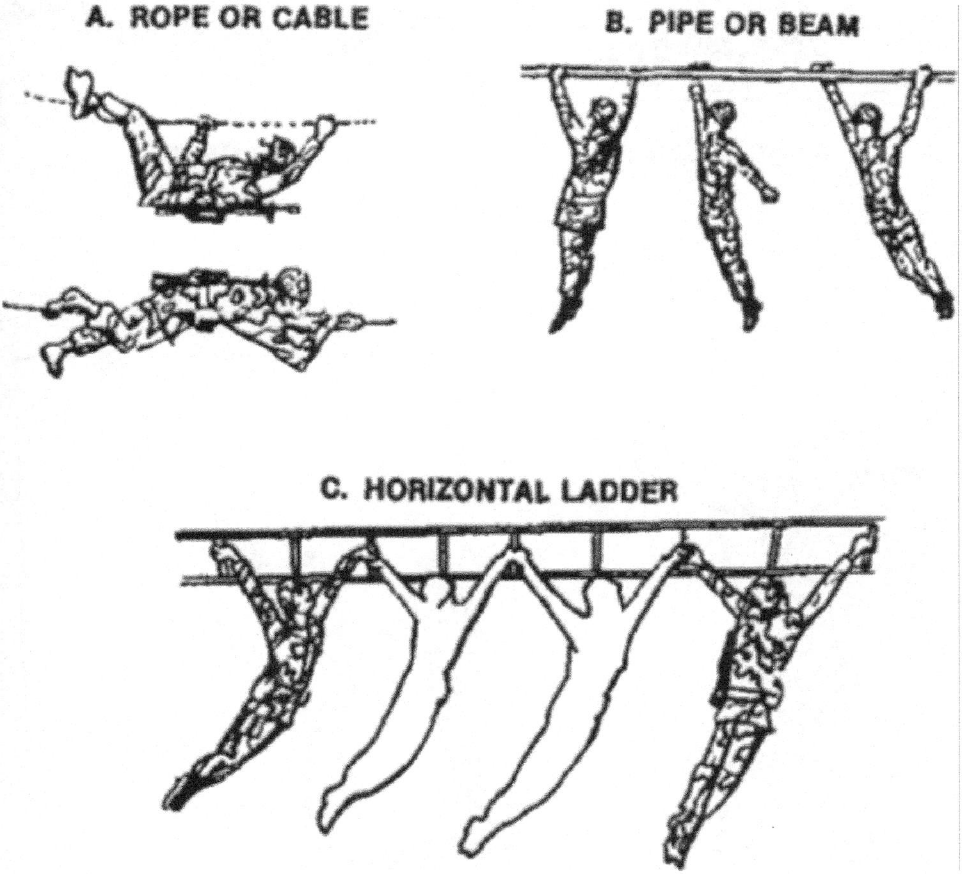

Figure 3-23. Traversing Horizontal Objects.

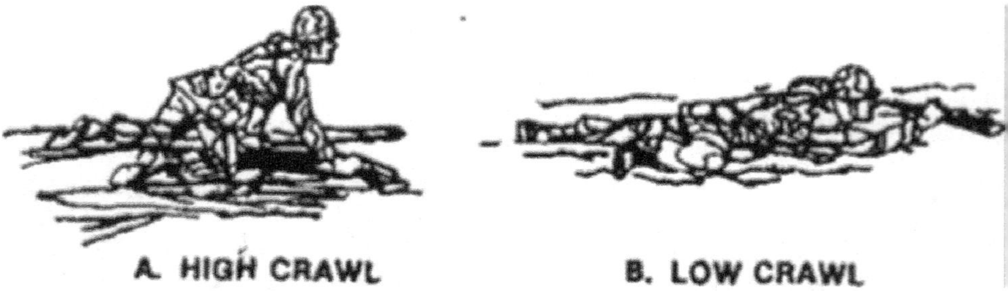

Figure 3-24.Crawling.

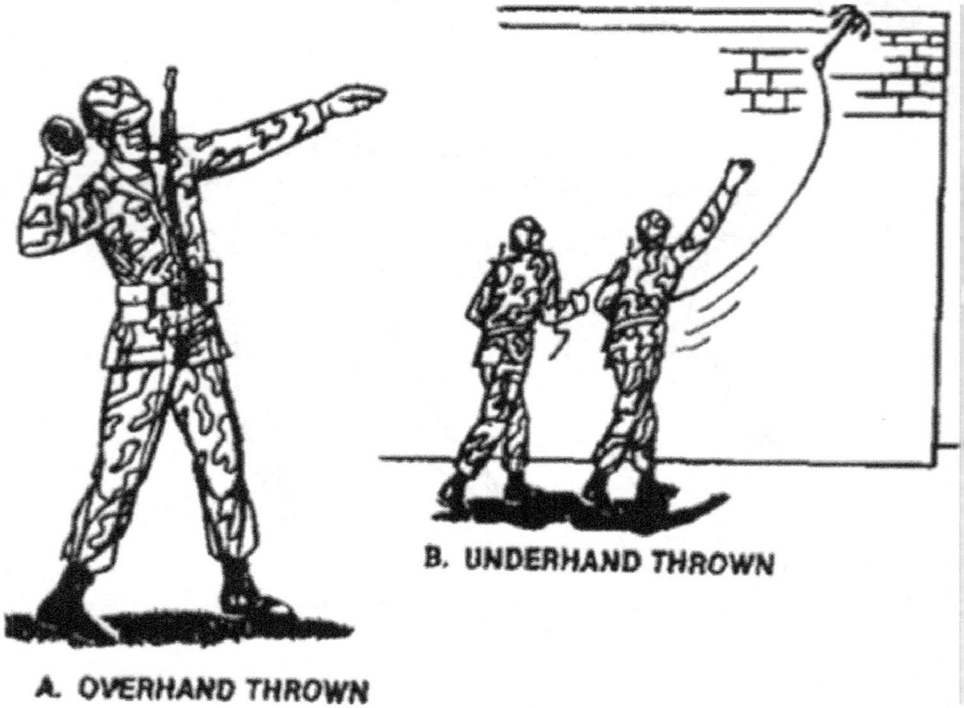

Figure 3-25.Throwing.

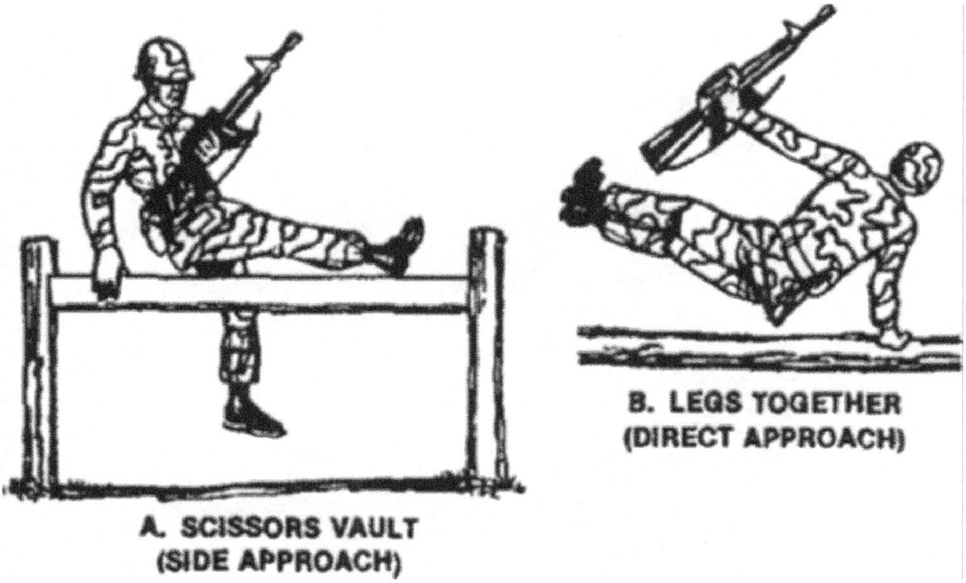

Figure 3-26. Vaulting

Figure 3-27 Balancing.

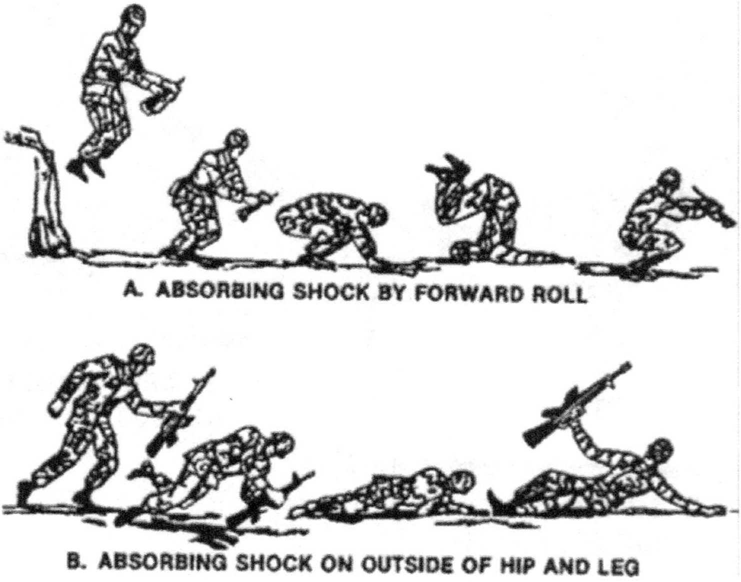

Figure 3-28.Falling.

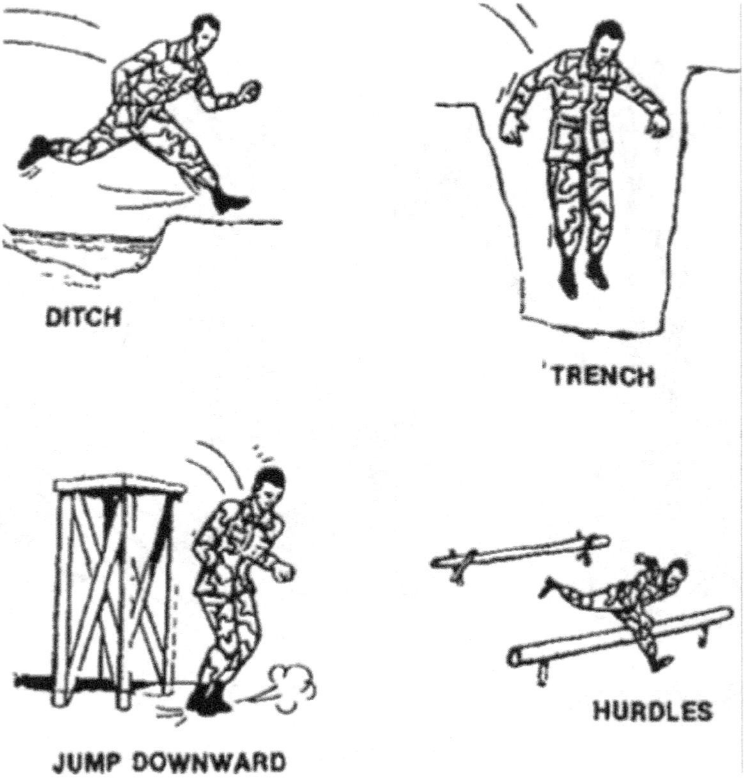

Figure 3-29.Jumping-Type Obstacles.

LANES TO GUIDE CHANGE OF DIRECTION

MAZES TO CAUSE CHANGE OF DIRECTION

Figure 3-30.Dodging-Type Obstacles.

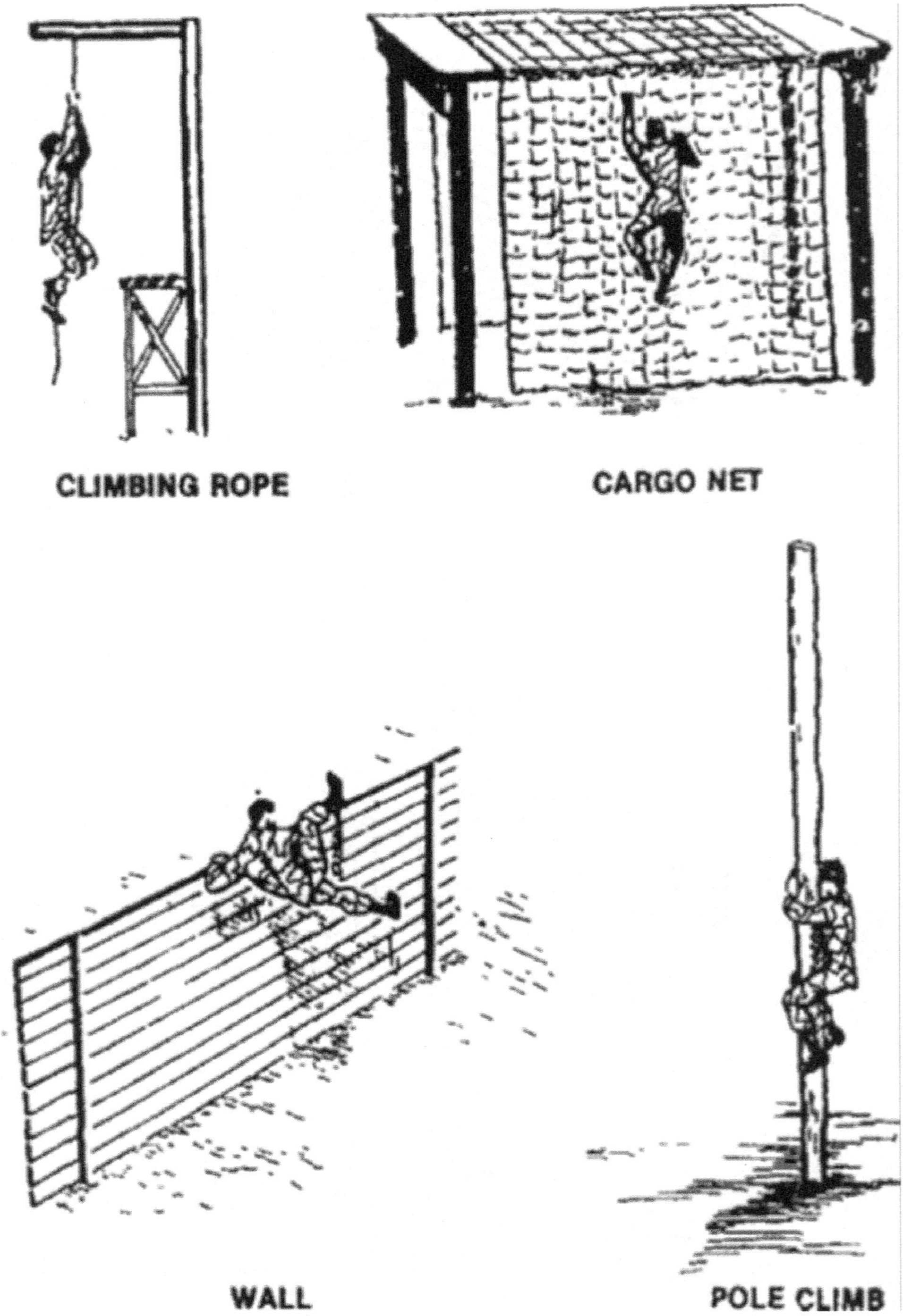

Figure 3-31. Vertical Climbing Obstacles.

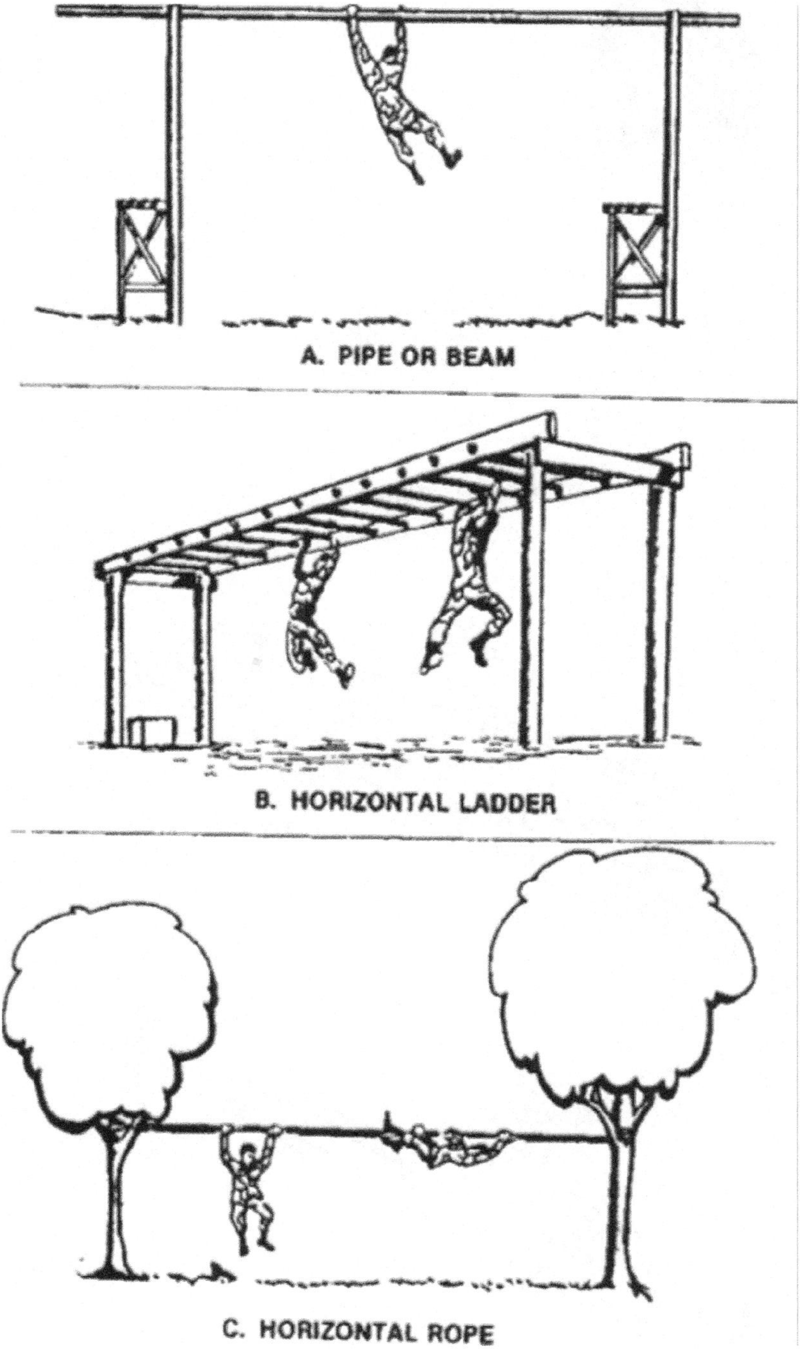

Figure 3-32.Horizontal Type Obstacles.

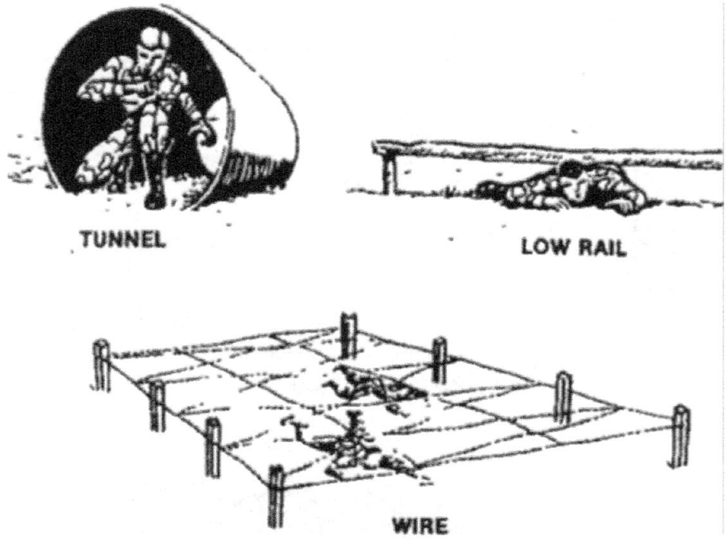

Figure 3-33.Crawling-Type Obstacles.

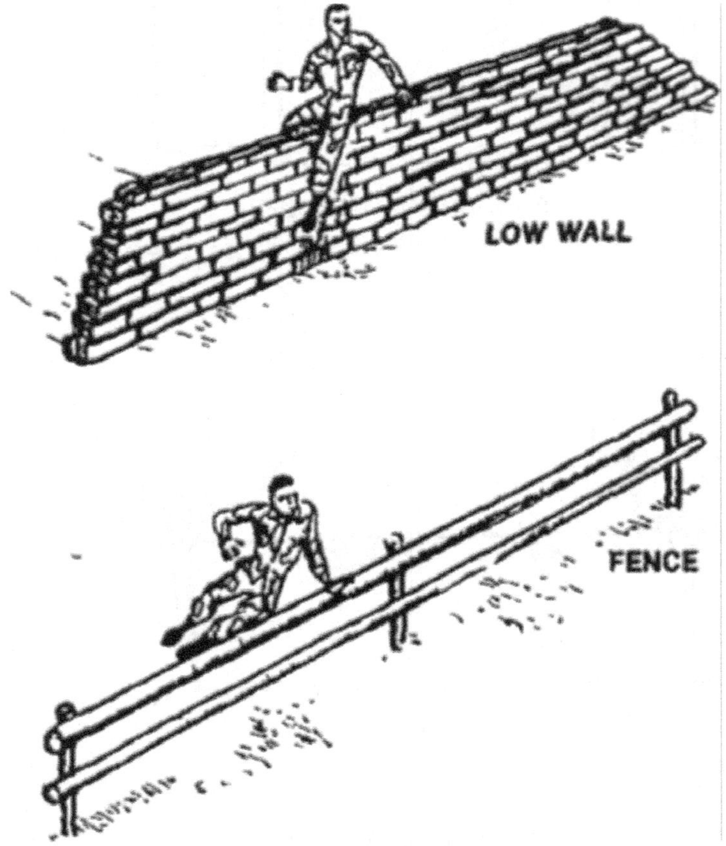

Figure 3-34.Vaulting Obstacles.

Figure 3-35.Balance-Type Obstacles.

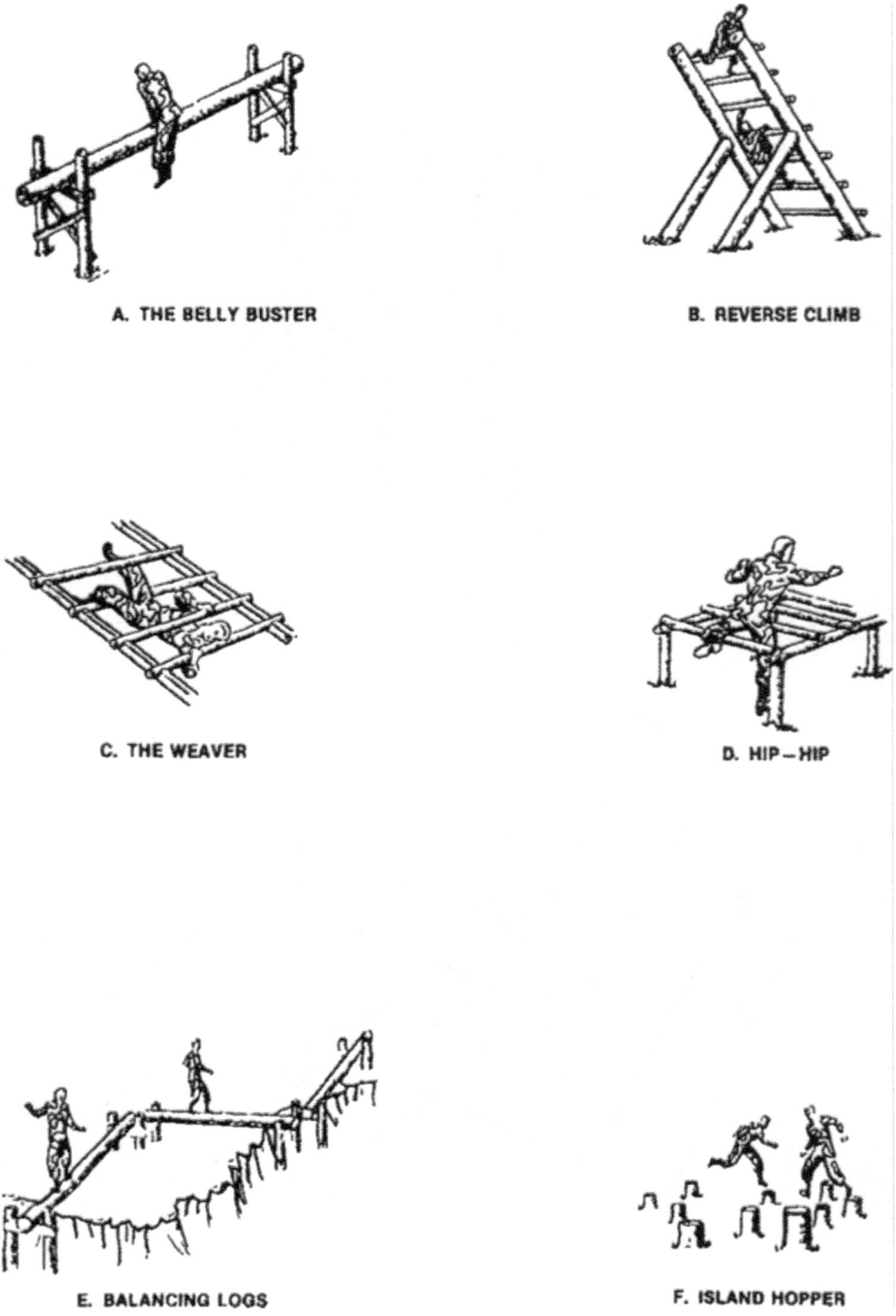

Figure 3-36.Red Group.

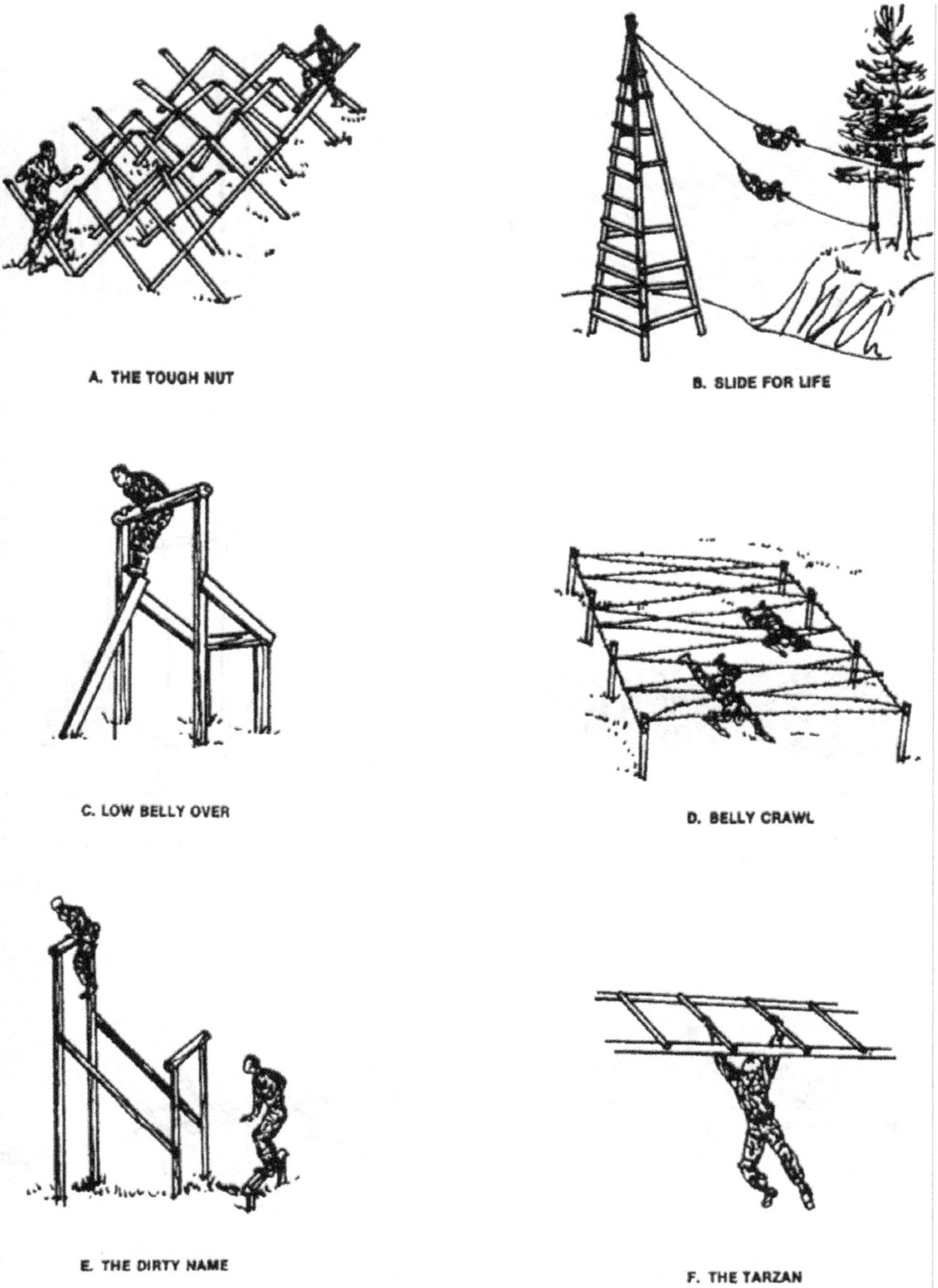

Figure 3-37.White Group.

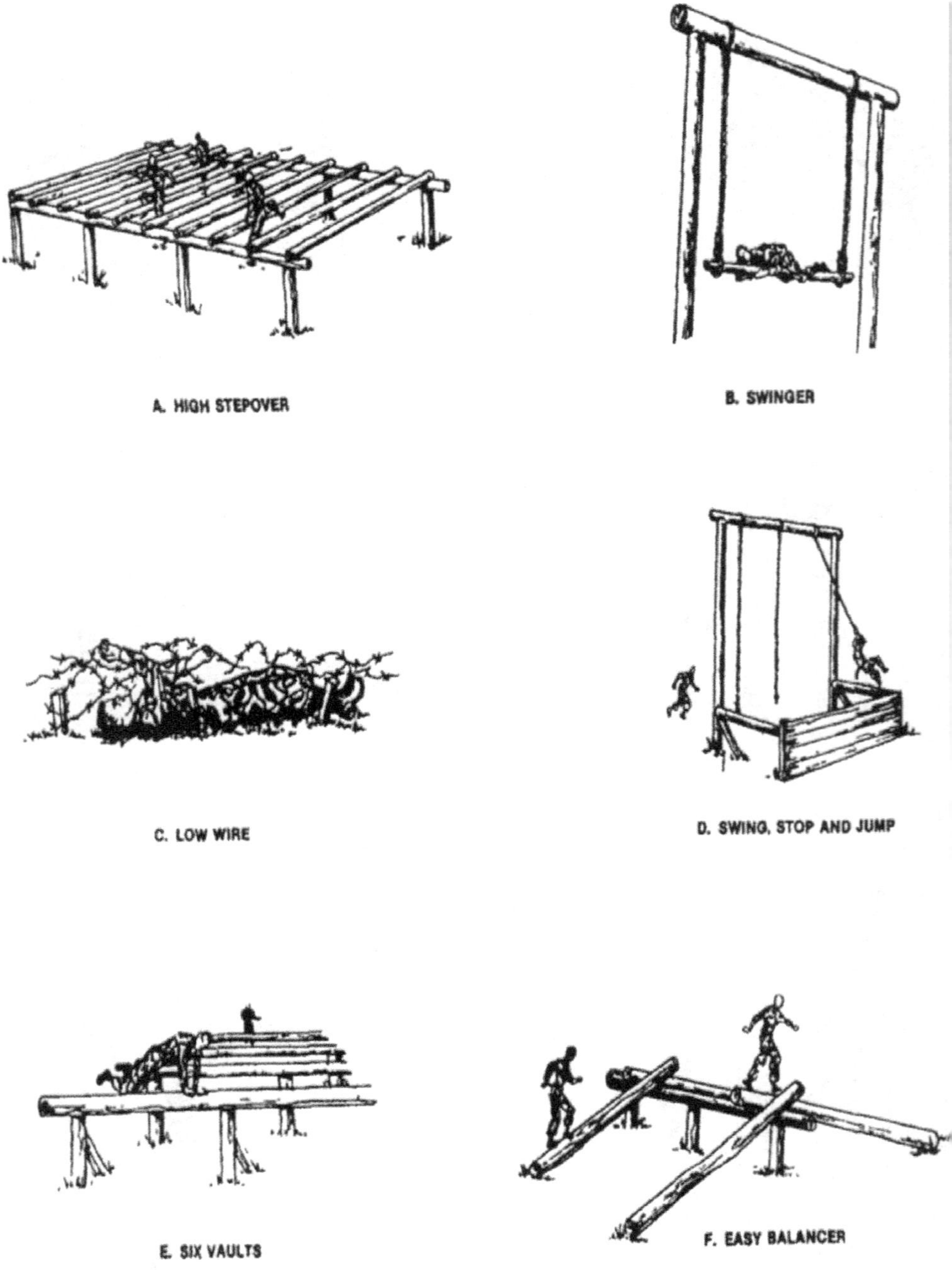

Figure 3-38.Blue Group.

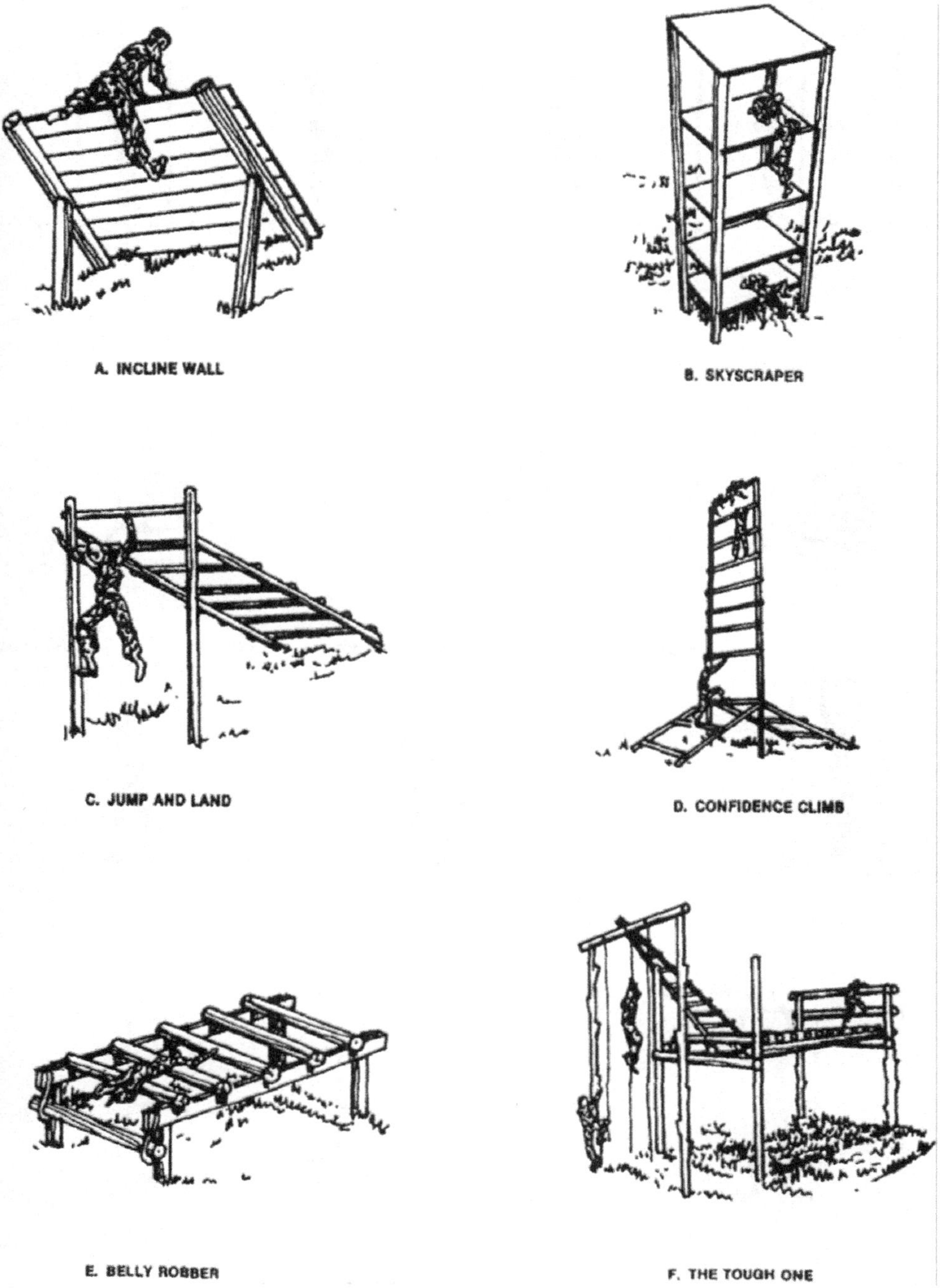

Figure 3-39.Black Group.

Figure 3-40. Bench Conditioning, Table I.

Figure 3-41. Bench Conditioning, Table II.

PROGRESSION GUIDE

AGE GROUP	LEVEL	EXERCISES 1	2	3	4	5	6
17	A	15	18	14	15	15	250
to	B	13	16	13	13	13	235
29	C	11	14	12	11	11	[illegible]
30	A	13	14	12	13	13	200
to	B	11	13	11	11	11	185
39	C	9	12	10	9	9	165
40	A	11	11	10	11	11	150
to	B	9	10	9	9	9	135
44	C	7	9	8	7	7	120
45	A	9	8	8	9	9	100
to	B	7	7	7	7	7	90
49	C	5	6	6	5	5	80
50	A	7	6	6	7	7	75
to	B	5	5	5	5	5	70
59	C	3	4	4	3	3	60
60	A	4	5	4	4	4	50
and	B	3	4	3	3	3	40
over	C	2	3	2	2	2	30
Minutes for each exercise		2	1	1	1	2	5

Table I. Progression Guide.

Table I. Exercises.

PROGRESSION GUIDE

AGE GROUP	LEVEL	EXERCISES 1	2	3	4	5	6
17	A	17	17	17	9	19	300
to	B	15	15	15	8	17	270
29	C	13	1[illegible]	13	7	15	245
30	A	15	15	15	8	17	235
to	B	13	13	13	7	15	210
39	C	11	11	11	6	13	190
40	A	13	13	13	7	15	175
to	B	11	11	11	6	13	155
44	C	9	10	9	5	11	135
45	A	11	11	11	6	13	125
to	B	9	9	9	5	11	110
49	C	7	7	7	4	9	100
50	A	9	9	9	5	11	95
to	B	7	7	7	4	9	85
59	C	5	5	5	3	7	75
60	A	6	7	7	4	9	70
and	B	5	5	5	3	7	60
over	C	4	4	4	2	5	50
Minutes for each exercise		1	1	1	1 1/2	1 1/2	6

Table II. Progression Guide.

Table II. Exercises.

PROGRESSION GUIDE

AGE GROUP	LEVEL	EXERCISES 1	2	3	4	5	6
17	A	10	19	19	16	10	350
to	B	9	17	17	15	9	315
29	C	8	15	15	14	8	280
30	A	9	17	17	14	9	270
to	B	8	15	15	13	8	240
39	C	7	13	13	12	7	210
40	A	8	15	15	12	8	200
to	B	7	13	13	11	7	180
44	C	6	11	11	10	6	160
45	A	7	13	13	10	7	150
to	B	6	11	11	9	6	135
49	C	5	9	9	8	5	120
50	A	6	11	11	8	6	115
to	B	5	9	9	7	5	105
59	C	4	7	7	6	4	95
60	A	5	9	9	7	5	90
and	B	4	7	7	6	4	80
over	C	3	5	5	4	3	70
Minutes for each exercise		1½	1	1	1½	1	6

Table III. Progression Guide.

Table III. Exercises.

PROGRESSION GUIDE

AGE GROUP	LEVEL	EXERCISES 1	2	3	4	5	6
17	A	12	9	12	24	25	400
to	B	11	8	11	22	23	380
29	C	10	7	10	21	21	360
30	A	11	8	11	23	23	305
to	B	10	7	10	21	21	290
39	C	9	6	9	20	20	275
40	A	10	7	10	20	21	225
to	B	9	6	9	18	18	215
44	C	8	5	8	16	16	205
45	A	8	6	8	16	14	175
to	B	7	5	7	14	14	165
49	C	6	4	6	12	12	155
50	A	6	5	6	13	13	135
to	B	5	4	5	11	11	130
59	C	4	3	4	10	10	120
60	A	5	4	5	10	10	100
and	B	4	3	4	9	9	95
over	C	3	2	3	8	8	90
Minutes for each exercise		1	2	1	1	1	6

Table IV. Progression Guide.

Table IV. Exercises.

PROGRESSION GUIDE

AGE GROUP	LEVEL	EXERCISES 1	2	3	4	5	6
17	A	14	13	28	14	30	450
to	B	13	12	27	13	28	430
29	C	12	11	26	12	26	410
30	A	12	12	25	12	26	350
to	B	11	11	24	11	24	330
39	C	10	10	23	10	22	310
40	A	11	11	23	11	23	250
to	B	10	10	21	10	21	240
44	C	9	9	19	9	19	230
45	A	9	9	20	9	20	200
to	B	8	8	18	8	18	190
49	C	7	7	16	7	16	180
50	A	7	7	16	7	16	170
to	B	6	6	14	6	14	155
59	C	5	5	12	5	12	140
60	A	6	6	12	6	12	115
and	B	5	5	11	5	10	110
over	C	4	4	9	4	9	105
Minutes for each exercise		2	1	1	2	1	5

Table V. Progression Guide.

Table V. Exercises.

PROGRESSION GUIDE

AGE GROUP	LEVEL	EXERCISES 1	2	3	4	5	6
17	A	17	15	32	32	35	500
to	B	16	14	30	30	33	480
29	C	15	13	28	28	31	460
30	A	15	13	30	30	31	400
to	B	14	12	28	28	29	380
39	C	13	11	26	26	27	360
40	A	13	10	27	27	27	310
to	B	12	9	25	25	25	285
44	C	11	8	23	23	23	265
45	A	11	9	23	23	23	250
to	B	10	8	21	21	21	230
49	C	9	7	19	19	19	210
50	A	9	8	19	19	19	200
to	B	8	7	17	17	17	190
59	C	7	6	15	15	15	175
60	A	8	7	15	15	17	140
and	B	7	6	13	13	15	130
over	C	5	5	10	10	12	120
Minutes for each exercise		2	1	1	1	1	6

Table VI. Progression Guide.

Table VI. Exercises.

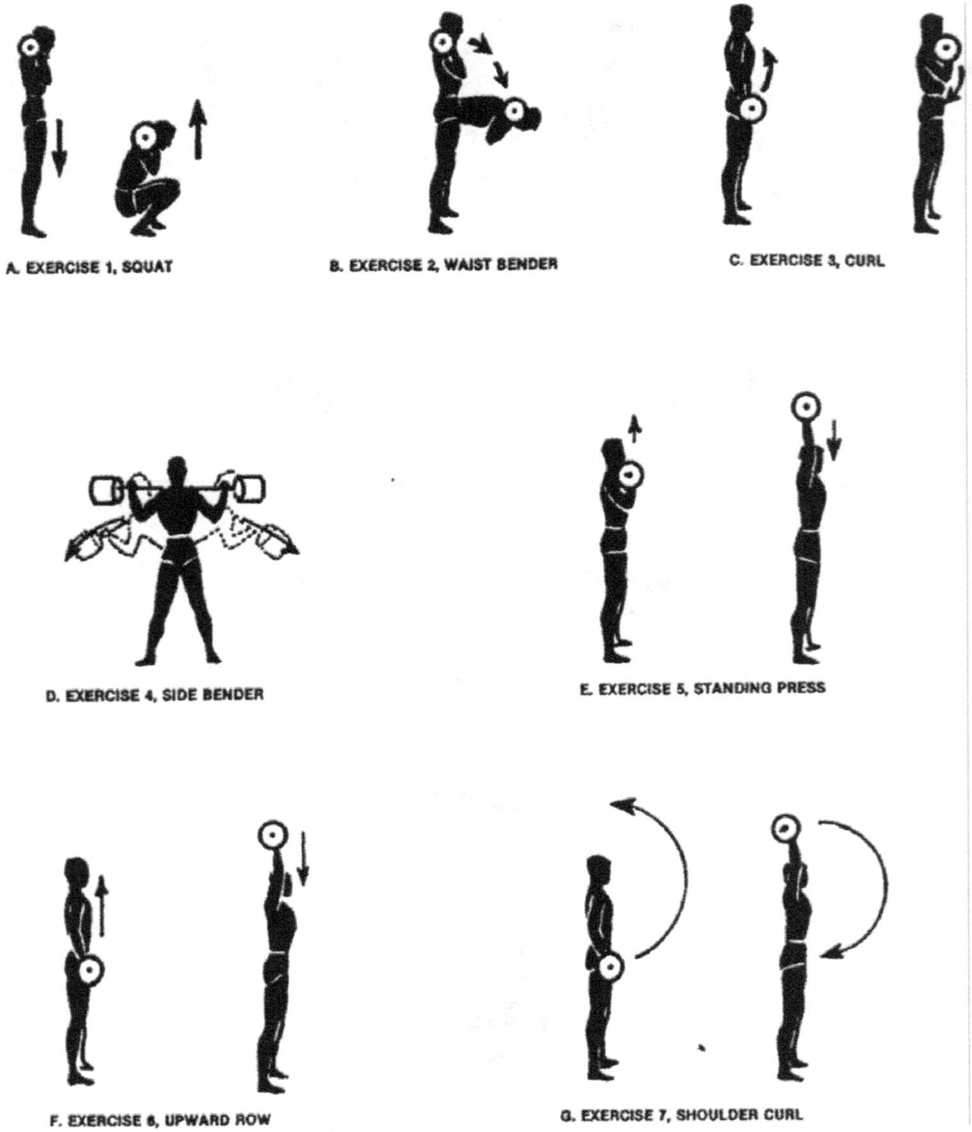

Figure 3-42. Weight Training.

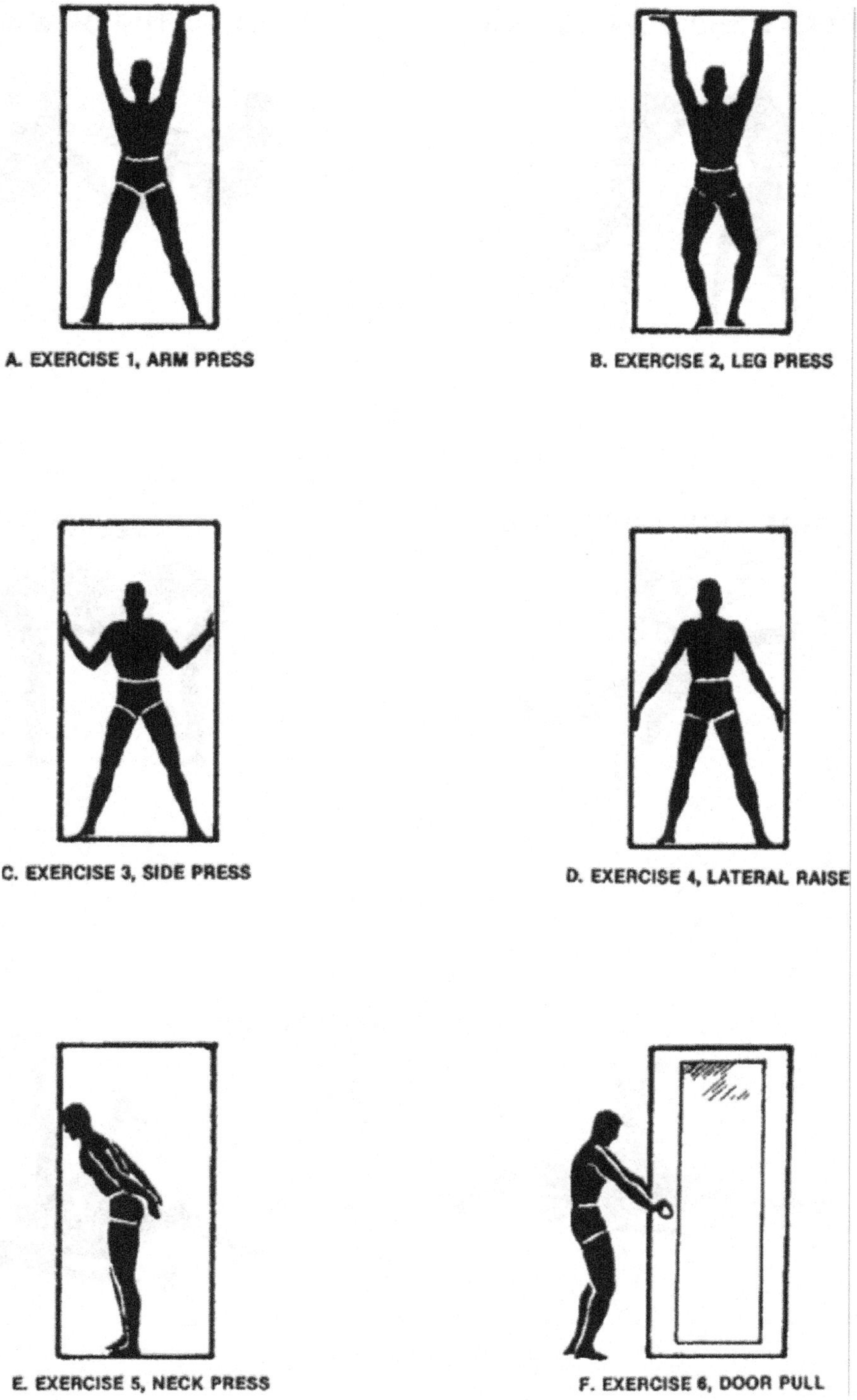

Figure 3-43. Isometric Exercises.

Figure 5-1.Combatives Table I.

Figure 5-2.Combatives Table II.

Figure. 5-3.Combatives Table III.

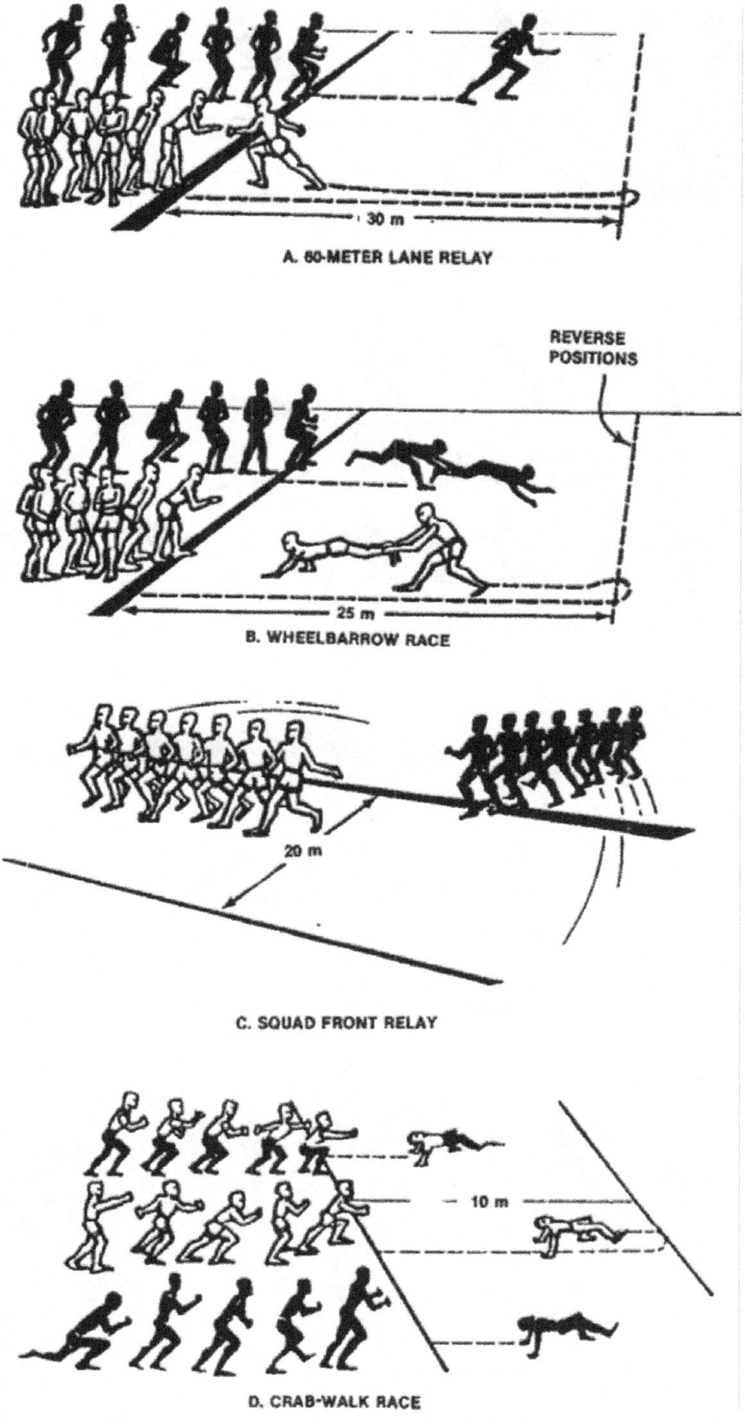

Figure 5-4.Relay Table I.

A. 100-METER LANE RELAY

B. FROG-JUMP RELAY

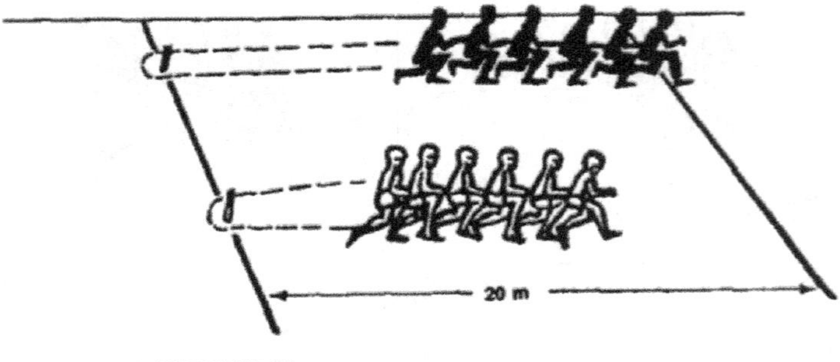

C. SIMPLE RELAY

D. FIREMAN'S CARRY RELAY

Figure 5-5.Relay Table II.

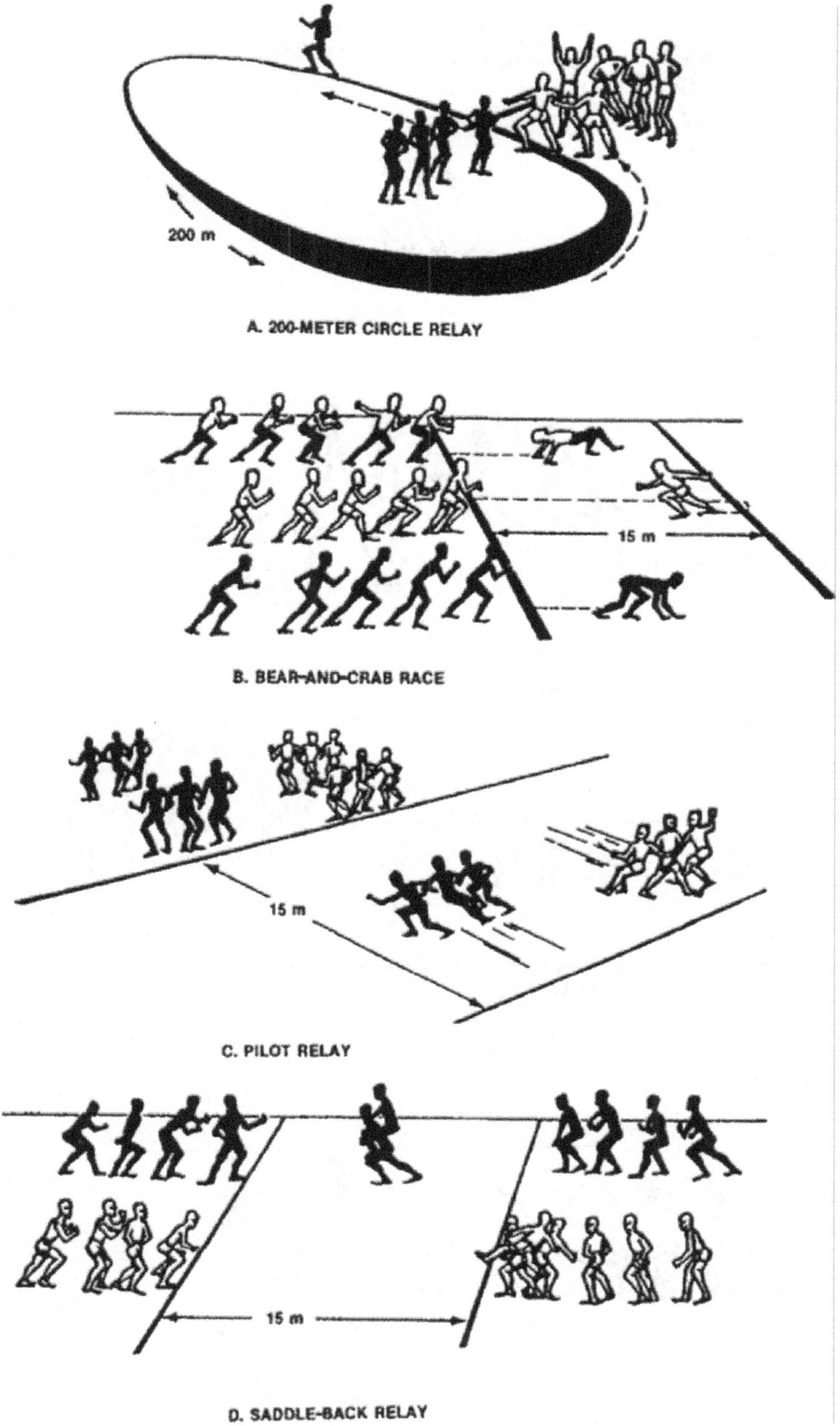

Figure 5-6.Relay Table III.

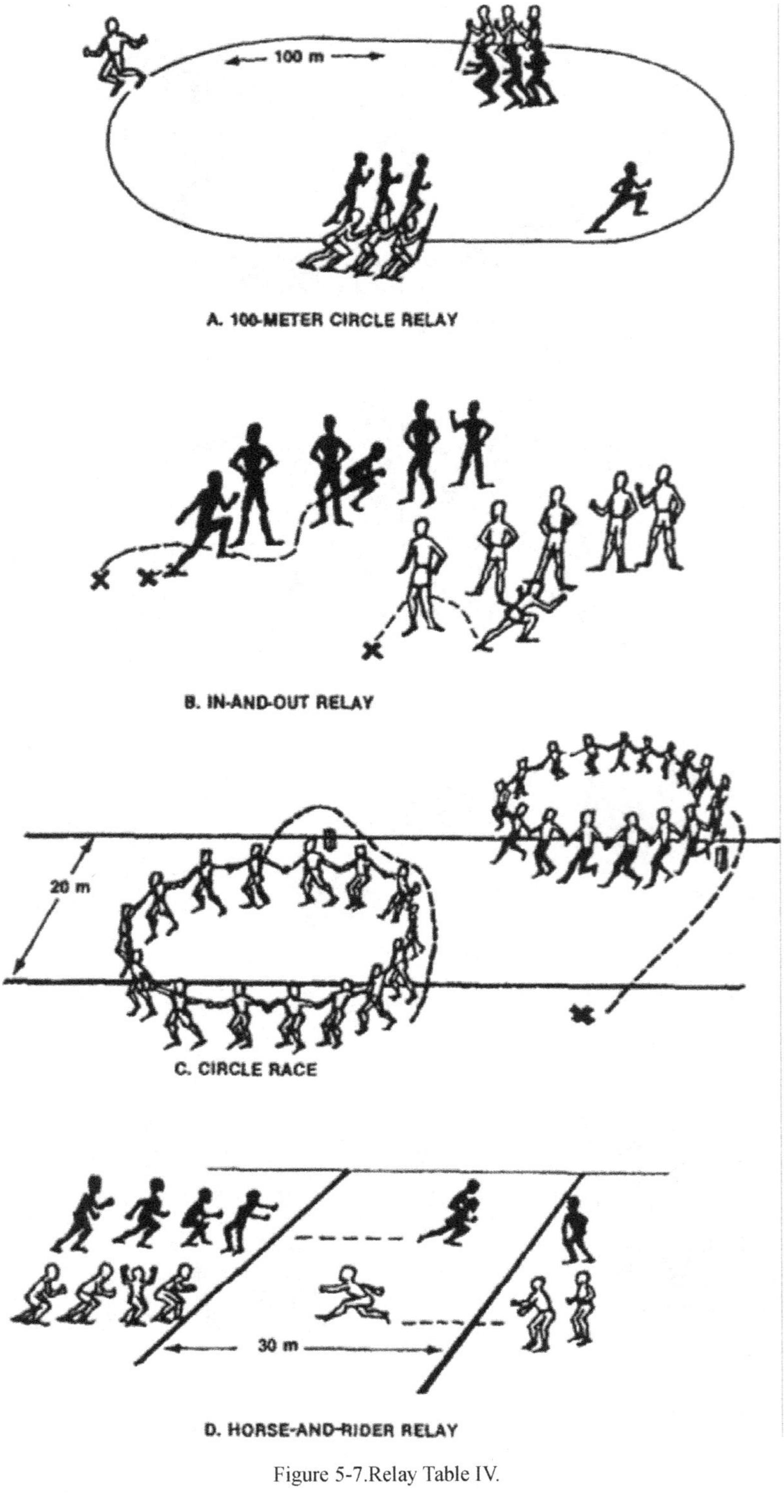

Figure 5-7.Relay Table IV.

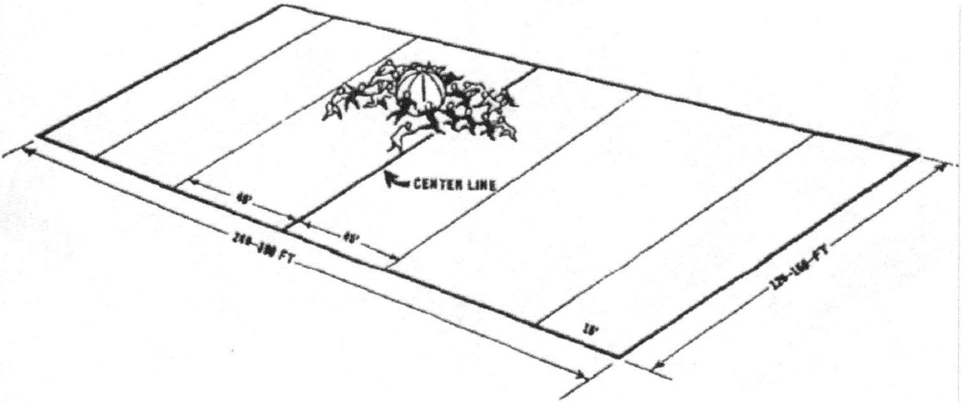

Figure 5-8.Pushball.

Figure 5-9.Line Rush.

Figure. 5-10. Human Tug-of-War.

Figure 5-11. Master of the Ring.

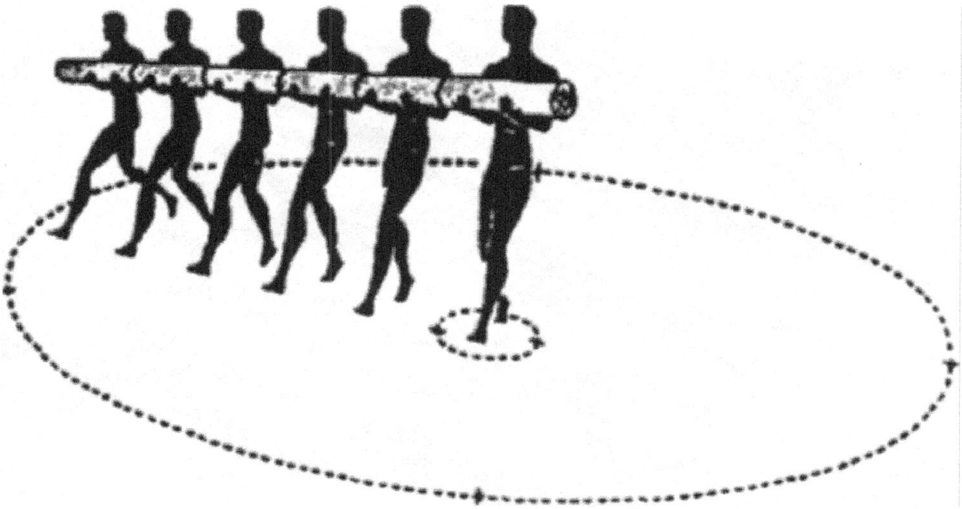

Figure 5-12.Log Pivot Circle.

Figure 5-13.Log-Rolling Race.

Figure 5-14. Prone Pushing Contest.

Figure 5-15. Shuttle Relay Race.

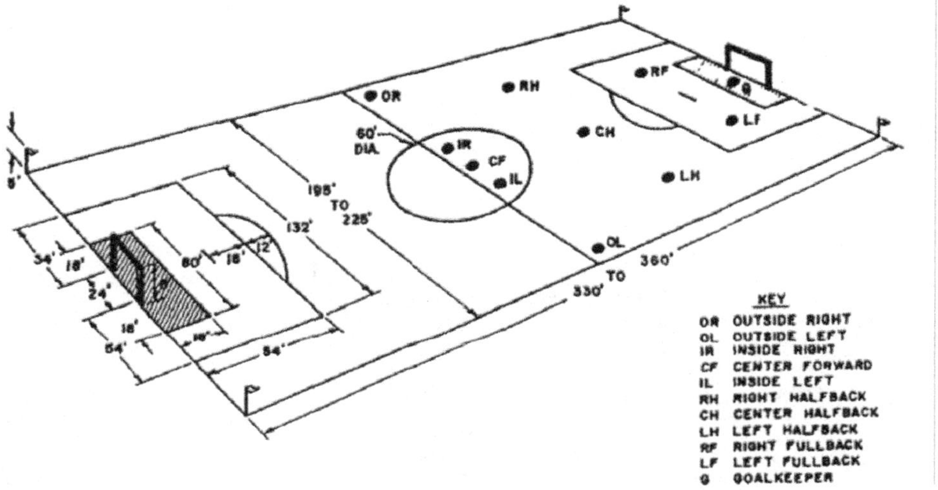

Figure 5-16 Soccer.

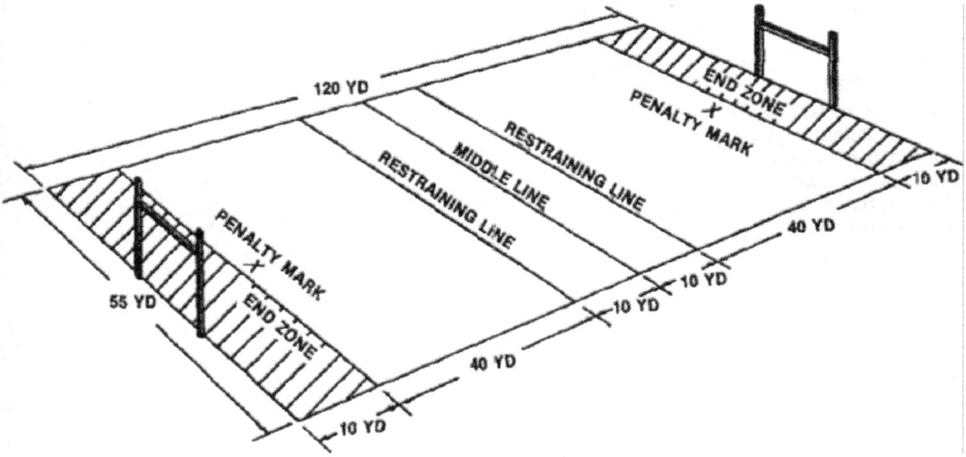

Figure 5-17.Speedball Field.

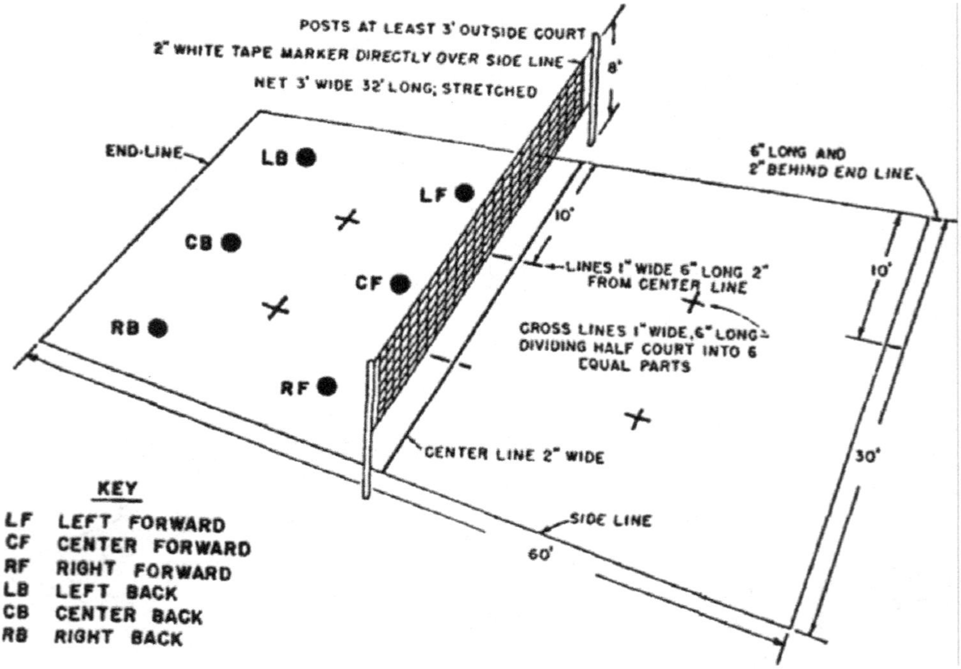

Figure 5-18. Volleyball.

PTS	UPHILL CLIMB	ROPE CLIMB	EVACUATION	FIRE AND MANEUVER	FORCED MARCH
100	1:20-BELOW	0:10-BELOW	0:25-BELOW	1:20-BELOW	26:00-BELOW
99	1:21-1:22	0:11		1:21-1:22	26:01-26:12
98	1:23-1:24	0:12	0:26	1:23-1:24	26:13-26:24
97	1:25-1:26	0:13		1:25-1:26	26:25-26:36
96	1:27-1:28	0:14	0:27	1:27-1:28	26:37-26:48
95	1:29-1:30	0:15		1:29-1:30	26:49-27:00
94	1:31-1:32	0:16	0:28	1:31-1:32	27:01-27:12
93	1:33-1:34	0:17		1:33-1.34	27:13-27:24
92	1:35-1:36	0:18	0:29	1:35-1.36	27:25-27:36
91	1:37-1:38	0:19		1:37-1:38	27:37-27:48
90	1:39-1:40	0:20	0:30	1:39-1:40	27:49-28:00
89	1:41-1:42	0:21	0:31	1:41-1:42	28:01-28:12
83	1:43-1:44	0:22	0:32	1:43-1:44	28:13-28:24
87	1:45-1:46	0:23	0:33	1:45-1:46	28:25-28:36
86	1:47-1:48	0:24	0:34	1.47-1:48	28:37-28:48
85	1:49-1:50	0:25	0:35	1.49-1:50	28:49-29:00
84	1:51-1:52	0:26	0:36	1:51-1:52	29:01-29:12
83	1:53-1:54	0:27	0.37	1:53-1:54	29:13-29:24
82	1:55-1:56	0:28	0:38	1:55-1:56	29:25-29:36
81	1:57-1:58	0:29	0:39	1:58-1:58	29:37-29:48
80	1:59-2.00	0:30	0:40	1:59-2:00	29:49-30:00
79	2.01-2:02	0:31	0:41	2:01-2:02	30:01-30:12
78	2:03-2:04	0:32	0:42	2:03-2.04	30:13-30:24
77	2:05-2:06	0.33	0:43	2:05-2:06	30:25-30:36
76	2:07-2:08	0.34	0.44	2:07-2:08	30:37-30:48
75	2:09-2:10	0:35	0:45	2:09-2:10	30:49-31:00
74	2:11-2:12	0:36	0:46	2:11-2 12	31.01-31:12
73	2:13-2:14	0:37	0.47	2.13-2:14	31:13-31:24
72	2:15-2:16	0:38	0:48	2.15-2:16	31:25-31:36
71	2:17-2:18	0:39	0:49	2:17-2:18	31:37-31:48
70	2:19-2:20	0:40	0:50	2:19-2.20	31:49-32.00
69	2.21-2:22	0:41	0:51	2:21-2:22	32:01-32:12
68	2:23-2:24	0:42	0:52	2:23-2:24	32:13-32:24
67	2:25-2.26	0:43	0:53	2:25-2:26	32:25-32:36
66	2:27-2:28	0:44	0:54	2.27-2:28	32.37-32:48
65	2:29-2:30	0.45	0.55	2.20-2 30	32.49-33:00

Figure 6-1.Physical Readiness Test Scoring Table for Male.

	EXAMPLE 1	EXAMPLE 2	EXAMPLE 3
Event #1	100 pts	100 pts	99 pts
Event #2	100 pts	99 pts	86 pts
Event #3	85 pts	75 pts (F)	80 pts
Event #4	90 pts	79 pts (F)	100 pts
Event #5	79 pts	85 pts	90 pts
Pts	454 pts	438 pts	455 pts
Final Score	79.0% (F)	69.0% (F)	91.0%
(F) = Failure			

Figure 6-2.Scoring Examples.

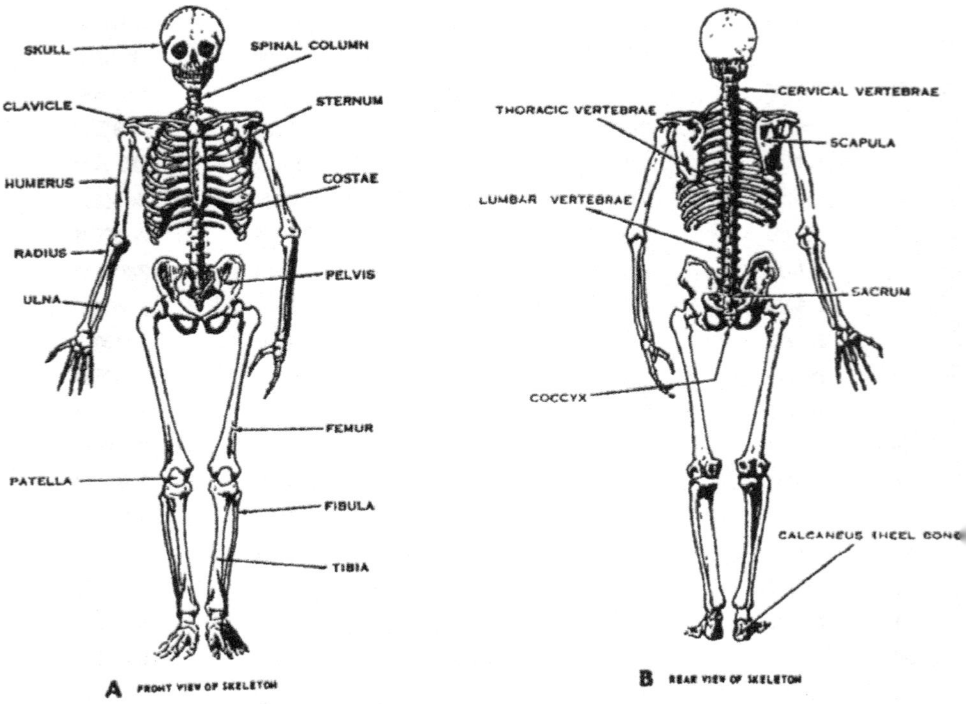

Figure 7-1.Front and Rear Views of Skeleton.

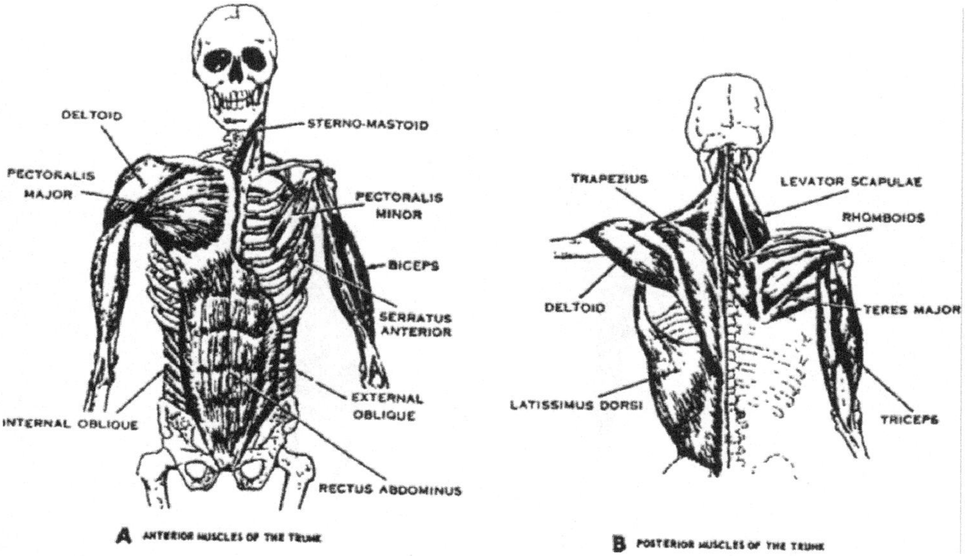

Figure 7-2.Muscles of the Trunk.

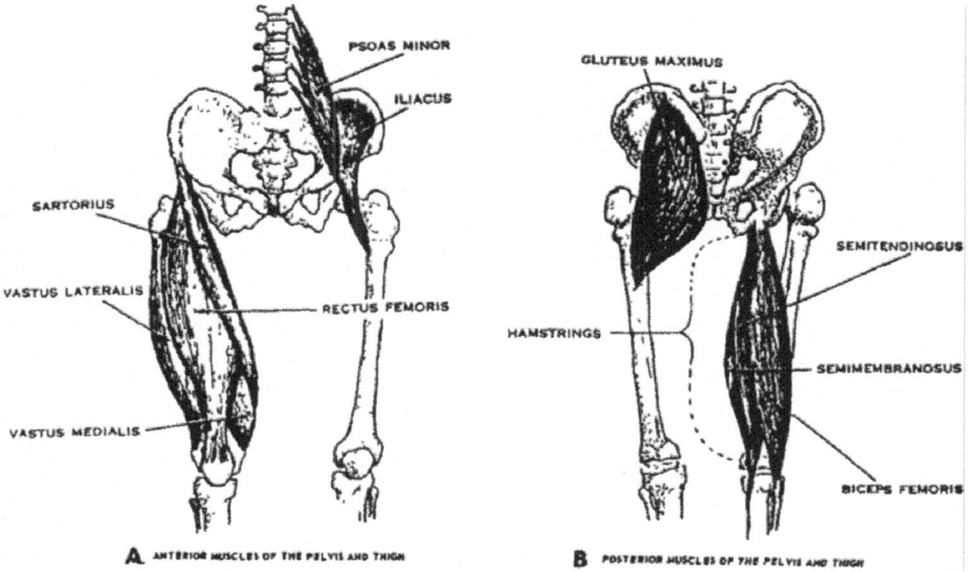

Figure 7-3.Anterior and Posterior Muscles of the Pelvisand Thigh.

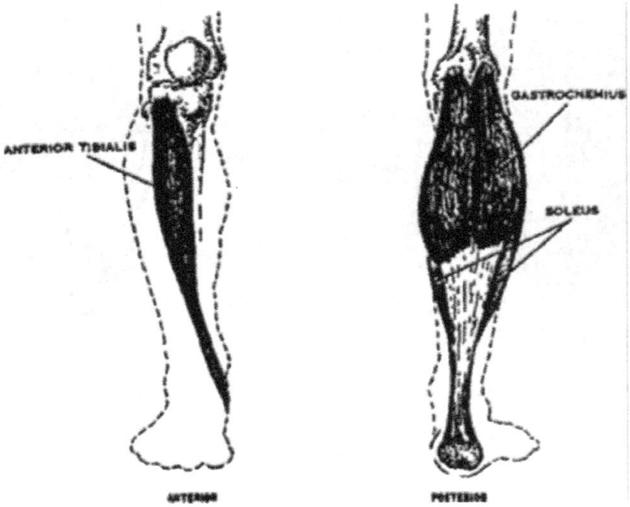

Figure 7-4. Muscles of Lower Leg.

Figure 7-5.Circulation of Blood from Heart to the Body.

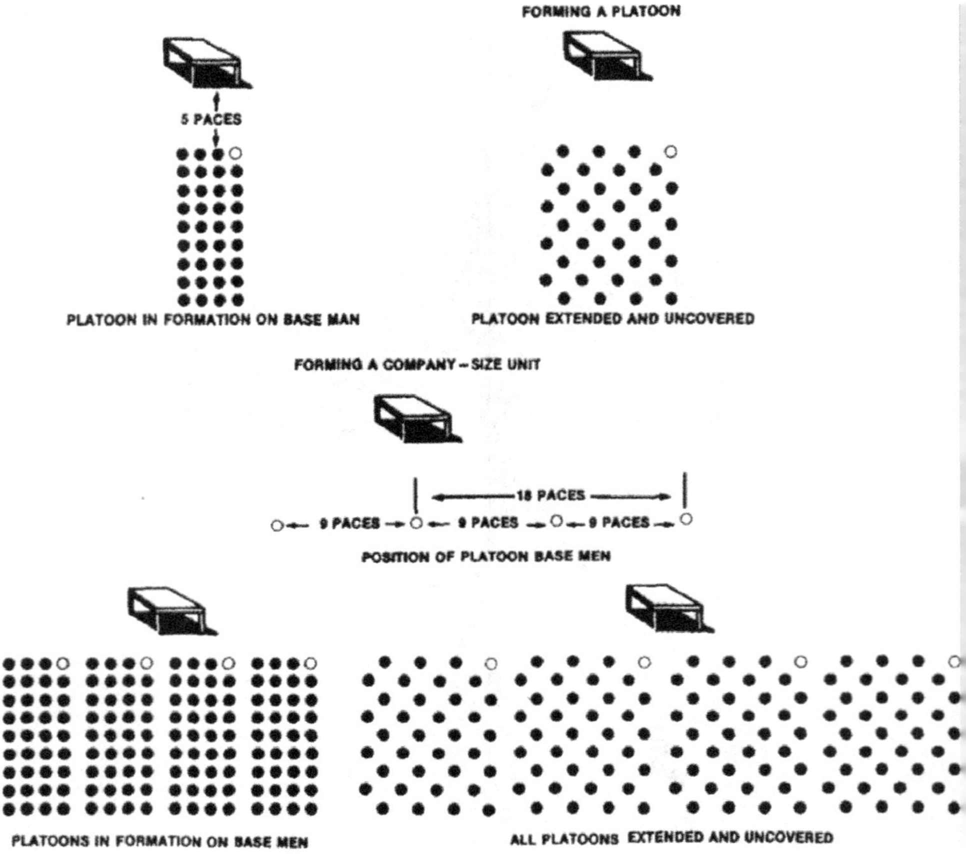

Figure A-l. Forming the Extended Rectangular Formation.

Figure A-2.Circle Formation.

Physical Fitness Training

Department of Defense

Pentagon
Publishing

CHANGE 1

HEADQUARTERS
DEPARTMENT OF THE ARMY
Washington, DC, 1 October 1998

PHYSICAL FITNESS TRAINING

1. Change FM 21-20, 30 September 1992, as follows:

REMOVE OLD PAGES	INSERT NEW PAGES
14-3 to 14-8	14-3 to 14-8.2
14-21 to 14-22	14-21 to 14-22

2. A star (*) marks new or changed material.

3. File this transmittal sheet in front of this publication.

DISTRIBUTION RESTRICTION: proved for public release; distribution is unlimited.

Approved for public release; distribution is unlimited.

FM 21-20
Ct

Change

No. 1

HEADQUARTERS
DEPARTMENT OF THE ARMY
Washington, OC, 1 October 1998

PHYSICAL FITNESS TRAINING

1. FM 21-20, 30 September 1992. 6 changed as follows:

Remove pages	*Insert pages*
14-3 th roug h 14-8	14-3 Ihroug h 14-5.2
14-21 and 14-22	U"21 anti 14"22

2 A star (*) Indicates new or changed material.

3, File this transmittal sheet in front of this publication.

By Ordar of the Secretary of the Army:

DENNIS J. REIMER
Genera/, ***United*** States ***Army***
CHtfof Smff

Official:

JOELB.HJ0SON
Adminislrsilve Assistant to fte
Secretary of the Army

Distribution:

Active Army, Army National Guard of the United States, and U.S. Army Reserve; To he distributed in accordance with Initial Distribution Number {IDN) 110166, requirements for FM 21-20.

Field Manual
No. 21-20

*FM 21-20
Headquarters
Department of the Army
Washington, DC, 30 September 1992

TABLE OF CONTENTS

DISTRIBUTION RESTRICTION: Approved for public release; distribution is unlimited

This publication supersedes FM 21-20, 28 August 1985.

TABLE OF CONTENTS (CONT.)

Preface

On 5 July 1950, U.S. troops, who were unprepared for the physical demands of war, were sent to battle. The early days of the Korean war were nothing short of disastrous, as U.S. soldiers were routed by a poorly equipped, but well-trained, North Korean People's Army. As American soldiers withdrew, they left behind wounded comrades and valuable equipment their training had not adequately prepared them to carry heavy loads.

The costly lessons learned by Task Force Smith in Korea are as important today as ever. If we fail to prepare our soldiers for their physically demanding wartime tasks, we are guilty of paying lip service to the principle of "Train as you fight." Our physical training programs must do more for our soldiers than just get them ready for the semiannual Army Physical Fitness Test (APFT).

FM 21 -20 is directed at leaders who plan and conduct physical fitness training. It provides guidelines for developing programs which will improve and maintain physical fitness levels for all Army personnel. These programs will help leaders prepare their soldiers to meet the physical demands of war. This manual can also be used as a source book by all soldiers. FM 21-20 was written to conform to the principles outlined in FM 25-100, Training the Force.

The benefits to be derived from a good physical fitness program are many. It can reduce the number of soldiers on profile and sick call, invigorate training, and enhance productivity and mental alertness. A good physical fitness program also promotes team cohesion and combat survivability. It will improve soldiers' combat readiness.

The proponent of this publication is HQ TRADOC. Send comments and recommendations on DA Form 2028 (Recommended Changes to Publications and Blank Forms) directly to Headquarters, US Army Infantry Center, US Army Physical Fitness School (ATZB-PF), Fort Benning, GA31905-5000.

Unless this publication states otherwise, masculine nouns and pronouns do not refer exclusively to men.

CHAPTER 1

Introduction

A soldier's level of physical fitness' has a direct impact on his combat readiness. The many battles in which American troops have fought underscore the important role physical fitness plays on the battlefield. The renewed nationwide interest in fitness has been accompanied by many research studies on the effects of regular participation in sound physical fitness programs. The overwhelming conclusion is that such programs enhance a person's quality of life, improve productivity, and bring about positive physical and mental changes. Not only are physically fit soldiers essential to the Army, they are also more likely to have enjoyable, productive lives.

This chapter provides an overview of fitness. It defines physical fitness, outlines the phases of fitness, and discusses various types of fitness programs and fitness evaluation. Commanders and leaders can use this information to develop intelligent, combat-related, physical fitness programs.

Components of physical fitness include weight control, diet, nutrition, stress management, and spiritual and ethical fitness.

Physical fitness, the emphasis of this manual, is but one component of total fitness. Some of the "others are weight control, diet and nutrition, stress management, dental health, and spiritual and ethical fitness, as well as the avoidance of hypertension, substance abuse, and tobacco use. This manual is primarily concerned with issues relating directly to the development and maintenance of the five components of physical fitness.

The Army's physical fitness training program extends to all branches of the total Army. This includes the USAR and ARNG and encompasses all ages and ranks and both sexes. Its purpose is to physically condition all soldiers throughout their careers beginning with initial entry training (IET). It also includes soldiers with limiting physical profiles who must also participate in physical fitness training.

Commanders and leaders must ensure that all soldiers in their units maintain the highest level of physical fitness in accordance with this manual and with AR 350-15 which prescribes policies, procedures, and responsibilities for the Army physical fitness program.

Leadership Responsibilities

Effective leadership is critical to the success of a good physical training program. Leaders, especially senior leaders, must understand and practice the new Army doctrine of physical fitness. They must be visible and active participants in physical training programs. In short, leaders must lead PT! Their example will emphasize the importance of physical fitness training and will highlight it as a key element of the unit's training mission.

Leaders must emphasize the value of physical training and clearly explain the objectives and benefits of the program. Master Fitness Trainers (MFTs), graduates of a special course taught by the U.S. Army Physical Fitness School, can help commanders do this. However, regardless of the level of technical experience MFTs have, the sole responsibility for good programs rests with leaders at every level.

A poorly designed and executed physical fitness program hurts morale. A good program is well planned and organized, has reasonable yet challenging requirements, and is competitive and progressive. It also has command presence at every level with leaders setting the example for their soldiers.

Leaders should also continually assess their units to determine which specific components of fitness they lack. Once they identify the shortcomings, they should modify their programs to correct the weaknesses.

Leaders should not punish soldiers who fail to perform to standard. Punishment, especially excessive repetitions or additional PT, often does more harm than good. Leaders must plan special training to help soldiers who need it. The application of sound leadership techniques is especially important in bringing physically deficient soldiers up to standard.

'COMMAND FUNCTIONS

Commanders must evaluate the effectiveness of physical fitness training and ensure that it is focused on the unit's missions. They can evaluate its effectiveness by participating in and observing training, relating their fitness programs to the unit's missions, and analyzing individual and unit APFT performance.

Leaders should regularly measure the physical fitness level of every soldier to evaluate his progress and determine the success of the unit's program.

Commanders should assure that qualified leaders supervise and conduct fitness training and use their MFTs, for they have received comprehensive training in this area.
Leaders can learn about fitness training in the following ways:

- Attend the four-week MFT course or one-week Exercise Leaders Course.
- Request a fitness workshop from the Army Physical Fitness School.
- Become familiar with the Army's fitness publications. Important examples include this manual, AR 350-15, and DA Pamphlets 350-15, 350-18, and 350-22.

Commanders must provide adequate facilities and funds to support a program which will improve each soldier's level of physical fitness. They must also be sure that everyone participates, since all individuals, regardless of rank, age, or sex, benefit from regular exercise. In some instances, leaders will need to make special efforts to overcome recurring problems which interfere with regular training.

Leaders must also make special efforts to provide the correct fitness training for soldiers who are physically substandard. "Positive profiling" (DA Form 3349) permits and encourages profiled soldiers to do as much as they can within the limits of their profiles. Those who have been away from the conditioning process because of leave, sickness, injury, or travel may also need special consideration.

Commanders must ensure that the time allotted for physical fitness training is used effectively.
Training times is wasted by the following:

Commanders must ensure that the time alloted for physical fitness training is used effectively.

- Unprepared or unorganized leaders.
- Assignment fo a group which us too large for one leader.
- Insufficient training intensity: it will result in no improvement.
- Rates of progression that are too slow or too fast.
- Extreme faomality that usually emphasizes form over substance. An example would be too many units runs at slow paces or "daily dozen" activities that look impressive but do not result in impovement.
- Inadequate facilities which cause long waiting periods between exercises during a workout and/or between workouts.
- Long rest periods which interfere with progress.

To foster a positive attitude, unit leaders and instructors must be knowledgeable, understanding, and fair, but demanding. They must recognize individual differences and motivate soldiers to put forth their best efforts. However, they must also emphasize training to standard. Attaining a high level of physical fitness cannot be done simply by going through the motions. Hard training is essential.

Commanders must ensure that leaders are familiar with approved techniques, directives, and publications and that they use them. The objective of every commander should be to incorporate the most effective methods of physical training into a balanced program. This program should result in the improved physical fitness of their soldiers and an enhanced ability to perform mission-related tasks.

MFTs can help commanders formulate sound programs that will attain their physical training goals, but commanders must know and apply the doctrine. However, since the responsibility for physical training is the commander's, programs must be based on his own training objectives. These he must develop from his evaluation of the unit's mission-essential task list (METL). Chapter 10 describes the development of the unit's program.

MASTER FITNESS TRAINERS

A Master Fitness Trainer (MFT) is a soldier who has completed either the four-week active-component, two-week reserve-component, or U.S. Military Academy's MFT course work. Although called "masters," MFTs are simply soldiers who know about all aspects of physical fitness training and how soldiers' bodies function. Most importantly, since MFTs are taught to design individual and unit programs, they should be used by commanders as special staff assistants for this purpose.

MFTs can do the following:

- Assess the physical fitness levels of individuals and units.
- Analyze the unit's mission-related tasks and develop sound fitness training programs to support those tasks.
- Train other trainers to conduct sound, safe physical training.
- Understand the structure and function of the human body, especially as it relates to exercise.

Components of Fitness

Physical fitness is the ability to function effectively in physical work, training, and other activities and still have enough energy left over to handle any emergencies which may arise.

The components of physical fitness are as follows:

- Cardiorespiratory (CR) endurance-the efficiency with which the body delivers oxygen and nutrients needed for muscular activity and transports waste products from the cells.
- Muscular strength - the greatest amount of force a muscle or muscle group can exert in a single effort.
- Muscular endurance - the ability of a muscle or muscle group to perform repeated movements with a sub-maximal force for extended periods of times.
- Flexibility-the ability to move the joints (for example, elbow, knee) or any group of joints through an entire, normal range of motion.
- Body composition-the amount of body fat a soldier has in comparison to his total body mass.

Improving the first three components of fitness listed above will have a positive impact on body composition and will result in less fat. Excessive body fat detracts from the other fitness components, reduces performance, detracts from appearance, and negatively affects one's health.

Factors such as speed, agility, muscle power, eye-hand coordination, and eye-foot coordination are classified as components of "motor" fitness. These factors affect a soldier's survivability on the battlefield. Appropriate training can improve these factors within the limits of each soldier's potential. The Army's fitness program seeks to improve or maintain all the components of physical and motor fitness through sound, progressive, mission-specific physical training for individuals and units.

Principles of Exercise

Adherence to certain basic exercise principles is important for developing an effective program. The principles of exercise apply to everyone at all levels of physical training, from the Olympic-caliber athlete to the week-end jogger. They also apply to fitness training for military personnel.

These basic principles of exercise must be followed:

- Regularity. To achieve a training effect, a person must exercise often. One should strive to exercise each of the first four fitness components at least three times a week. Infrequent exercise can do more harm than good. Regularity is also important in resting, sleeping, and following a good diet.
- Progression. The intensity (how hard) and/or duration (how long) of exercise must gradually increase to improve the level of fitness.
- Balance. To be effective, a program should include activities that address all the fitness components, since overemphasizing any one of them may hurt the others.
- Variety. Providing a variety of activities reduces boredom and increases motivation and progress.
- Specificity. Training must be geared toward specific goals. For example, soldiers become better runners if their training emphasizes running. Although swimming is great exercise, it does not improve a 2-mile-run time as much as a running program does.
- Recovery. A hard day of training for a given component of fitness should be followed by an easier training day or rest day for that component and/or muscle group(s) to help permit recovery. Another way to allow recovery is to alternate the muscle groups exercised every other day, especially when training for strength and/or muscle endurance.
- Overload. The work load of each exercise session must exceed the normal demands placed on the body in order to bring about a training effect.

FITT Factors

Factors for a successful training program are Frequency, Intensity, Time, and Type; "FUT".

Certain factors must be part of any fitness training program for it to be successful. These factors are Frequency, Intensity, Time, and Type. The acronym FITT makes it easier to remember them. (See Figure 1- 1.)

FREQUENCY

Army Regulation 350-15 specifies that vigorous physical fitness training will be conducted 3 to 5 times per week. For optimal results, commanders must strive to conduct 5 days of physical training per week. Ideally, at least three exercise sessions for CR fitness, muscle endurance, muscle strength, and flexibility should be performed each week to improve fitness levels. Thus, for example, to obtain maximum gains in muscular strength, soldiers should have at least three strength-training sessions per week. Three physical activity periods a week, however, with only one session each of cardiorespiratory, strength, and flexibility training will not improve any of these three components.

With some planning, a training program for the average soldier can be developed which provides fairly equal emphasis on all the components of physical fitness. The following training program serves as an example.

In the first week, Monday, Wednesday, and Friday are devoted to CR fitness, and Tuesday and Thursday are devoted to muscle endurance and strength. During the second week, the

FITT Factors Applied to Physical Conditioning Program	Cardiorespiratory Endurance	Muscular Strength	Muscular Endurance	Muscular Strength and Muscular Endurance	Flexibility
F — Frequency	3-5 times/week	3 times/week	3-5 times/week	3 times/week	Warm-up and Cool-down: Stretch before and after each exercise session Developmental Stretching: To improve flexibility, stretch 2-3 times/week
I — Intensity	60-90% HRR*	3-7 RM*	12+ RM	8-12 RM	Tension and slight discomfort, NOT PAIN
T — Time	20 minutes or more	The time required to do 3-7 repetitions of each exercise	The time required to do 12+ repetitions of each exercise	The time required to do 8-12 repetitions of each exercise	Warm-up and Cool-down Stretches: 10-15 seconds/stretch Developmental Stretches: 30-60 seconds/stretch
T — Type	Running Swimming Cross-Country Skiing Rowing Bicycling Jumping Rope Walking/Hiking Stair Climbing	Free Weights Resistance Machines Partner-Resisted Exercises Body-Weight Exercises (Pushups/Situps/Pullups/Dips, etc.)			Stretching: Static Passive P.N.F.

* HRR = Heart Rate Reserve * RM = Repetition Maximum

Figure 1-1

training days are flip-flopped: muscle endurance and strength are trained on Monday, Wednesday, and Friday, and CR fitness is trained on Tuesday and Thursday. Stretching exercises are done in every training session to enhance flexibility. By training continuously in this manner, equal emphasis can be given to developing muscular endurance and strength and to CR fitness while training five days per week.

If the unit's mission requires it, some muscular and some CR training can be done during each daily training session as long as a "hard day/recovery day" approach is used. For example, if a unit has a hard run on Monday, Wednesday, and Friday, it may also choose to run on Tuesday and Thursday. However, on Tuesday and Thursday the intensity and/or distance/time should be reduced to allow recovery. Depending on the time available for each session and the way training sessions are conducted, all components of fitness can be developed using a three-day-per-week schedule. However, a five-day-per-week program is much better than three per week. (See Training Program in Chapter 10.)

Numerous other approaches can be taken when tailoring a fitness program to meet a unit's mission as long as the principles of exercise are not violated. Such programs, when coupled with good nutrition, will help keep soldiers fit to win.

INTENSITY

Training at the right intensity is the biggest problem in unit programs. The intensity should vary with the type of exercise being done. Exercise for CR development must be strenuous enough to elevate the heart rate to between 60 and 90 percent of the heart rate reserve (HRR). (The calculation of percent HRR is explained in Chapter 2.) Those with low fitness levels should start exercising at a lower training heart rate (THR) of about 60 percent of HRR.

For muscular strength and endurance, intensity refers to the percentage of the maximum resistance that is used for a given exercise. When determining intensity in a strength-training program, it is easier to refer to a "repetition maximum" or "RM." For example, a IO-RM is the maximum weight that can be correctly lifted 10 times. An 8-12 RM is the weight that can be lifted 8 to 12 times correctly. Doing an exercise "correctly" means moving the weight steadily and with proper form without getting help from other muscle groups by jerking, bending, or twisting the body. For the average person who wants to improve both muscular strength and endurance, an 8-12 RM is best.

The person who wants to concentrate on muscular strength should use weights which let him do three to seven repetitions before his muscles fatigue. Thus, for strength development, the weight used should be a 3-7 RM. On the other hand, the person who wants to concentrate on muscular endurance should use a 12+ RM. When using a 12+ RM as the training intensity, the more repetitions performed per set, over time, the greater will be the improvement in muscular endurance. Conversely, the greater the number of repetitions performed, the smaller will be the gains in strength. For example, a person who regularly trains with a weight which lets him do 100 repetitions per exercise (a 1OO-RM) greatly increases his muscular endurance but minimally improves his muscular strength. (See Chapter 3 for information on resistance training.)

All exercise sessions should include stretching during the warm-up and cool-down. One should stretch so there is slight discomfort, but no pain, when the movement is taken beyond the normal range of motion. (See Chapter 4 for information on stretching.)

All exercises sessions should include stretching during the warm-up and cool-down.

TIME

Like intensity, the time spent exercising depends on the type of exercise being done. At least 20 to 30 continuous minutes of intense exercise must be used in order to improve cardiorespiratory endurance.

For muscular endurance and strength, exercise time equates to the number of repetitions done. For the average soldier, 8 to 12 repetitions with enough resistance to cause muscle failure improves both muscular endurance and strength. As soldiers progress, they will make better strength gains by doing two or three sets of each resistance exercise.

Flexibility exercises or stretches should be held for varying times depending on the objective of the session. For warming-up, such as before a run, each stretch should be held for 10 to 15 seconds. To improve flexibility, it is best to do stretching during the cool-down, with each stretch held for 30 to 60 seconds. If flexibility improvement is a major goal, at least one session per week should be devoted to developing it.

TYPE

Type refers to the kind of exercise performed. When choosing the type, the commander should consider the principle of specificity. For example, to improve his soldiers' levels of CR fitness (the major fitness component in the 2-mile run), he should have them do CR types of exercises. These are discussed in Chapter 2.

Ways to train for muscular strength and endurance are addressed in Chapter 3, while Chapter 4 discusses flexibility. These chapters will help commanders design programs which are tailor-made to their soldiers' needs. The basic rule is that to improve performance, one must practice the particular exercise, activity, or skill he wants to improve. For example, to be good at push-ups, one must do push-ups. No other exercise will improve push-up performance as effectively.

Warm-up and Cool-Down

One must prepare the body before taking part in organized PT, unit sports competition, or vigorous physical activity. A warm-up may help prevent injuries and maximize performance. The warm-up increases the body's internal temperature and the heart rate. The chance of getting injured decreases when the heart, muscles, ligaments, and tendons are properly prepared for exertion. A warm-up should include some running-in-place or slow jogging, stretching, and calisthenics. It should last five to seven minutes and should occur just before the CR or muscular endurance and strength part of the workout. After a proper warm-up, soldiers are ready for a more intense conditioning activity.

Soldiers should cool down properly after each exercise period, regardless of the type of workout. The cool-down serves to gradually slow the heart rate and helps prevent pooling of the blood in the legs and feet. During exercise, the muscles squeeze the blood through the veins. This helps return the blood to the heart. After exercise, however, the muscles relax and no longer do this, and the blood can accumulate in the legs and feet. This can cause a person to faint. A good cool-down will help avoid this possibility.

Soldiers should walk and stretch until their heart rates return to less than 100 beats per minute (BPM) and heavy sweating stops. This usually happens five to seven minutes after the conditioning session.

Phases of Fitness Conditioning

The physical fitness training program is divided into three phases: preparatory, conditioning, and maintenance. The starting phases for different units or individuals vary depending on their age, fitness levels, and previous physical activity.

Young, healthy persons may be able to start with the conditioning phase, while those who have been exercising regularly may already be in the maintenance phase. Factors such as extended field training, leave time, and illness can cause soldiers to drop from a maintenance to a conditioning phase.

Persons who have not been active, especially if they are age 40 or older, should start with the preparatory phase. Many soldiers who fall into this category may be recovering from illness or injury, or they may be just out of high school. Most units will have soldiers in all three phases of training at the same time.

PREPARATORY PHASE

The preparatory phase helps both the cardiorespiratory and muscular systems get used to exercise, preparing the body to handle the conditioning phase. The work load in the beginning must be moderate. Progression from a lower to a higher level of fitness should be achieved by gradual, planned increases in frequency, intensity, and time.

Initially, poorly conditioned soldiers should run, or walk if need be, three times a week at a comfortable pace that elevates their heart rate to about 60 percent HRR for 10 to 15 minutes. Recovery days should be evenly distributed throughout the week, and training should progress slowly. Soldiers should continue at this or an appropriate level until they have no undue fatigue or muscle soreness the day following the exercise. They should then lengthen their exercise session to 16 to 20 minutes and/or elevate their heart rate to about 70 percent HRR by increasing their pace. To be sure their pace is faster, they should run a known distance and try to cover it in less time. Those who feel breathless or whose heart rate rises beyond their training heart rate (THR) while running should resume walking until the heart rate returns to the correct training level. When they can handle an intensity of 70 percent HRR for 20 to 25 minutes, they should be ready for the next phase. Chapter 2 shows how to determine the THR, that is, the right training level during aerobic training.

The preparatory phase for improving muscular endurance and strength through weight training should start easily and progress gradually. Beginning weight trainers should select about 8 to 12 exercises that work all the body's major muscle groups. They should use only very light weights the first week (that is, the first two to three workouts). This is very important, as they must first learn the proper form for each exercise. Light weights will also help minimize muscle soreness and decrease the likelihood of injury to the muscles, joints, and ligaments. During the second week, they should use progressively heavier weights on each resistance exercise. By the end of the second week (four to six workouts), they should know how much weight will let them do 8 to 12 repetitions to muscle failure for each exercise. At this point the conditioning phase be-

CONDITIONING PHASE

To reach the desired level of fitness, soldiers must increase the amount of exercise and/or the workout intensity as their strength and/or endurance increases.

To improve cardiorespiratory endurance, for example, they must increase the length of time they run. They should start with the preparatory phase and gradually increase the running time by one or two minutes each week until they can run continuously for 20 to 30 minutes. At this point, they can increase the intensity until they reach the desired level of fitness. They should train at least three times a week and take no more than two days between workouts.

For weight trainers, the conditioning phase normally begins during the third week. They should do one set of 8 to 12 repetitions for each of the selected resistance exercises. When they can do more than 12 repetitions of any exercise, they should increase the

Soldiers and units should be encouraged to progress beyond minimum requirements.

weight used on that exercise by about five percent so they can again do only 8 to 12 repetitions. This process continues throughout the conditioning phase. As long as they continue to progress and get stronger while doing only one set of each exercise, it is not necessary for them to do more than one set per exercise. When they stop making progress with one set, they should add another set on those exercises in which progress has slowed. As training progresses, they may want to increase the sets to three to help promote further increases in strength and/or muscle mass.

For maximum benefit, soldiers should do strength training three times a week with 48 hours of rest between workouts for any given muscle group. It helps to periodically do a different type of exercise for a given muscle or muscle group. This adds variety and ensures better strength development.

The conditioning phase ends when a soldier is physically mission-capable and all personal, strength-related goals and unit-fitness goals have been met.

MAINTENANCE PHASE

The maintenance phase sustains the high level of fitness achieved in the conditioning phase. The emphasis here is no longer on progression. A well-designed, 45- to 60-minute workout (including warm-up and cool-down) at the right intensity three times a week is enough to maintain almost any appropriate level of physical fitness. These workouts give soldiers time to stabalize their flexibility, CR endurance, and muscular endurance and strength. However, more frequent training may be needed to reach and maintain peak fitness levels.

Soldiers and units should always be encouraged to progress beyond minimum requirements. Maintaining an optimal level of fitness should become part of every soldier's life-style and should be continued throughout his life.

An effective program uses a variety of activities to develop muscular endurance and strength, CR endurance, and flexibility, and to achieve good body composition. It should also promote the development of coordination as well as basic physical skills. (See Chapter 10 for guidance in constructing a unit program.)

Types of Fitness Programs

The Army has too many types of units with different missions to have one single fitness program for everyone. Therefore, only broad categories of programs and general considerations are covered here. They are classified as unit, individual, and special programs.

UNIT PROGRAMS

Unit programs must support unit missions. A single unit may require several types of programs. Some units, such as infantry companies, have generally the same types of soldiers and MOSS. On the other hand, certain combat--service-support units have many different types of soldiers, each with unique needs. Commanders can develop programs for their own unit by following the principles in this chapter. MFTs know how to help commanders develop programs for their units/soldiers.

Commanders of units composed of both men and women must also understand the physiological differences between the sexes. These are summarized in Appendix A. Although women are able to participate in the same fitness programs as men, they must work harder to perform at the same absolute level of work or exercise. The same holds true for poorly-conditioned soldiers running with well-conditioned soldiers.

To overcome this problem in the case of running, for example, the unit should use ability group runs rather than unit runs. Soldiers in a given ability group will run at a set pace, with groups based on each soldier's most recent 2-mile-run time. Three to six groups per company-sized unit are usually enough. Within each group, each soldier's heart rate while running should be at his own THR. When the run is not intense enough to bring one or more of the soldiers to THR, it is time for those soldiers to move up to the next ability group.

Ability group running does two things more effectively than unit runs: 1) it lets soldiers improve to their highest attainable fitness level; and, 2) it more quickly brings subpar performers up to minimum standards.

It also allows soldiers to train to excel on the APFT which, in turn, helps promotion opportunities. Holding a fit soldier back by making him run at a slow, unit-run pace (normally less than his minimum pace for the 2-mile run on the APFT) hurts his morale and violates the principle of training to challenge.

initial Entry Training (IET)

The training program in basic training (BT) brings soldiers up to the level of physical fitness they need to do their jobs as soldiers. However, the program requires good cadre leadership to ensure that it is appropriate, demanding, and challenging.

Trainees report to active duty at various levels of physical fitness and ability. During basic training they pass through the preparatory into the conditioning phase. During "fill" periods and the first week of training, the focus is on learning and developing the basics of physical fitness.

Training emphasizes progressive conditioning of the whole body. To minimize the risk of injury, exercises must be done properly, and the intensity must progress at an appropriate rate. Special training should be considered for soldiers who fail to maintain the unit's or group's rate of progression. Commanders should evaluate each basic trainee who falls below standard and give him individualized, special assistance to improve his deficiencies.

Additional training should not be used as punishment for a soldier's inability to perform well.

More PT is not necessarily better. Chapter 11 describes how to develop physical training programs in IET units.

Advanced Incividual Training (AIT)

Although AIT focuses on technical and MOS-oriented subjects, physical fitness must be emphasized throughout. Most soldiers arriving from basic training are already well into the conditioning phase. Therefore, AIT unit training should focus on preparing soldiers to meet the physical requirements of their initial duty assignments. (See TRADOC Reg. 350-6, Chapter 4.)

Walking, running, and climbing during unit training contribute to physical fitness, but they are not enough. Physical training in AIT requires continued, regular, vigorous exercise which stresses the whole body and addresses all the components of fitness.

By the end of AIT, soldiers must meet APFT standards. With good programs and special training, all healthy AIT graduates should easily be able to demonstrate that they, possess the required level of physical fitness.

By the end of AIT, soldiers must meet APFT standards.

There must be a positive approach to all special fitness training.

TOE and TDA Units-Active Component

There are many types of units in the Army, and their missions often require different levels of fitness. TOE and TDA units must emphasize attaining and maintaining the fitness level required for the mission.

The unit's standards may exceed the Army's minimums. By regulation (AR 350- 15), the unit's standards can be established by the unit's commander, based on mission requirements.

TOE and TDA Units-Reserve Components

The considerations for the active component also apply to reserve components (RCS). However, since members of RC units cannot participate together in collective physical training on a regular basis, RC unit programs must focus on the individual's fitness responsibilities and efforts. Commanders, however, must still ensure that the unit's fitness level and individual PT programs are maintained. MFTs can give valuable assistance to RC commanders and soldiers.

INDIVIDUAL PROGRAMS

Many soldiers are assigned to duty positions that offer little opportunity to participate in collective unit PT programs. Examples are HQDA, MACOM staffs, hospitals, service school staff and faculty, recruiting, and ROTC. In such organizations, commanders must develop leadership environments that encourage and motivate soldiers to accept individual responsibility for their own physical fitness. Fitness requirements are the same for these personnel as for others. Section chiefs and individual soldiers need to use the fundamental principles and techniques outlined in this manual to help them attain and maintain a high level of physical fitness. MFTs can help develop individual fitness programs.

SPECIAL PROGRAMS

The day-to-day unit PT program conducted for most soldiers may not be appropriate for all unit members. Some of them may not be able to exercise at the intensity or duration best suited to their needs.

At least three groups of soldiers may need special PT programs. They are as follows:

- Those who fail the APFT and do not have medical profiles.
- Those who are overweight/overfat according to AR 600-9
- Those who have either permanent or temporary medical profiles.

Leaders must also give special consideration to soldiers who are age 40 or older and to recent arrivals who cannot meet the standards of their new unit.

Special programs must be tailored to each soldier's needs, and trained, knowledgeable leaders should develop and conduct them. This training should be conducted with the unit, If this is impossible, it should at least occur at the same time.

There must be a positive approach to all special fitness training. Soldiers who lack enough upper body strength to do a given number of push-ups or enough stamina to pass the 2-mile run should not be ridiculed. Instead, their shortcomings should be assessed and the information used to develop individualized programs to help them remedy their specific shortcomings. A company-sized unit may have as many as 20 soldiers who need special attention. Only smart planning will produce good programs for all of them.

Commanders must counsel soldiers, explaining that special programs are being developed in their best interests. They must make it clear that standards will be enforced. Next, they should coordinate closely with medical personnel to develop programs that fit the capabilities of soldiers with medical limitations. Each soldier should then begin an individualized program based on his needs.

MFTs know how to assess CR endurance, muscular strength and endurance, flexibility, and body composition. They can also develop thorough, tailor-made programs for all of a unit's special population.

APFT Failures

Although it is not the heart of the Army's physical fitness program, the APFT is the primary instrument for evaluating the fitness level of each soldier. It is structured to assess the muscular endurance of specific muscle groups and the functional capacity of the CR system.

Soldiers with reasonable levels of overall physical fitness should easily pass the APFT. Those whose fitness levels are substandard will fail. Soldiers who fail the APFT must receive special attention. Leaders should analyze their weaknesses and design programs to overcome them. For example, if the soldier is overweight, nutrition and dietary counseling may be needed along with a special exercise program. DA Pam 350-22 outlines several ways to improve a soldier's performance on each of the APFT events.

When trying to improve APFT performances, leaders must ensure that soldiers are not overloaded to the point where the fitness training becomes counterproductive. They should use ability groups for their running program and, in addition to a total-body strength-training program, should include exercises designed for push-up and sit-up improvement. When dealing with special populations, two very important principles are overload and recovery. The quality, not just the quantity, of the workout should be emphasized. Two-a-day sessions, unless designed extremely well, can be counter-productive. More PT is not always better.

Overweight Soldiers

Designers of weight loss and physical training programs for overweight soldiers should remember this: even though exercise is the key to sensible weight loss, reducing the number of calories consumed is equally important. A combination of both actions is best.

The type of exercise the soldier does affects the amount and nature of the weight loss. Both running and walking burn about 100 calories per mile. One pound of fat contains 3,500 calories. Thus, burning one pound of fat through exercise alone requires a great deal of running or walking. On the other hand, weight lost through dieting alone includes the loss of useful muscle tissue. Those who participate in an exercise program that emphasizes the development of strength and muscular endurance, however, can actually increase their muscle mass while losing body fat. These facts help explain why exercise and good dietary practices must be combined.

Unit MFTs can help a soldier determine the specific caloric requirement he needs to safely and successfully lose excess fat. They can devise a sound, individualized plan to arrive at that reduced caloric intake. Likewise, unit MFTs can also develop training programs which will lead to fat loss without the loss of useful muscle tissue.

Generally, overweight soldiers should strive to reduce their fat weight by two pounds per week. When a soldier loses weight, either by diet or exercise or both, a large initial weight loss is not unusual. This may be due to water loss associated with the using up of the body's carbohydrate stores. Although these losses may be encouraging to the soldier, little of this initial weight loss is due to the loss of fat.

Soldiers should be weighed under similar circumstances and at the same time each day. This helps avoid false measurements due to normal fluctuations in their body weight during the day. As a soldier develops muscular endurance and strength, lean muscle mass generally increases. Because muscle weighs more per unit of volume than fat, caution is advised in assessing his progress. Just because a soldier is not losing weight rapidly does not necessarily mean he is not losing fat. In fact, a good fitness program often results in gaining muscle mass while simultaneously losing fat weight. If there is reasonable doubt, his percentage of body fat should be determined.

Soldiers with Profiles

This manual stresses what soldiers can do while on medical profile rather than what they cannot do.

DOD Directive 1308.1 requires that, 'Those personnel identified with medically limiting defects shall be placed in a physical fitness program consistent with their limitations as advised by medical authorities."

AR 350-15 states, "For individuals with limiting profiles, commanders will develop physical fitness programs in cooperation with health care personnel."

The Office of the Surgeon General has developed DA Form 3349 to ease the exchange of information between health care personnel and the units. On this form, health care personnel list, along with limitations, those activities that the profiled soldier can do to maintain his fitness level. With this information, the unit should direct profiled soldiers to participate in the activities they can do. (An example of DA Form 3349 is in Appendix B.)

All profiled soldiers should take part in as much of the regular fitness program as they can. Appropriate activities should be substituted to replace those regular activities in which they cannot participate.

Chapter 2 describes some aerobic activities the soldier can do to maintain cardiorespiratory fitness when he cannot run. Chapter 3 shows how to strengthen each body part. Applying this information should allow some strength training to continue even when body parts are injured. The same principle applies to flexibility (Chapter 4).

Medical treatment and rehabilitation should be aimed at restoring the soldier to a suitable level of physical fitness. Such treatment should use appropriate, progressive physical activities with medical or unit supervision.

MFTs can help profiled soldiers by explaining alternative exercises and how to do them safely under the limitations of their profile. MFTs are not, however, trained to diagnose injuries or prescribe rehabilitative exercise programs. This is the domain of qualified medical personnel.

The activity levels of soldiers usually decrease while they are recovering from sickness or injury. As a result, they should pay special attention to their diets to avoid gaining body fat. This guidance becomes more important as soldiers grow older. With medical supervision, proper diet, and the right PT programs, soldiers should be able to overcome their physical profiles and quickly return to their normal routines and fitness levels.

All profiled soldiers hould do as much of the regular fitness rogram as they can, long with substitute activities.

Age as a Factor in Physical Fitness

Soldiers who are age 40 and older represent the Army's senior leadership. On the battlefield, they must lead other soldiers under conditions of severe stress. To meet this challenge and set a good example, these leaders must maintain and demonstrate a high level of physical fitness. Since their normal duties may be stressful but nonphysical, they must take part regularly in a physical fitness program. The need to be physically fit does not decrease with increased age.

People undergo many changes as they grow older. For example, the amount of blood the heart can pump per beat and per minute decreases during maximal exercise, as does the maximum heart rate. This lowers a person's physical ability, and performance suffers. Also, the percent of body weight composed of fat generally increases, while total muscle mass decreases. The result is that muscular strength and endurance, CR endurance, and body composition suffer. A decrease in flexibility also occurs.

Men tend to maintain their peak levels of muscular strength and endurance and CR fitness until age 30. After 30 there is a gradual decline throughout their lives. Women tend to reach their peak in physical capability shortly after puberty and then undergo a progressive decline.

Although a decline in performance normally occurs with aging, those who stay physically active do not have the same rate of decline as those who do not. Decreases in muscular strength and endurance, CR endurance, and flexibility occur to a lesser extent in those who regularly train these fitness components.

Soldiers who are fit at age 40 and continue to exercise show a lesser decrease in many of the physiological functions related to fitness than do those who seldom exercise. A trained 60-year-old, for example, may have the same level of CR fitness as a sedentary 20-year-old. In short, regular exercise can help add life to your years and years to your life.

The assessment phase of a program is especially important for those age 40 and over. However, it is not necessary or desirable to develop special fitness programs for these soldiers. Those who have been exercising regularly may continue to exercise at the same level as they did before reaching age 40. A program based on the principles of exercise and the training concepts in this manual will result in a safe, long-term conditioning program for all soldiers. Only those age 40 and over who have not been exercising regularly may need to start their exercise program at a lower level and progress more slowly than younger soldiers. Years of inactivity and possible abuse of the body cannot be corrected in a few weeks or months.

As of 1 January 1989, soldiers reaching age 40 are no longer required to get clearance from a cardiovascular screening program before taking the APFT. Only a medical profile will exempt them from taking the biannual record APFT. They must, however, have periodic physical examinations in accordance with AR 40-501 and NGR 40-501. These include screening for cardiovascular risk factors.

Evaluation

To evaluate their physical fitness and the effectiveness of their physical fitness training programs, all military personnel are tested biannually using the APFT in accordance with AR 350-15. (Refer to Chapter 14.) However, commanders may evaluate their physical fitness programs more frequently than biannually.

SCORING CATEGORIES

There are two APFT categories of testing for all military personnel Initial Entry Training (IET) and the Army Standard.

Safety is a major consideration when planning and evaluating physical training programs

IET Standard

The APFT standard for basic training is a minimum of 50 points per event and no less than 150 points overall by the end of basic training. Graduation requirements for AIT and One Station Unit Training (OSUT) require 60 points per event.

Army Standard

All other Army personnel (active and reserve) who are non-IET soldiers must attain the minimum Army standard of at least 60 points per event. To get credit for a record APFT, a medically profiled soldier must, as a minimum, complete the 2-mile run or one of the alternate aerobic events.

SAFETY

Safety is a major consideration when planning and evaluating physical training programs. Commanders must ensure that the programs do not place their soldiers at undue risk of injury or accident. They should address the following items:

"Environmental conditions (heat/cold/traction).

• Soldiers' levels of conditioning (low/high/age/sex).

•Facilities (availability/instruction/repair).

•Traffic (routes/procedures/formations).

•Emergency procedures (medical/communication/transport).

The objective of physical training in the Army is to enhance soldiers' abilities to meet the physical demands of war. Any physical training which results in numerous injuries or accidents is detrimental to this goal. As in most training, common sense must prevail. Good, sound physical training should challenge soldiers but should not place them at undue risk nor lead to situations where accidents or injuries are likely to occur.

CHAPTER 2

Cardiorespiratory Fitnfiss

Cardiorespiratory (CR) fitness, sometimes called CR endurance, aerobic fitness, or aerobic capacity, is one of the five basic components of physical fitness. CR fitness is a condition in which the body's cardiovascular (circulatory) and respiratory systems function together, especially during exercise or work, to ensure that adequate oxygen is supplied to the working muscles to produce energy. CR fitness is needed for prolonged, rhythmic use of the body's large muscle groups. A high level of CR fitness permits continuous physical activity without a decline in performance and allows for rapid recovery following fatiguing physical activity.

Activities such as running, road marching, bicycling, swimming, cross-country skiing, rowing, stair climbing, and jumping rope place an extra demand on the cardiovascular and respiratory systems. During exercise, these systems attempt to supply oxygen to the working muscles. Most of this oxygen is used to produce energy for muscular contraction. Any activity that continuously uses large muscle groups for 20 minutes or longer taxes these systems. Because of this, a wide variety of training methods is used to improve cardiorespiratory endurance.

Physiology of Aerobic Training

Aerobic exercise uses oxygen to produce most of the body's energy needs. It also brings into play a fairly complex set of physiological events.

To provide enough energy-producing oxygen to the muscles, the following events occur:

- Greater movement of air through the lungs.
- Increased movement of oxygen from the lungs into the blood stream.
- Increased delivery of oxygen-laden blood to the working muscles by the heart's accelerated pumping action.
- Regulation of the blood vessel's size to distribute blood away from inactive tissue to working muscle.
- Greater movemen t of oxygen from the blood into the muscle tissue.
- Accelerated return of veinous blood to the heart.

Correctly performed aerobic exercise, over time, causes positive changes in the body's CR system. These changes allow the heart and vascular systems to deliver more oxygen-rich blood to the working muscles during exercise. Also, those muscles regularly used during aerobic exercise undergo positive changes. By using more oxygen, these changes let the muscles make and use more energy during exercise and, as a result, the muscles can work longer and harder.

During maximum aerobic exercise, the trained person has an increased^maximum oxygen consumption (VC»2max). He is better able to process oxygen and fuel and can therefore provide more energy to the working muscles.

VO^nax, also called aerobic capacity, is the most widely accepted single indicator of one's CR fitness level.

CR fitness is needed for prolonged, rhythmic use of the body's large muscle groups.

Aerobic exercise is the best type of activity for attaining and maintaining a low percentage of body fat.

The best way to determine aerobic capacity is to measure it in the laboratory. It is much easier, however, to estimate maximum oxygen uptake by using other methods.

It is possible to determine a soldier's CR fitness level and get an accurate estimate of his aerobic capacity by using his APFT 2-mile-run time. (Appendix F explains how to do this.) Other tests - the bicycle, walk, and step tests - may also be used to estimate one's aerobic capacity and evaluate one's CR fitness level.

In the presence of oxygen, muscle cells produce energy by breaking down carbohydrates and fats. In fact, fats are only used as an energy source when oxygen is present. Hence, aerobic exercise is the best type of activity for attaining and maintaining a low percentage of body fat.

A person's maximum aerobic capacity can be modified through physical training. To reach very high levels of aerobic fitness, one must train hard. The best way to improve CR fitness is to participate regularly in a demanding aerobic exercise program.

Many factors can negateively affect one's ability to perform well aerobically. These include the following:

- Age.
- Anemia,
- Carbon monoxide from tobacco smoke or pollution.
- High altitude (reduced oxygen pressure).
- Illness (heart disease).
- Obesity.
- Sedentary life-style.

Any condition that reduces the body's ability to bring in, transport, or use oxygen reduces a person's ability to perform aerobically. Inactivity causes much of the decrease in physical fitness that occurs with increasing age. Some of this decrease in aerobic fitness can be slowed by taking part in a regular exercise program.

Certain medical conditions also impair the transport of oxygen. They include diseases of the lungs, which interfere with breathing, and disabling heart conditions. Another is severe blocking of the arteries which inhibits blood flow to the heart and skeletal muscles.

Smoking can lead to any or all of the above problems and can, in the long and short term, adversely affect one's ability to do aerobic exercise.

FITT Factors

As mentioned in Chapter 1, a person must integrate several factors into any successful fitness training program to improve his fitness level. These factors are summarized by the following words which form the acronym FITT. Frequency, Intensity, Time, and Type. They are described below as they pertain to cardiorespiratory fitness. A warm-up and cool-down should also be part of each workout. Information on warming up and cooling down is given in Chapters 1 and 4.

FREQUENCY

Frequency refers to how often one exercises. It is related to the intensity and duration of the exercise session. Conditioning the CR system can best be accomplished by three adequately intense workouts per week. Soldiers should do these on alternate days. By building up gradually, soldiers can get even greater benefits from working out five times a week. However, leaders should recognize the need for recovery between hard exercise periods and should adjust the training intensity accordingly. They must also be aware of the danger of overtraining and recognize that the risk of injury increases as the intensity and duration of training increases.

INTENSITY

Intensity is related to how hard one exercises. It represents the degree of effort with which one trains and is probably the single most important factor for improving performance. Unfortunately, it is the factor many units ignore.

Changes in CR fitness are directly related to how hard an aerobic exercise is performed. The more energy expended per unit of time, the greater the intensity of the exercise. Significant changes in CR fitness are brought about by sustaining training heart rates in the range of 60 to 90 percent of the heart rate reserve (HRR). Intensities of less than 60 percent HRR are generally inadequate to produce a training effect, and those that exceed 90 percent HRR can be dangerous.

Soldiers should gauge the intensity of their workouts for CR fitness by determining and exercising at their training heart rate (THR). Using the THR method lets them find and prescribe the correct level of intensity during CR exercise. By determining one's maximum heart rate, resting heart rate, and relative conditioning level, an appropriate THR or intensity can be prescribed.

One's ability to monitor the heart rate is the key to success in CR training. (Note: Ability-group running is better than unit running because unit running does not accommodate the individual soldier's THR. For example, some soldiers in a formation may be training at 50 percent HRR and others at 95 percent HRR. As a result, the unit run will be too intense for some and not intense enough for others.)

The heart rate during work or exercise is an excellent indicator of how much effort a person is exerting. Keeping track of the heart rate lets one gauge the intensity of the CR exercise being done. With this information, one can be sure that the intensity is enough to improve his CR fitness level.

Following are two methods for determining training heart rate (THR). The first method, percent maximum heart rate (% MHR), is simpler to use, while the second method, percent heart rate reserve (% HRR), is more accurate. Percent HRR is the recommended technique for determining THR.

Percent MHR **Method**

With this method, the THR is figured using the estimated maximal heart rate. A soldier determines his estimated maximum heart rate by subtracting his age from 220. Thus, a 20-year-old would have an estimated maximum heart rate (MHR) of 200 beats per minute (220 -20 = 200).

To figure a THR that is 80 percent of the estimated MHR for a 20-year-old soldier in good physical condition, multiply 0.80 times the MHR of 200 beats per minute (BPM). This example is shown below.

FORMULA

% x MHR ⇒ THR

CALCULATION

0.80 x 200 BPM = 160 BPM

When using the MHR method, one must compensate for its built-in weakness. A person using this method may exercise at an intensity which is not high enough to cause a training effect. To compensate for this, a person who is in poor shape should exercise at 70 percent of his MHR; if he is in relatively good shape, at 80 percent MHR; and, if he is in excellent shape, at 90 percent MHR.

Intensity is probably the single most important factor for improving performance.

By determining one's maximum heart rate, resting rate, and conditioning level, an appropriate THR can be prescribed.

Percent HRR Method

A more accurate way to calculate THR is the percent HRR method. The range from 60 to 90 percent HRR is the THR range in which people should exercise to improve their CR fitness levels. If a soldier knows his general level of CR fitness, he can determine which percentage of HRR is a good starting point for him. For example, if he is in excellent physical condition, he could start at 85 percent of his HRR; if he is in reasonably good shape, at 70 percent HRR; and, if he is in poor shape, at 60 percent HRR.

Most CR workouts should be conducted with the heart rate between 70 to 75 percent HRR to attain, or maintain, an adequate level of fitness. Soldiers who have reached a high level of fitness may derive more benefit from working at a higher percentage of HRR, particularly if they cannot find more than 20 minutes for CR exercise. Exercising at any lower percentage of HRR does not give the heart, muscles, and lungs an adequate training stimulus.

Before anyone begins aerobic training, he should know his THR (the heart rate at which he needs to exercise to get a training effect).

The example below shows how to figure the THR by using the resting heart rate (RHR) and age to estimate heart rate reserve (HRR). A 20-year-old male soldier in reasonably good physical shape is the example.

STEP 1: Determine the MHR by subtracting the soldier's age from 220.

FORMULA
220 - age = MHR
(GIVEN)

CALCULATION
220 - 20 = 200 BPM

STEP 2: Determine the RHR in beats per minute (DPM) by counting the resting pulse for 30 seconds, and multiply the count by two. A shorter period can be used, but a 30-second count is more accurate. This count should be taken while the soldier is completely relaxed and rested. How to determine heart rate is described below. Next, determine the heart rate reserve (HRR) by subtracting the RHR from the estimated MHR. If the soldier's RHR is 69 BPM, the HRR is calculated as shown here.

FORMULA
MHR - RHR = HRR

CALCULATION
200 BPM - 69 BPM = 131 BPM

STEP 3: Calculate the THR based on 70 percent of HRR (a percentage based on a good level of CR fitness).

FORMULA
(% x HRR) + RHR = THR

CALCULATION
(0.70x131 BPM)+69 BPM=160.7 BPM

As shown, the percentage (70 percent in this example) is converted to the decimal form (0.70) before it is multiplied by the HRR. The result is then added to the resting heart rate (RHR) to get the THR. Thus, the product obtained by multiplying 0.70 and 131 is 91.7. When 91.7 is added to the RHR of 69, a THR of 160.7 results. When the calculations produce a fraction of a heart beat, as in the example, the value is rounded off to the nearest whole number. In this case, 160.7 BPM is rounded off to give a THR of 161 BPM. In summary, a reasonably fit 20-year-old soldier with a resting heart rate of 69 BPM has a training heart rate goal of 161 BPM. To determine the RHR, or to see if one is within the THR during and right after exercise, place the tip of the third finger lightly over one of the carotid arteries in the neck. These arteries are located to the left and right of the Adam's apple. (See Figure 2-1 A.) Another convenient spot from which to monitor the pulse is on the radial artery on the wrist just above the base of the thumb. (See Figure 2-1B.) Yet another way is to place the hand over the heart and count the number of heart beats. (See Figure 2-1 C.)

During aerobic exercise, the body will usually have reached a "Steady State" after five minutes of exercise, and the heart rate will have leveled off. At this time, and immediately after exercising, the soldier should monitor his heart rate.

He should count his pulse for 10 seconds, then multiply this by six to get his heart rate for one minute. This will let him determine if his training intensity is high enough to improve his CR fitness level.

For example, use the THR of 161 BPM figured above. During the 10-second period, the soldier should get a count of 27 beats (161/6= 26.83 or 27) if he is exercising at the right intensity. If his pulse rate is below the THR, he must exercise harder to increase his pulse to the THR. If his pulse is above the THR, he should normally exercise at a lower intensity to reduce the pulse rate to the prescribed THR. He should count as accurately as possible, since one missed beat during the 10-second count, multiplied by six, gives an error of six BPM.

Figure 2-1

A soldier who maintains his THR throughout a 20-30-minute exercise period is doing well and can expect improvement in his CR fitness level.

A soldier who maintains his THR throughout a 20- to 30-minute exercise period is doing well and can expect improvement in his CR fitness level. He should check his exercise and post-exercise pulse rate at least once each workout. If he takes only one pulse check, he should do it five minutes into the workout.

Figure 2-2 is a chart that makes it easy to determine what a soldier's THR should be during a 10-second count. Using this figure, a soldier can easily find his own THR just by knowing his age and general fitness level. For example, a 40-year-old soldier with a low fitness level should, during aerobic exercise, have a THR of 23 beats in 10 seconds. He can determine this from the table by locating his age and then tracking upward until he reaches the percent HRR for his fitness level. Again, those with a low fitness level should work at about 60 percent HRR and those with a good fitness level at **70** percent HRR. Those with a high level of fitness may benefit most by training at 80 to 90 percent HRR.

Another way to gauge exercise intensity is "perceived exertion." This method relies on how difficult the exercise seems to be and is described in Appendix G.

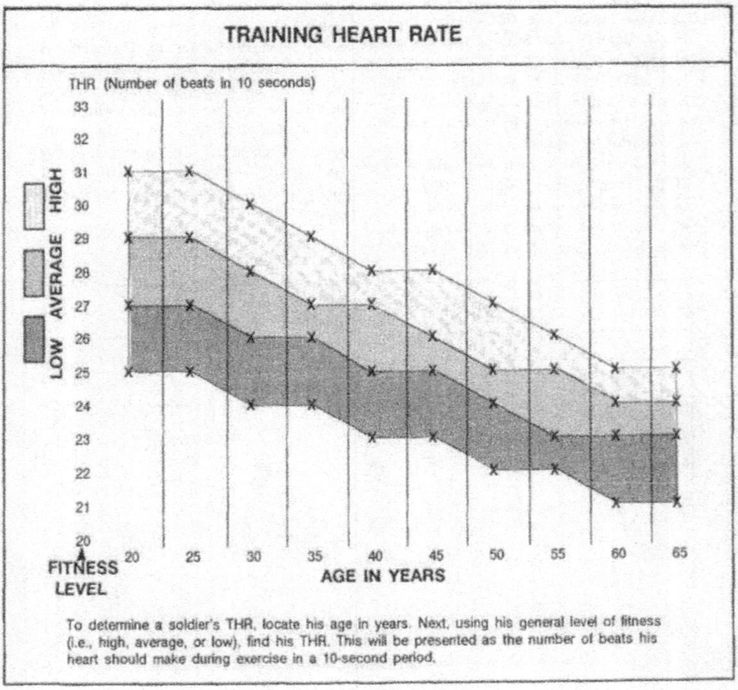

Figure 2-2

TIME

Time, or duration, refers to how long one exercises. It is inversely related to intensity. The more intense the activity, the shorter the time needed to produce or maintain a training effect; the less intense the activity, the longer the required duration. To improve CR fitness, the soldier must train for at least 20 to 30 minutes at his THR.

TYPE

Only aerobic exercises that require breathing in large volumes of air improve CR fitness. Worthwhile aerobic activities must involve the use of large muscle groups and must be rhythmic. They must also be of sufficient duration and intensity (60 to 90 percent HRR). Examples of primary and secondary exercises for improving CR fitness are as follows:

PRIMARY

- Running.
- Rowing.
- Jogging.
- Skiing (cross-country).
- Walking (vigorous).
- Exercising to music.
- Road marching.
- Rope skipping.
- Bicycling (stationary).
- Swimming.
- Bicycling (road/street).
- Stair climbing.

SECONDARY (Done with partners or opponents of equal or greater ability.)

- Racquetball (singles).
- Basketball (full court).
- Handball (singles).
- Tennis (singles).

The primary exercises are more effective than the secondary exercises in producing positive changes in CR fitness. The secondary activities may briefly elevate the heart rate but may not keep it elevated to the THR throughout the entire workout.

Every activity has its advantages and disadvantages. Trainers must weigh these and design programs that fit the unit's needs.

Every activity has its advantages and disadvantages. Trainers must design programs that fit the unit's needs.

Running

Running enables the body to improve the transport of blood and oxygen to the working muscles and brings about positive changes in the muscles' ability to produce energy. Running fits well into any physical training program 'because a training effect can be attained with only three 20-minute workouts per week.

Some soldiers may need instruction to improve their running ability. The following style of running is desired. The head is erect with the body in a straight line or slightly bent forward at the waist. The elbows are bent so the forearms are relaxed and held loosely at waist level. The arms swing naturally from front to rear in straight lines. (Cross-body arm movements waste energy. The faster the run, the faster the arm action.) The toes point straight ahead, and the feet strike on the heel and push off at the big toe.

Besides learning running techniques, soldiers need information on ways to prevent running injuries. The most common injuries associated with PT in the Army result from running and occur to the feet, ankles, knees, and legs. Proper warm-up and cool-down, along with stretching exercises and wearing appropriate clothing and well-fitting running shoes, help prevent injuries. Important information on safety factors and common running injuries is presented in Chapter 13 and Appendix E.

Important information on safety factors and common running injuries is presented in Chapter 13 and Appendix E.

Failure to allow recovery between hard bouts of running cannot only lead to overtraining, but can also be a major cause of injuries. A well-conditioned soldier can run five to six times a week. However, to do this safely, he should do two things: 1) gradually buildup to running that frequently; and, 2) vary the intensity and/or duration of the running sessions to allow recovery between them.

ABILITY GROUP RUNNING

Traditionally, soldiers have run in unit formations at a pace prescribed by the PT leader. Commanders have used unit runs to improve unit cohesion and fitness levels. Unfortunately, too many soldiers are not challenged enough by the intensity or duration of the unit run, and they do not receive a training benefit. For example, take a company that runs at a nine-minute-per-mile pace for two miles. Only soldiers who cannot run two miles in a time faster than 18 minutes will receive a significant training effect. Therefore, in terms of conditioning, most soldiers who can pass the 2-mile-run test are wasting their time and losing the chance to train hard to excel. Ability group running (AGR) is the best way to provide enough intensity so each soldier can improve his own level of CR fitness.

AGR lets soldiers train in groups of near-equal ability. Each group runs at a pace intense enough to produce a training effect for that group and each soldier in it. Leaders should program these runs for specific lengths of time, not miles to be run. This procedure lets more-fit groups run a greater distance than the less-fit groups in the same time period thus enabling every soldier to improve.

*The best way to assign soldiers to ability groups is to, make a list, in order, of the unit's most recent APFT **2-mile-run times.***

The best way to assign soldiers to ability groups is to make a list, in order, of the unit's most recent APFT 2-mile-run times. The number of groups depends on the unit size, number of leaders available to conduct the runs, and range of 2-mile-run times. A company-sized unit broken down into four to six ability groups, each with a leader, is best for aerobic training, For activities like circuits, strength training, and competitive events, smaller groups are easier to work with than one large group.

Because people progress at different rates, soldiers should move to faster groups when they are ready. To help them train at their THR and enhance their confidence, those who have a hard time keeping up with a group should be placed in a slower group. As the unit's fitness level progresses, so should the intensity at which each group exercises. Good leadership will prevent a constant shifting of soldiers between groups due to lack of effort.

AGR is best conducted at the right intensity at least three times a week. As explained, the CR system should not be exercised "hard" on consecutive days. If AGR is used on hard CR-training days, unit runs at lower intensities are good for recovery days. Using this rotation, soldiers can gain the desired benefits of both unit and ability-group runs. The problem comes when units have a limited number of days for PT and there is not enough time for both. In this case, unit runs should seldom, if ever, be used and should be recognized for what they are — runs to build unit cohesion.

Leaders can use additional methods to achieve both goals. The unit can begin in formation and divide into ability groups at a predetermined release point. The run can also begin with soldiers divided into ability groups which join at a link-up point. Alternately, ability groups can be started over the same route in a stagger, with the slowest group first. Link-ups occur as each faster group overtakes slower groups.

With imagination and planning, AGR will result in more effective training workouts for each soldier. The argument that ability-group running detracts from unit cohesion is invalid. Good leadership and training in all areas promote unit cohesion and team spirit; training that emphasizes form over substance does not.

INTERVAL TRAINING

Interval training also works the cardiorespiratory system. It is an advanced form of exercise training which helps a person significantly improve his fitness level in a relatively short time and increase his running speed.

In interval training, a soldier exercises by running at a pace that is slightly faster than his race pace for short periods of time. This may be faster than the pace he wants to maintain during the next APFT 2-mile run. He does this repeatedly with periods of recovery placed between periods of fast running. In this way, the energy systems used are allowed to recover, and the exerciser can do more fast-paced running in a given workout than if he ran continuously without resting. This type of intermittent training can also be used with activities such as cycling, swimming, bicycling, rowing, and road marching.

The following example illustrates how the proper work-interval times and recovery times can be calculated for interval training so that it can be used to improve a soldier's 2-mile-run performance.

The work-interval time (the speed at which a soldier should run each 440-yard lap) depends on his actual race pace for one mile. If a soldier's actual 1-mile-race time is not known, it can be estimated from his last APFT by taking one half of his 2-mile-run time. Using a 2-mile-run time of 1600 minutes as an example, the pace for an interval training workout is calculated as follows:

Step 1. Determine (or estimate) the actual 1-mile-race pace. The soldier's 2-mile-run time is 16:00 minutes, and his estimated pace for 1 mile is one half of this or 8:00 minutes.
Step 2. Using the time from Step 1. determine the time it took to run 440 yards by dividing the 1-mile-race pace by four. (8:00 minutes/4 = 2:00 minutes per 440 yards.)
Step 3. Subtract one to four seconds from the 440-yard time in Step 2 to find the time each 440-yard lap should be run during an interval training session. (2:00 minutes - 1 to 4 seconds = 1:59 to 1:56.)

Thus, each 440-yard lap should be run in 1 munute, 56 seconds to 1 minute, 59 seconds during interval training based on the soldier's 16:00, 2-mile run time. Recovery periods, twice the length of the work-interval periods. These recovery periods, therefore, will be 3 minutes, 52 seconds long (1:56 + 1:56 = 3:52).

Using the work-interval time for each 440-yard lap from Step 3, the soldier can run six to eight repetitions of 440 yards at a pace of 1 minute, 56 seconds (1:56) for each 440-yard run. This can be done on a 440-yard track (about 400 meters) as follows:

1. Run six to eight 440-yard repetitions with each interval run at a 1:56 pace.
2. Follow each 440-yard run done in 1 minute, 56 secons by an easy jog of 440 yards for recovery. Each 440-yard jog should take twice as much time as the work interval (that is, 3:52). For each second of work, there are two seconds of recovery. Thus, the work-to-rest ratio is 1:2.

440-YARD TIMES FOR INTERVAL TRAINING	
1-Mile Time	**440- Yard Time**
4:45 - 5:00*	1:05 - 1:09*
5:01 - 5:59	1:14 - 1:25
6:00 - 6:59	1:25 - 1:40
7:00 - 7:59	1:41 - 1:55
8:00 - 8:59	1:55 - 2:10
9:00 9:59	2:10 2.25
10:00 - 10:59	2:25 +

* The slower 1-mile run times correspond to the slower 440-yard times, as do the faster 1-mile times with the faster 440-yard times.

Table 2-1

In Fartlek training, the soldier varies the intensity (speed) of the running throughout the workout.

To help determine the correct time intervals for a wide range of fitness levels, refer to Table 2-1. It shows common 1 -mile times and the corresponding 440-yard times.

Monitoring the heart-rate response during interval training is not as important as making sure that the work intervals are run at the proper speed. Because of the intense nature of interval training, during the work interval the heart rate will generally climb to 85 or 90 percent of HRR. During the recovery interval, the heart rate usually falls to around 120 to 140 beats per minute. Because the heart rate is not the major concern during interval training, monitoring THR and using it as a training guide is not necessary.

As the soldier becomes more conditioned, his recovery is quicker. As a result, he should either shorten the recovery interval (jogging time) or run the work interval a few seconds faster.

After a soldier has reached a good CR fitness level using the THR method, he should be ready for interval training. As with any other new training method, interval training should be introduced into his training program gradually and progressively. At first, he should do it once a week. If he responds well, he may do it twice a week at the most, with at least one recovery day in between. He may also do recovery workouts of easy jogging on off days. It is recommended that interval training be done two times a week only during the last several weeks before an APFT. Also, he should rest the few days before the test by doing no, or very easy, running.

As with any workout, soldiers should start interval workouts with a warm-up and end them with a cool-down.

FARTLEK TRAINING

In Fartlek training, another type of CR training sometimes called speed play, the soldier varies the intensity (speed) of the running during the workout. Instead of running at a constant speed, he starts with veryslow jogging. When ready, he runs hard for a few minutes until he feels the need to slow down. At this time he recovers by jogging at an easy pace. This process of alternating fast and recovery running (both of varying distances) gives the same results as interval training. However, neither the running *nor* recovery interval is timed, and the running is not done on a track. For these reasons, many runners prefer Fartlek training to interval training.

LAST-MAN-UP RUNNING

This type of running, which includes both sprinting and paced running, improves CR endurance and conditions the legs. It consists of 40- to 50-yard sprints at near-maximum effort. This type of running is best done by squads and sections. Each squad leader places the squad in an evenly-spaced, single-file line on a track or a smooth, flat course. During a continuous 2- to 3-mile run of moderate intensity, the squad leader, running in the last position, sprints to the front of the line and becomes the leader. When he reaches the front, he resumes the moderate pace of the whole squad. After he reaches the front, the next soldier, who is now at the rear, immediately sprints to the front. The rest of the soldiers continue to run at a moderate pace. This pattern of sprinting by the last person continues until each soldier has resumed his original position in line. This pattern of sprinting and running is repeated several times during the run. The distance run and number of sprints performed should increase as the soldiers' conditioning improves.

CROSS-COUNTRY RUNNING

Cross-country running conditions the leg muscles and develops CR endurance. It consists of running a certain distance on a course laid out across fields, over hills, through woods, or on any other irregular terrain. It can be used as both a physical conditioning activity and a competitive event. The object is to cover the distance in the shortest time.

The unit is divided into ability groups using 2-mile-run times. Each group starts its run at the same time. This lets the better-conditioned groups run farther and helps ensure that they receive an adequate training stimulus.

The speed and distance can be increased gradually as the soldiers' conditioning improves. At first, the distance should be one mile or less, depending on the terrain and fitness level. It should then be gradually increased to four miles. Cross-country runs have several advantages: they provide variety in physical fitness training, and they can accommodate large numbers of soldiers. Interest can be stimulated by competitive runs after soldiers attain a reasonable level of fitness. These runs may also be combined with other activities such as compass work (orienteering).

Cross-country runs can accommodate large numbers of soldiers.

Road Marches

The road or foot march is one of the best ways to improve and maintain fitness. Road marches are classified as either administrative or tactical, and they can be conducted in garrison or in the field. Soldiers must be able to move quickly, carry a load (rucksack) of equipment, and be physically able to perform their missions after extended marching.

BENEFITS OF ROAD MARCHES

Road marches are an excellent aerobic activity. They also help develop endurance in the muscles of the lower body when soldiers carry a heavy load. Road marches offer several benefits when used as part of a fitness program. They are easy to organize, and large numbers of soldiers can participate. In addition, when done in an intelligent, systematic, and progressive manner, they produce relatively few injuries. Many soldier-related skills can be integrated into road marches. They can also help troops acclimatize to new environments. They help train leaders to develop skills in planning, preparation, and supervision and let leaders make first-hand observations of the soldiers' physical stamina. Because road marches are excellent fitness-training activities, commanders should make them a regular part of their unit's PT program.

Road marches help troops acclimatize to new environments,

Soldiers should receive advance notcie before going on a march, to help morale and give them time to prepare.

TYPES OF MARCHES

The four types of road marches - day, limited visibility, forced, and shuttle - are described below. For more information on marches, see FM 21-18.

Day Marches

Day marches, which fit easily into the daily training plan, are most conducive to developing physical fitness. They are characterized by dispersed formations and ease of control and reconnaissance.

Limited Visibility Marches

Limited visibility marches require more detailed planning and supervision and are harder to control than day marches. Because they move more slowly and are in tighter formations, soldiers may not exercise hard enough to obtain a conditioning effect. Limited visibility marches do have some advantages, however. They protect soldiers from the heat of the day, challenge the ability of NCOS and officers to control their soldiers, and provide secrecy and surprise in tactical situations.

Forced Marches

Forced marches require more than the normal effort in speed and exertion. Although they are excellent conditioners, they may leave soldiers too fatigued to do other required training tasks.

Shuttle Marches

Shuttle marches alternate riding and marching, usually because there are not enough vehicles to carry the entire unit. These marches may be modified and used as fitness activities. A shuttle march can be planned to move troops of various fitness levels from one point to another, with all soldiers arriving at about the same time. Soldiers who have high fitness levels can generally march for longer stretches than those who are less fit.

PLANNING A ROAD MARCH

Any plan to conduct a road march to improve physical fitness should consider the following:

- Load to be carried.
- Discipline and supervision.
- Distance to be marched.
- Route reconnaissance.
- Time allotted for movement.
- Water stops.
- Present level of fitness.
- Rest stops.
- Intensity of the march.
- Provisions for injuries.
- Terrain an weather conditions.
- Safety precautions.

Soldiers should usually receive advance notice before going on a march. This helps morale and gives them time to prepare. The leader should choose an experienced soldier as a pacesetter to lead the march. The pacesetter should carry the same load as the other soldiers and should be of medium height to ensure normal strides. The normal stride for a foot march, according to FM 21-18, is 30 inches. This stride, and a cadence of 106 steps per minute, results in a speed of 4.8 kilometers per hour (kph). When a 10-minute rest is taken each hour, a net speed of 4 kph results.

The pacesetter should keep in mind that ground slope and footing affect stride length. For example, the length decreases when soldiers march up hills or down steep slopes. Normal stride and cadence are maintained easily on moderate, gently rolling terrain unless the footing is muddy, slippery, or rough.

Personal hygiene is important in preventing unnecessary injuries. Before the march, soldiers should cut their toenails short and square them

off, wash and dry their feet, and lightly apply foot powder. They should wear clean, dry socks that fit well and have no holes. Each soldier should take one or more extra pair of socks depending on the length of the march. Soldiers who have had problems with blisters should apply a thin coating of petroleum jelly over susceptible areas. Leaders should check soldiers' boots before the march to make sure that they fit well, are broken in and in good repair, with heels that are even and not worn down.

During halts soldiers should lie down and elevate their feet. If time permits, they should massage their feet, apply powder, and change socks. Stretching for a few minutes before resuming the march may relieve cramps and soreness and help prepare the muscles to continue exercising. To help prevent lower back strain, soldiers should help each other reposition the rucksacks and other loads following rest stops. Soldiers can relieve swollen feet by slightly loosening the laces across their arches.

After marches, soldiers should again care for their feet, wash and dry their socks, and dry their boots.

PROGRAMS TO IMPROVE LOAD-CARRYING ABILITY

The four generalized programs described below can be used to improve the soldiers' load-carrying ability. Each program is based on a different number of days per week available for a PT program.

If only two days are available for PT, both should include exercises for improving CR fitness and muscular endurance and strength. Roughly equal emphasis should be given to each of these fitness components.

If there are only three days available for PT, they should be evenly dispersed throughout the week. Two of the days should stress the development of muscular endurance and strength for the whole body. Although all of the major muscle groups of the body should be trained, emphasis should be placed on the leg (hamstrings and quadriceps), hip (gluteal and hip flexors), low back (spinal erector), and abdominal (rectus abdominis) muscles. These two days should also include brief (2-mile) CR workouts of light to moderate intensity (65 to 75 percent HRR). On the one CR fitness day left, soldiers should take a long distance run (4 to 6 miles) at a moderate pace (70 percent HRR), an interval workout, or an aerobic circuit. They should also do some strength work of light volume and intensity. If four days are available, a road march should be added to the three-day program at least twice monthly. The speed, load, distance, and type of terrain should be varied.

If there are five days, leaders should devote two of them to muscular strength and endurance and two of them to CR fitness. One CR fitness day will use long distance runs; the other can stress more intense workouts including interval work, Fartlek running, or last-man-up running. At least two times per month, the remaining day should include a road march.

Leaders must train and march with their units as much as possible.

Soldiers can usually begin road-march training by carrying a total load equal to 20 percent of their body weight. This includes all clothing and equipment. However, the gender make-up and/or physical condition of a unit may require using a different starting load. Beginning distances should be between five and six miles, and the pace should be at 20 minutes per mile over flat terrain with a hard surface. Gradual increases should be made in speed, load, and distance until soldiers can do the anticipated, worst-case, mission-related scenarios without excessive difficulty or exhaustion. Units should take maintenance marches at least twice a month. Distances should vary from six to eight miles, with loads of 30 to 40 percent of body weight. The pace should be 15 to 20 minutes per mile.

Units should do maintenance marches at least twice a month.

A recent Army study showed that road-march training two times a month and four times a month produced similar improvements in road-marching performance. Thus, twice-monthly road marches appear to produce a favorable improvement in soldiers' abilities to road march if they are supported by a sound PT program (five days per week)

Commanders must establish realistic goals for road marching based on assigned missions. They should also allow newly assigned soldiers and those coming off extended profiles to gradually build up to the unit's fitness level before making them carry maximum loads. This can be done with ability groups.

Road marching should be integrated into all other training. Perhaps the best single way to improve load-earring capacity is to have a regular training program which systematically increases the load and distance. It must also let the soldier regularly practice carrying heavy loads over long distances.

As much as possible, leaders at all levels must train and march with their units. This participation enhances leaders' fitness levels and improves team spirit and confidence, both vital elements in accomplishing difficult and demanding road marches.

Alternate Forms of Aerobic Exercise

Some soldiers cannot run. In such cases, they may use other activities as supplements *or* alternatives. Swimming, bicycling, and cross-country skiing are all excellent endurance exercises and are good substitutes for running. Their drawback is that they require special equipment and facilities that are not always available. As with all exercise, soldiers should start slowly and progress gradually. Those who use non-running activities to

such training may not improve running ability. To prepare a soldier for the APFT 2-mile run, there is no substitute for running.

SWIMMING

Swimming is a good alternative to running. Some advantages of swimming include the following:

- o Involvement of all the major muscle groups.
- o Body position that enhances the blood's return to the heart.
- o Partial support of body weight by the water, which minimizes lower body stress in overweight soldiers.

Swimming may be used to improve one's CR fitness level and to maintain and improve CR fitness during recovery from an injury. It is used to supplement running and develop upper body endurance and limited strength. The swimmer should start slowly with a restful stroke. After five minutes, he should stop to check his pulse, compare it with his THR and, if needed, adjust the intensity.

Compared with all the other modes of aerobic exercise presented in this manual (e.g., running, walking, cycling, cross-country skiing, rope jumping, etc.) in swimming alone, one's THR should be lower than while doing the other forms of aerobic exercise. This is because, in swimming, the heart does not beat as fast as when doing the other types of exercise at the same work rate. Thus, in order to effectively train the CR system during swimming, a soldier should set his THR about 10 bpm lower than while running. For example, a soldier whose THR while running is 150 bpm should have a THR of about 140 bpm while swimming. By modifying their THRs in this manner while swimming, soldiers will help to ensure that they are working at the proper intensity.

Non-swimmers can run in waist-to chest-deep water, tread water, and do pool-side kicking for an excellent aerobic workout. They can also do calisthenics in the water. Together these activities combine walking and running with moderate resistance work for the upper body.

For injured soldiers, swimming and aerobic water-training are excellent for improving CR fitness without placing undue stress on injured weight-bearing parts of the body.

CYCLING

Cycling is an excellent exercise for developing CR fitness. Soldiers can bicycle outdoors or on a stationary cycling machine indoors. Road cycling should be intense enough to allow the soldier to reach and maintain THR at least 30 minutes.

Soldiers can alter the cycling intensity by changing gears, adding hill work, and increasing velocity. Distance can also be increased to enhance CR fitness, but the distance covered is not as important as the amount of time spent training at THR. The intensity of a workout can be increased by increasing the resistance against the wheel or increasing the pedaling cadence (number of RPM). For interval training, the soldier can vary the speed and resistance and use periods of active recovery at low speed and/or low resistance.

Cycling should be intense enough to let the soldier reach and maintain THR at least 30 minutes.

For swimming, a soldier should set his THR at about 10 beats per minute lower then when running.

WALKING

Walking is another way to develop cardiorespiratory fitness. It is enjoyable, requires no equipment, and causes few injuries. However, unless walking is done for a long time at the correct intensity, it will not produce any significant CR conditioning.

Sedentary soldiers with a low degree of fitness should begin slowly with 12 minutes of walking at a comfortable pace. The heart rate should be monitored to determine the intensity. The soldier should walk at least four times a week and add two minutes each week to every workout until the duration reaches 45 to 60 minutes per workout. He can increase the intensity by adding hills or stairs.

As the walker's fitness increases, he should walk 45 to 60 minutes at a faster pace. A simple way to increase walking speed is to carry the arms the same way as in running. With this technique the soldier has a shorter arm swing and takes steps at a faster rate. Swinging the arms faster to increase the pace is a modified form of race walking (power walking) which allows for more upper-body work. This method may also be used during speed marches. After about three months, even the most unfit soldiers should reach a level of conditioning that lets them move into a running program.

CROSS-COUNTRY SKIING

Cross-country skiing requires vigorous movement of the arms and legs, developing muscular and CR endurance.

Cross-country or Nordic skiing is another excellent alternative to the usual CR activities. It requires vigorous movement of the arms and legs which develops muscular and CR endurance and coordination. Some of the highest levels of aerobic fitness ever measured have been found in cross-country skiers.

Although some regions lack snow, one form or another of cross-country skiing can be done almost anywhere—on country roads, golf courses, open fields, and in parks and forests.

Cross-country skiing is easy to learn. The action is similar to that used in brisk walking, and the intensity may be varied as in running. The work load is determined by the difficulty of terrain, the pace, and the frequency and duration of rest periods. Equipment is reasonably priced, with skis, boots, and poles often obtainable from the outdoor recreation services.

ROPE SKIPPING

Rope skipping is also a good exercise for developing CR fitness. It requires little equipment, is easily learned, may be done almost anywhere, and is not affected by weather. Some runners use it as a substitute for running during bad weather.

A beginner should select a jump rope that, when doubled and stood on, reaches to the armpits. Weighted handles or ropes may be used by better-conditioned soldiers to improve upper body strength. Rope skippers should begin with five minutes of jumping rope and then monitor their heart rate. They should attain and maintain their THR to ensure a training effect, and the time spent jumping should be increased as the fitness level improves.

Rope jumping, however, may be stressful to the lower extremities and therefore should be limited to no more than three times a week. Soldiers should skip rope on a cushioned surface such as a mat or carpet and should wear cushioned shoes.

HANDBALL AND RACQUET SPORTS

Handball and the racquet sports (tennis, squash, and racquetball) involve bursts of intense activity for short periods. They do not provide the same degree of aerobic training as exercises of longer duration done at lower intensities. However, these sports are good supplements and can provide excellent aerobic benefits depending on the skill of the players. If played vigorously each day, they may be an adequate substitute for low-level aerobic training. Because running increases endurance, it helps improve performance in racket sports, but the reverse is not necessarily true.

EXERCISE TO MUSIC

Aerobic exercise done to music is another excellent alternative to running. It is a motivating, challenging activity that combines exercise and rhythmic movements. There is no prerequisite skill, and it can be totally individualized to every fitness level by varying the frequency, intensity, and duration. One can move to various tempos while jogging or doing jumping jacks, hops, jumps, or many other calisthenics.

Workouts can be done in a small space by diverse groups of varying fitness levels. Heart rates should be taken during the conditioning phase to be sure the workout is sufficiently intense. If strengthening exercises are included, the workout addresses every component of fitness. Holding relatively light dumbbells during the workout is one way to increase the intensity for the upper body and improve muscular endurance. Warm-up and cool-down stretches should be included in the aerobic workout.

CHAPTER 3

Muscular Endurance and Strength

On today's battlefield, in addition to cardiorespiratory fitness, soldiers need a high level of muscular endurance and strength. In a single day they may carry injured comrades, move equipment, lift heavy tank or artillery rounds, push stalled vehicles, or do many other strength-related tasks. For example, based on computer-generated scenarios of an invasion of Western Europe, artillerymen may have to load from 300 to 500, 155mm-howitzer rounds (95-lb rounds) while moving from 6 to 10 times each day over 8 to 12 days. Infantrymen may need to carry loads exceeding 100 pounds over great distances, while supporting units will deploy and displace many times. Indeed, survival on the battlefield may, in large part, depend on the muscular endurance and strength of the individual soldier.

Muscular Fitness

Muscular fitness has two components: muscular strength and muscular endurance.

Muscular strength is the greatest amount of force a muscle or muscle group can exert in a single effort.

Muscular endurance is the ability of a muscle or muscle group to do repeated contractions against a less-than-maximum resistance for a given time.

Although muscular endurance and strength are separate fitness components, they are closely related. Progressively working against resistance will produce gains in both of these components.

Muscular Contractions

Isometric, isotonic, and isokinetic muscular endurance and strength are best produced by regularly doing each specific kind of contraction. They are described here.

Isometric contraction produces contraction but no movement, as when pushing against a wall. Force is produced with no change in the angle of the joint.

Isotonic contraction causes a joint to move through a range of motion against a constant resistance. Common examples are push-ups, sit-ups, and the lifting of weights.

Isokinetic contraction causes the angle at the joint to change at a constant rate, for example, at 180 degrees per second. To achieve a constant speed of movement, the load or resistance must change at different joint angles to counter the varying forces produced by the muscle(s) at different angles. This requires the use of isokinetic machines. There are other resistance-training machines which, while not precisely controlling the speed of movement, affect it by varying the resistance throughout the range of motion. Some of these devices are classified as pseudo-isokinetic and some as variable-resistance machines.

Isotonic and isokinetic contractions have two specific phases - the concentric or "positive" phase and the eccentric or "negative" phase. In the concentric phase (shortening) the muscle contracts, while in the eccentric phase (elongation) the muscle returns to its normal length. For example, on the upward phase of the biceps curl, the biceps are shortening. This is a concentric (positive) contraction. During the lowering phase of the curl the biceps are lengthening. This is an eccentric (negative) contraction.

A muscle can control more weight in the eccentric phase of contraction than it can lift concentrically. As a result, the muscle may be able to handle more of an overload eccentrically. This greater overload, in return, may produce greater strength gains.

The nature of the eccentric contraction, however, makes the muscle and connective tissue more susceptible to damage, so there is more muscle soreness following eccentric work.

When a muscle is overloaded, whether by isometric, isotonic, or isokinetic contractions, it adapts by becoming stronger. Each type of contraction has advantages and disadvantages, and each will result in strength gains if done properly.

The above descriptions are more important to those who assess strength than to average people trying to develop strength and endurance. Actually, a properly designed weight training program with free weights or resistance machines will result in improvements in all three of these categories.

Principles of Muscular Training

To have a good exercise program, the seven principles of exercise, described in Chapter 1, must be applied to all muscular endurance and strength training. These principles are overload, progression, specificity, regularity, recovery, balance, and variety.

OVERLOAD

The overload principle is the basis for all exercise training programs. For a muscle to increase in strength, the workload to which it is subjected during exercise must be increased beyond what it normally experiences. In other words, the muscle must be overloaded. Muscles adapt to increased workloads by becoming larger and stronger and by developing greater endurance.

To understand the principle of overload, it is important to know the following strength-training terms:

- Full range of motion. To obtain optimal gains, the overload must be applied throughout the full range of motion. Exercise a joint and its associated muscles through its complete range starting from the prestretched position (stretched past the relaxed position) and ending in a fully contracted position. This is crucial to strength development.
- Repetition. When an exercise has progressed through one complete range of motion and back to the beginning, one repetition has been completed.
- One-repetition maximum (1-RM). This is a repetition performed against the greatest possible resistance (the maximum weight a person can lift one time). A 10-RM is the maximum weight one can lift correctly 10 times. Similarly, an 8-12 RM is that weight which allows a person to do from 8 to 12 correct repetitions. The intensity for muscular endurance and strength training is often expressed as a percentage of the 1-RM.
- Set. This is a series of repetitions done without rest.
- Muscle Failure. This is the inability of a person to do another correct repetition in a set.

The minimum resistance needed to obtain strength gains is 50 percent of the 1-RM. However, to achieve enough overload, programs are designed to require sets with 70 to 80 percent of one's 1-RM. (For example, if a soldier's 1-RM is 200 pounds, multiply 200 pounds by 70 percent [200 X 0.70 = 140 pounds] to get 70 percent of the 1-RM.)

When a muscle is overloaded by isometric, isotonic, or isokinetic contractions, it adapts by becoming stronger.

A better and easier method is the repetition maximum (RM) method. The exerciser finds and uses that weight which lets him do the correct number of repetitions. For example, to develop both muscle endurance and strength, a soldier should choose a weight for each exercise which lets him do 8 to 12 repetitions to muscle failure. (See Figure 3-1.) The weight should be heavy enough so that, after doing from 8 to 12 repetitions, he momentarily cannot correctly do another repetition. This weight is the 8-12 RM for that exercise.

FITT Factors Applied to Conditioning Programs for Muscular Endurance and/or Strength		
Muscular Strength	Muscular Endurance	Muscular Strength and Muscular Endurance
3 times/week	3-5 times/week	3 times/week
3-7 RM*	12+ RM	8-12 RM
The time required to do 3-7 repetitions of each resistance exercise	The time required to do 12+ repetitions of each resistance exercise	The time required to do 8-12 repetitions of each resistance exercise
Free Weights Resistance Machines Partner-Resisted Exercises Body-Weight Exercises (Push-ups/Sit-ups/Pull-ups/Dips, etc.)		

* RM = Repetition Maximum

Figure 3-1

MUSCULAR ENDURANCE/ STRENGTH DEVELOPMENT

To develop muscle strength, the weight selected should be heavier and the RM will also be different. For example, the soldier should find that weight for each exercise which lets him do 3 to 7 repetitions correctly. This weight is the 3-7 RM for that exercise. Although the greatest improvements seem to come from resistances of about 6-RM, an effective range is a 3-7 RM. The weight should be heavy enough so that an eighth repetition would be impossible because of muscle fatigue.

The weight should also not be too heavy. If one cannot do at least three repetitions of an exercise, the resistance is too great and should be reduced. Soldiers who are just beginning a resistance-training program should not start with heavy weights. They should first build an adequate foundation by training with an 8-12 RM or a 12+ RM.

To develop muscular endurance, the soldier should choose a resistance that lets him do more than 12 repetitions of a given exercise. This is his 12+ repetition maximum (12+ RM). With continued training, the greater the number of repetitions per set, the greater will be the improvement in muscle endurance and the smaller the gains in strength. For example, when a soldier trains with a 25-RM weight, gains in muscular endurance will be greater than when using a 15-RM weight, but the gain in strength will not be as great. To optimize a soldier's performance, his RM should be determined from an analysis of the critical tasks of his mission. However, most soldiers will benefit most from a resistance-training program with an 8-12 RM.

Whichever RM range is selected, the soldier must always strive to overload his muscles. The key to overloading a muscle is to make that muscle exercise harder than it normally does.

An overload may be achieved by any of the following methods:

- Increasing the resistance.
- Increasing the number of repetitions per set.
- Increasing the number of sets.
- Reducing the rest time between sets.
- Increasing the speed of movement in the concentric phase. (Good form is more important than the speed of movement.)
- Using any combination of the above.

PROGRESSION

When an overload is applied to a muscle, it adapts by becoming stronger and/or by improving its endurance. Usually significant increases in strength can be made in three to four weeks of proper training depending on the individual. If the workload is not progressively increased to keep pace with newly won strength, there will be no further gains. When a soldier can correctly do the upper limit of repetitions for the set without reaching muscle failure, it is usually time to increase the resistance. For most soldiers, this upper limit should be 12 repetitions.

For example, if his plan is to do 12 repetitions in the bench press, the soldier starts with a weight that causes muscle failure at between 8 and 12 repetitions (8- 12 RM). He should continue with that weight until he can do **12** repetitions correctly. He then should increase the weight by about 5 percent but no more than 10 percent. In a multi-set routine, if his goal is to do three sets of eight repetitions of an exercise, he starts with a weight that causes muscle failure before he completes the eighth repetition in one or more of the sets. He continues to work with that weight until he can complete all eight repetitions in each set, then increases the resistance by no more than 10 percent.

SPECIFICITY

A resistance-training program should provide resistance to the specific muscle groups that need to be strengthened. These groups can be identified by doing a simple assessment. The soldier slowly does work-related movements he wants to improve and, at the same time, he feels the muscles on each side of the joints where motion occurs. Those muscles that are contracting or becoming tense during the movement are the muscle groups involved. If the soldier's performance of a task is not adequate or if he wishes to improve, strength training for the identified muscle(s) will be beneficial. To improve his muscular endurance and strength, in a given task, the soldier must do resistance movements that are as similar as possible to those of doing the task. In this way, he ensures maximum carryover value to his soldiering tasks.

REGULARITY

Exercise must be done regularly to produce a training effect. Sporadic exercise may do more harm than good. Soldiers can maintain a moderate level of strength by doing proper strength workouts only once a week, but three workouts per week are best for optimal gains. The principle of regularity also applies to the exercises for individual muscle groups. A soldier can work out three times a week, but when different muscle groups are exercised at each workout, the principle of regularity is violated and gains in strength are minimal.

Exercise must be done regularly to produce a training effect.

RECOVERY

There should be at least a 48-hour recovery period between workouts for the same muscle group.

Consecutive days of hard resistance training for the same muscle group can be detrimental. The muscles must be allowed sufficient recovery time to adapt. Strength training can be done every day only if the exercised muscle groups are rotated, so that the same muscle or muscle group is not exercised on consecutive days. There should be at least a 48-hour recovery period between workouts for the same muscle groups. For example, the legs can be trained with weights on Monday, Wednesday, and Friday and the upper body muscles on Tuesday, Thursday, and Saturday.

Recovery is also important within a workout. The recovery time between different exercises and sets depends, in part, on the intensity of the workout. Normally, the recovery time between sets should be 30 to 180 seconds.

BALANCE

It is important to include exercises that work all the major muscle groups in both the upper and lower body.

When developing a strength training program, it is important to include exercises that work all the major muscle groups in both the upper and lower body. One should not work just the upper body, thinking that running will strengthen the legs.

Most muscles are organized into opposing pairs. Activating one muscle results in a pulling motion, while activating the opposing muscle results in the opposite, or pushing, movement. When planning a training session, it is best to follow a pushing exercise with a pulling exercise which results in movement at the same joint(s). For example, follow an overhead press with a lat pull-down exercise. This technique helps ensure good strength balance between opposing muscle groups which may, in turn, reduce the risk of injury. Sequence the program to exercise the larger muscle groups first, then the smaller muscles. For example, the lat pull-down stresses both the larger latissimus dorsi muscle of the back and the smaller biceps muscles of the arm. If curls are done first, the smaller muscle group will be exhausted and too weak to handle the resistance needed for the lat pull-down. As a result, the soldier cannot do as many repetitions with as much weight as he normally could in the lat pull-down. The latissimus dorsi muscles will not be overloaded and, as a result, they may not benefit very much from the workout.

The best sequence to follow for a total-body strength workout is to first exercise the muscles of the hips and legs, followed by the muscles of the upper back and chest, then the arms, abdominal, low back, and neck. As long as all muscle groups are exercised at the proper intensity, improvement will occur.

VARIETY

A major challenge for all fitness training programs is maintaining enthusiasm and interest. A poorly designed strength- training program can be very boring. Using different equipment, changing the exercises, and altering the volume and intensity are good ways to add variety, and they may also produce better results. The soldier should periodically substitute different exercises for a given muscle group(s). For example, he can do squats with a barbell instead of leg presses on a weight machine. Also, for variety or due to necessity (for example, when in the field), he can switch to partner-resisted exercises or another form of resistance training. However, frequent wholesale changes should be avoided as soldiers may become frustrated if they do not have enough time to adapt or to see improvements in strength.

Workout Techniques

Workouts for improving muscular endurance or strength must follow the principles just described. There are also other factors to consider, namely, safety, exercise selection, and phases of conditioning.

SAFETY FACTORS

Major causes of injury when strength training are improper lifting techniques combined with lifting weights that are too heavy. Each soldier must understand how to do each lift correctly before he starts his strength training program.

The soldier should always do weight training with a partner, or spotter, who can observe his performance as he exercises. To ensure safety and the best results, both should know how to use the equipment and the proper spotting technique for each exercise.

A natural tendency in strength training is to see how much weight one can lift. Lifting too much weight forces a compromise in form and may lead to injury. All weights should be selected so that proper form can be maintained for the appropriate number of repetitions.

Correct breathing is another safety factor in strength training. Breathing should be constant during exercise. The soldier should never hold his breath, as this can cause dizziness and even loss of consciousness. As a general rule, one should exhale during the positive (concentric) phase of contraction as the weight or weight stack moves away from the floor, and inhale during the negative (eccentric) phase as the weight returns toward the floor.

EXERCISE SELECTION

When beginning a resistance-training program, the soldier should choose about 8 to 16 exercises that work all of the body's major muscle groups. Usually eight well-chosen exercises will serve as a good starting point. They should include those for the muscles of the leg, low back, shoulders, and so forth. The soldier should choose exercises that work several muscle groups and try to avoid those that isolate single muscle groups. This will help him train a greater number of muscles in a given time. For example, doing lat pull-downs on the "lat machine" works the latissimus dorsi of the back and the biceps muscles of the upper arm. On the other hand, an exercise like concentration curls for the biceps muscles of the upper arm, although an effective exercise, only works the arm flexor muscles. Also, the concentration curl requires twice as much time as lat pull-downs because only one arm is worked at a time.

Perhaps a simpler way to select an exercise is to determine the number of joints in the body where movement occurs during a repetition. For most people, especially beginners, most of the exercises in the program should be "multi-joint" exercises. The exercise should provide movement at more than one joint. For example, the pull-down exercise produces motion at both the shoulder and elbow joints. The concentration curl, however, only involves the elbow joint.

PHASES OF CONDITIONING

There are three phases of conditioning: preparatory, conditioning, and maintenance. These are also described in Chapter 1.

Preparatory Phase

The three phases of conditioning are preparatory, conditioning, and maintenance.

The soldier should use very light weights during the first week (the preparatory phase) which includes the first two to three workouts. This is very important, because the beginner must concentrate at first on learning the proper form for each exercise. Using light weights also helps minimize muscle soreness and decreases the likelihood of injury to the muscles, joints, and ligaments. During the second week, he should use progressively heavier weights. By the end of the second week (4 to 6 workouts), he should know how much weight on each exercise will allow him to do 8 to 12 repetitions to muscle failure. If he can do only seven repetitions of an exercise, the weight must be reduced; if he can do more than 12, the weight should be increased.

Conditioning Phase

The third week is normally the start of the conditioning phase for the beginning weight trainer. During this phase, the soldier should increase the amount of weight used and/or the intensity of the workout as his muscular strength and/or endurance increases. He should do one set of 8 to 12 repetitions for each of the heavy-resistance exercises. When he can do more than 12 repetitions of any exercise, he should increase the weight until he can again do only 8 to 12 repetitions. This usually involves an increase in weight of about five percent. This process continues indefinitely. As long as he continues to progress and get stronger, he does not need to do more than one set per exercise. If he stops making progress with one set of 8 to 12 repetitions per exercise, he may benefit from adding another set of 8 to 12 repetitions on those exercises in which progress has slowed. As time goes on and he progresses, he may increase the number to three sets of an exercise to get even further gains in strength and/or muscle mass. Three sets per exercise is the maximum most soldiers will ever need to do.

Maintenance Phase

Once the soldier reaches a high level of fitness, the maintenance phase is used to maintain that level. The emphasis in this phase is no longer on progression but on retention. Although training three times a week for muscle endurance and strength gives the best results, one can maintain them by training the major muscle groups properly one or two times a week. More frequent training, however, is required to reach and maintain peak fitness levels. Maintaining the optimal level of fitness should become part of each soldier's life-style and training routine. The maintenance phase should be continued throughout his career and, ideally, throughout his life.

As with aerobic training, the soldier should do strength training three times a week and should allow at least 48 hours of rest from resistance training between workouts for any given muscle group.

TIMED SETS

Timed sets refers to a method of physical training in which as many repetitions as possible of a given exercise are performed in a specified period of time. After an appropriate period of rest, a second, third, and so on, set of that exercise is done in an equal or lesser time period. The exercise period, recovery period, and the number of sets done should be selected to make sure that an overload of the involved muscle groups occurs.

The use of timed sets, unlike exercises performed in cadence or for a specific number of repetitions, helps to ensure that each soldier does as many repetitions of an exercise as possible within a period of time. It does not hold back the more capable performer by restricting the number of repetitions he may do. Instead, soldiers at all levels of fitness can individually do the number of repetitions they are capable of and thereby be sure they obtain an adequate training stimulus.

In this FM, timed sets will be applied to improving soldier's sit-up and push-up performance. (See Figures 3-2 and 3-3.) Many different but equally valid approaches can be taken when using timed sets to improve push-up and sit-up performance. Below, several of these will be given.

It should first be stated that improving sit-up and push-up performance, although important for the APFT, should not be the main goal of an Army physical training program. It must be to develop an optimal level of physical fitness which will help soldiers carry out their mission during combat. Thus, when a soldier performs a workout geared to develop muscle endurance and strength, the goal should be to develop sufficient strength and/or muscle endurance in all the muscle groups he will be called upon to use as he performs his mission. To meet this goal, and to be assured that all emergencies can be met, a training regimen which exercises all the body's major muscle groups must be developed and followed. Thus, as a general rule, a muscle endurance or strength training workout should not be designed to work exclusively, or give priority to, those muscle groups worked by the sit-up or push-up event.

For this reason, the best procedure to follow when doing a resistance exercise is as follows. First, perform a workout to strengthen all of the body's major muscles. Then, do timed sets to improve push-up and sit-up performance. Following this sequence ensures that all major muscles are worked. At the same time, it reduces the amount of time and work that must be devoted to push-ups and sit-ups. This is because the muscles worked by those two exercises will already be pre-exhausted.

The manner in which timed sets for push-ups and sit-ups are conducted should occasionally be varied. This ensures continued gains and minimizes boredom. This having been said, here is a very time-efficient way of conducting push-up/sit-up improvement. Alternate timed sets of push-ups and timed sets of sit-ups with little or no time between sets allowed for recovery. In this way, the muscle groups used by the push-up can recover while the muscles used in the sit-up are exercised, and vice versa. The following is an example of this type of approach:

TIMED SETS

SET NO.	ACTIVITY	TIME PERIOD	REST INTERVAL
1	Push-ups	45 seconds	0
2	Sit-ups	45 seconds	0
3	Push-ups	30 seconds	0
4	Sit-ups	30 seconds	0
5	Push-ups	30 seconds	0
6	Sit-ups	30 seconds	0

Figure 3-2

If all soldiers exercise at the same time, the above activity can be finished in about 3.5 minutes. As the soldiers' levels of fitness improve, the difficulty of the activity can be increased. This is done by lengthening the time period of any or all timed sets, by decreasing any rest period between timed sets, by increasing the number of timed sets performed, or by any combination of these.

To add variety and increase the overall effectiveness of the activity, different types of push-ups (regular, feet-elevated, wide-hand, close-hand, and so forth) and sit-ups (regular, abdominal twists, abdominal curls, and so forth) can be done. When performing this type of workout, pay attention to how the soldiers are responding, and make adjustments accordingly. For example, the times listed in the chart above may prove to be too long or too short for some soldiers. In the same way, because of the nature of the sit-up, it may become apparent that some soldiers can benefit by taking slightly more time for timed sets of sit-ups than for push-ups.

When using timed sets for push-up and sit-up improvement, soldiers can also perform all sets of one exercise before doing the other. For example, several timed sets of push-ups can be done followed by several sets of sit-ups, or vice versa. With this approach, rest intervals must be placed between timed sets. The following example can be done after the regular strength workout and is reasonable starting routine for most soldiers.

During a timed set of push-ups, a soldier may reach temporary muscle failure at any time before the set is over. If this happens, he should immediately drop to his knees and continue doing modified push-ups on his knees.

Finally, as in any endeavor, soldiers must set goals for themselves. This applies when doing each timed set and when planning for their next and future APFTs.

Major Muscle Groups

In designing a workout it is important to know the major muscle groups, where they are located, and their primary action. (See Figure 3-4.)

To ensure a good, balanced workout, one must do at least one set of exercises for each of the major muscle groups.

TIMED SETS

SET NO.	ACTIVITY	TIME PERIOD	REST INTERVAL
1	Regular Push-ups	30 seconds	30 seconds
2	Wide-hand Push-ups	30 seconds	30 seconds
3	Close-hand Push-ups	30 seconds	30 seconds
4	Regular Push-ups	20 seconds	30 seconds
5	Regular Push-ups done on knees	30 seconds	30 seconds
6	Regular Sit-ups	60 seconds	30 seconds
7	Abdominal Twists	40 seconds	30 seconds
8	Curl-ups	30 seconds	30 seconds
9	Abdominal Crunches	30 seconds	End

Figure 3-3

MAJOR MUSCLE GROUPS

The Major Skeletal Muscles of the Human Body

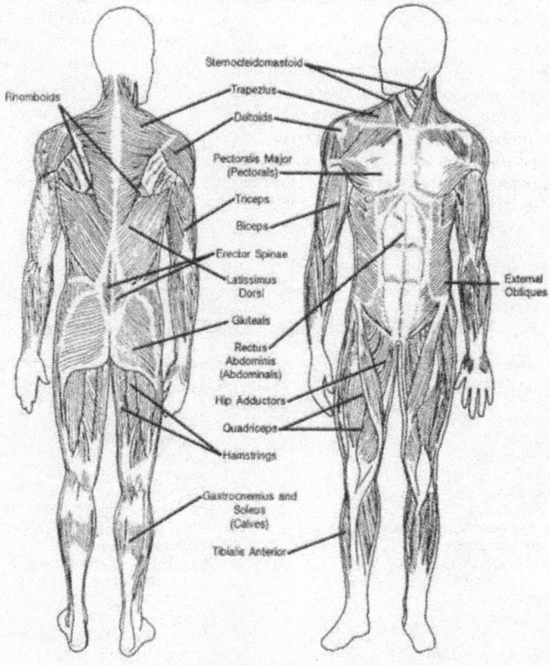

The iliopsoas muscle (a hip flexor) cannot be seen as it lies beneath other muscles. It attaches to the lumbar, the pelvis, the vertebrae and the femur.

Figure 3-4

BEGINNING EXERCISE PROGRAM

NAME OF EXERCISE	MAJOR MUSCLE GROUP(S) WORKED*
1. Leg press or squat	--Quadriceps, Gluteals
2. Leg curl	--Hamstrings
3. Heel raise	--Gastrocnemius
4. Bench press	--Pectorals, Triceps, Deltoids
5. Lat pull-down or pull-up	--Latissimus Dorsi, Biceps
6. Overhead press	--Deltoids, Triceps
7. Sit-up	--Rectus Abdominus, Iliopsoas, oblique muscles
8. Bent-leg dead lift	--Erector Spinae, Quadriceps, Gluteals

Figure 3-5

The beginning weight-training program shown at Figure 3-5 will work most of the important, major muscle groups. It is a good program for beginners and for those whose time is limited. The exercises should be done in the order presented.

The weight-training program shown at Figure 3-6 is a more comprehensive program that works the major muscle groups even more thoroughly. It has some duplication with respect to the muscles that are worked. For example, the quadriceps are worked by the leg press/squat and leg extensions, and the biceps are worked by the seated row, lat pull-down, and biceps curl. Thus, for the beginner, this program may overwork some muscle groups. However, for the more advanced lifter, it will make the muscles work in different ways and from different angles thereby providing a better over-all development of muscle strength. This program also includes exercises to strengthen the neck muscles.

When doing one set of each exercise to muscle failure, the average soldier should be able to complete this routine and do a warm-up and cool-down within the regular PT time.

MORE ADVANCED EXERCISE PROGRAM

NAME OF EXERCISE	MAJOR MUSCLE GROUP(S) WORKED
1. Leg press or squat	--Quadriceps, Gluteals
2. Leg raises	--Iliopsoas (hip flexors)
3. Leg extension	--Quadriceps
4. Leg curl	--Hamstrings
5. Heel raise	--Gastrocnemius, Soleus
6. Bench press	--Pectorals, Triceps, Deltoids
7. Seated row	--Rhomboids, Latissimus dorsi, Biceps
8. Overhead press	--Deltoids, Triceps
9. Lat pull-down or pull-up	--Latissimus dorsi, Biceps
10. Shoulder shrug	--Upper trapezius
11. Triceps extension	--Triceps
12. Biceps curl	--Biceps
13. Sit-up	--Rectus abdominus, iliopsoas
14. Bent-leg dead lift	--Erector spinae, Quadriceps, Gluteals
15. Neck flexion	--Sternocleidomastoid
16. Neck extension	--Upper trapezius

Figure 3-6

Key Points to Emphasize

Some key points to emphasize when doing resistance training tire as follows

- Train with a partner if possible. This helps to increase motivation, the intensity of the workout, and safety.
- Always breathe when lifting. Exhale during the concentric (positive] phase of contraction, and inhale during the eccentric (negative) phase.
- Accelerate the weight through the concentric phase of contraction, and return the weight to the starting position in a controlled manner during the eccentric phase.
- Exercise the large muscle groups first, then the smaller ones.
- Perform all exercises through their full range of motion. Begin from a fully extended, relaxed position (prestretched), and end the concentric phase in a fully contracted position.
- Always use strict form. Do not twist, lurch, lunge, or arch the body. This can cause serious injury. These motions also detract from the effectiveness of the exercise because they take much of the stress off the targeted muscle groups and place it on other muscles.
- Rest from 30 to 180 seconds between different exercises and sets of a given exercise.
- Allow at least 48 hours of recovery between workouts, but not more than 96 hours, to let the body recover and help prevent over training and injury.
- Progress slowly. Never increase the resistance used by more than 10 percent at a time.
- Alternate pulling and pushing exercises. For example, follow triceps extensions with biceps curls.
- Ensure that every training program is balanced. Train the whole body, not just specific areas. Concentrating on weak areas is all right, but the rest of the body must also be trained.

Exercise Programs

When developing strength programs for units, there are limits to the type of training that can be done. The availability of facilities is always a major concern. Although many installations have excellent strength-training facilities, it is unreasonable to expect that all units can use them on a regular basis. However, the development of strength does not require expensive equipment. All that is required is for the soldier, three times a week, to progressively overload his muscles.

TRAINING WITHOUT SPECIAL EQUIPMENT

Muscles do not care what is supplying the resistance. Any regular resistance exercise that makes the muscle work harder than it is used to causes it to adapt and become stronger. Whether the training uses expensive machines, sandbags, or partners, the result is largely the same.

Sandbags are convenient for training large numbers of soldiers, as they are available in all military units. The weight of the bags can be varied depending on the amount of fill. Sandbag exercises are very effective in strength-training circuits. Logs, ammo boxes, dummy rounds, or other equipment that is unique to a unit can also be used to provide resistance for strength training. Using a soldier's own body weight as the resistive force is another excellent alternative method of strength training. Pull-ups, push-ups, dips, sit-ups, and single-leg squats are examples of exercises which use a person's body weight. They can improve an untrained soldier's level of strength.

Partner-resisted exercises (PREs) are another good way to develop muscular strength without equipment, especially when training large numbers of soldiers at one time. As with all training, safety is a critical factor. Soldiers should warm up, cool down, and follow the principles of exercise previously outlined.

PARTNER-RESISTED EXERCISE

In partner-resisted exercises (PREs) a person exercises against a partner's opposing resistance. The longer the partners work together, the more effective they should become in providing the proper resistance for each exercise. They must communicate with each other to ensure that neither too much nor too little resistance is applied. The resister must apply enough resistance to bring the exerciser to muscle failure in 8 to 12 repetitions. More resistance usual] y can and should be applied during the eccentric (negative) phase of contraction (in other words, the second half of each repetition as the exerciser returns to the starting position). The speed of movement for PREs should always be slow and controlled. As a general rule, the negative part of each exercise should take at least as long to complete as the positive part. Proper exercise form and regularity in performance are key ingredients when using PREs for improving strength.

Following are descriptions and illustrations of several PREs. They should be done in the order given to ensure that the exercising soldier is working his muscle groups from the largest to the smallest. More than one exercise per muscle group may be used. The PT leader can select exercises which meet the unit's specific goals while considering individual limitations:

A 36-to 48-inch stick or bar one inch in diameter may be used for some of the exercises. This gives the resister a better grip and/or leverage and also provides a feel similar to that of free weights and exercise machines.

SPLIT-SQUAT

This exercise is for beginning trainees' quadriceps and gluteal muscles.

Exerciser

Position: Stand erect with both feet pointed straight ahead, the left foot placed in a forward position and the right foot placed about 2.5 feet behind the left foot.

Action: Keeping the back straight and the head up, bend both legs at the same time, and lower yourself slowly until the right knee barely touches the floor. Return to the starting position. This is one repetition. After 8 to 12 repetitions to muscle failure, repeat the action with the opposite leg forward.

Resister

Position: Stand directly behind the exerciser with the fleshy portion of your forearms resting squarely on the exerciser's shoulders. You may clasp your hands to gain extra leverage as long as you do not squeeze the exerciser's neck. Be sure to place the same foot forward as the exerciser.

Action: As the exerciser lowers himself, apply a steady, forceful pressure downward against his shoulders. A slightly lesser pressure should be applied as the exerciser returns to the starting position.

SINGLE-LEG SQUAT

This exercise is for advanced trainees' quadriceps and gluteal muscles.

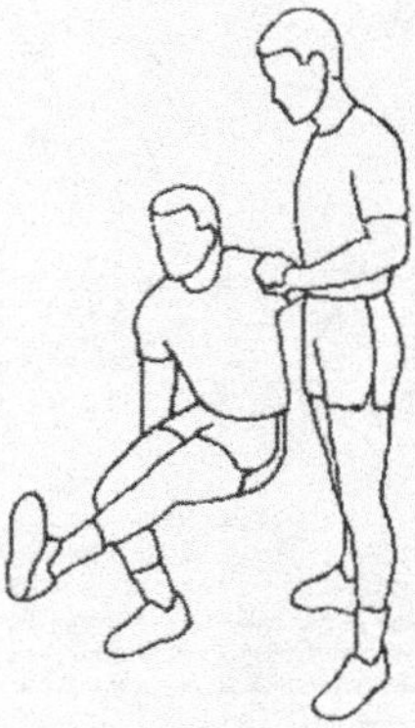

Exerciser

Position: Face your partner and grasp his wrists. Extend your right leg in front; keep it straight but do not let it contact your partner.

Action: Lower yourself in a controlled manner. Next, return to the upright position. After 8-12 repetitions to muscle failure, repeat this exercise with the other leg.

Resister

Position: Face the exerciser with your arms extended obliquely forward.

Action: Provide stability to the exerciser along with resistance or assistance as needed. When the exerciser can do more than 12 repetitions, apply an appropriate resistance that results in muscle failure in 8-12 repetitions.

LEG EXTENSION

This exercise is for the quadriceps muscles.

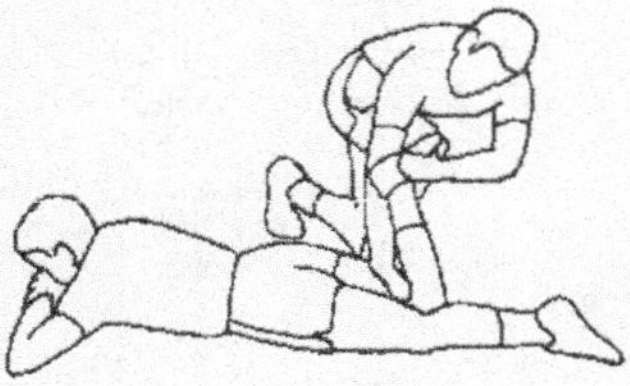

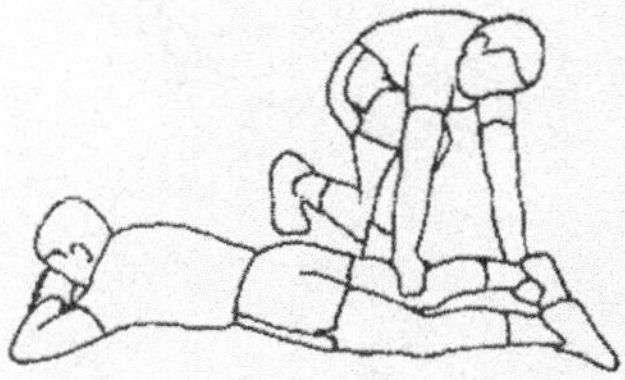

Exerciser

Position: Lie face down with one leg straight and the other flexed at the knee. Move your heel as close to your buttocks as possible.

Action: Extend your knee against the partner's resistance. Next, resist as your partner returns you to the starting position. Do 8 to 12 repetitions to muscle failure. Repeat this exercise with the other leg.

Resister

Position: Support the leg being exercised by placing your foot under the exerciser's thigh just above his knee.

Action: Resist while exerciser extends his leg. Next, apply upward pressure to return the exerciser to the starting position.

LEG CURL

This exercise is for the hamstring muscles.

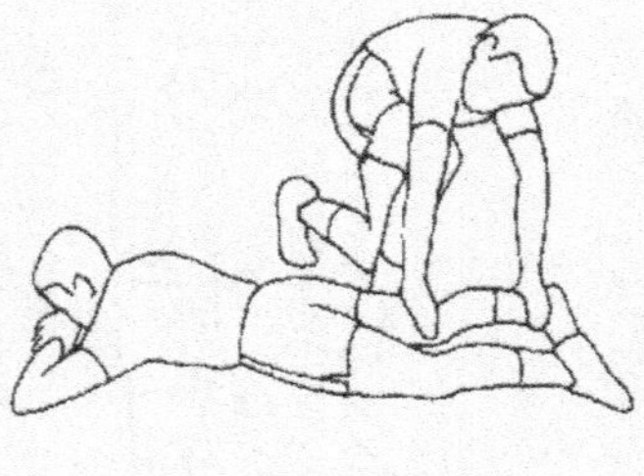

Exerciser

Position: Lie face down with your legs extended.
Action: Flex one leg against your partner's resistance until your heel is as close to your buttocks as possible. Next, resist your partner's efforts as he returns you to the starting position. Do 8 to 12 repetitions to muscle failure. Repeat this exercise with the other leg.

Resister

Position: Support the exerciser's leg as in the Leg Extension exercise.
Action: Resist the exerciser's movement with your hand(s) placed on his heel. Next, apply downward pressure to return the exerciser to the starting position.

HEEL RAISE (BENT OVER)

This exercise is for the gastrocnemius and soleus muscles.

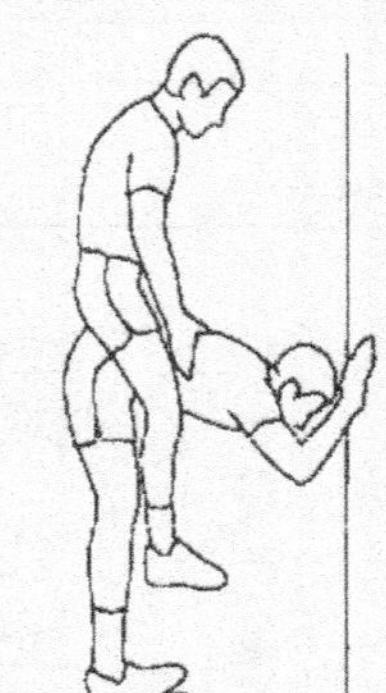

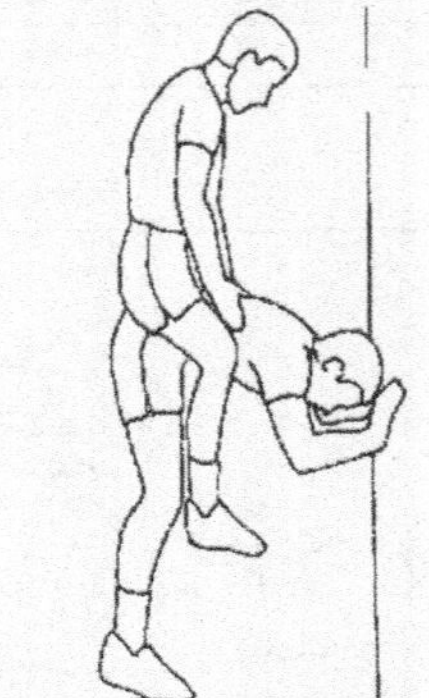

Exerciser

Position: Form a 90-degree angle between your upper body and legs by bending over at the hips. Use an additional partner or a fixed object for support.
Action: Keep your legs straight and rise up on the balls of your feet. Do 8 to 12 repetitions to muscle failure. If possible, perform the exercise by placing the balls of your feet firmly on a 4" x 4" board or the edge of a curb. Be sure to lower and raise your heels as far as possible.

Resister

Position: Sit on the upper part of the exerciser's buttocks; DO NOT SIT ON THE EXERCISER'S LOW BACK. (Properly positioning your body places less pressure on the exerciser's back and helps him better work his gastrocnemius and soleus muscles.)
Action: Provide resistance to the exerciser with your body weight.

TOE RAISE

This exercise is for the tibialis anterior muscle.

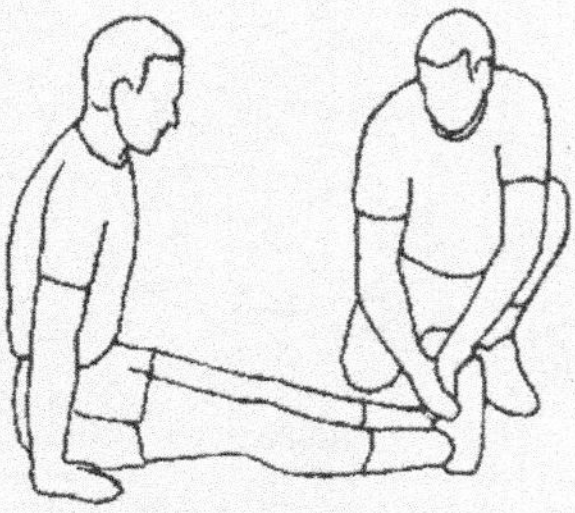

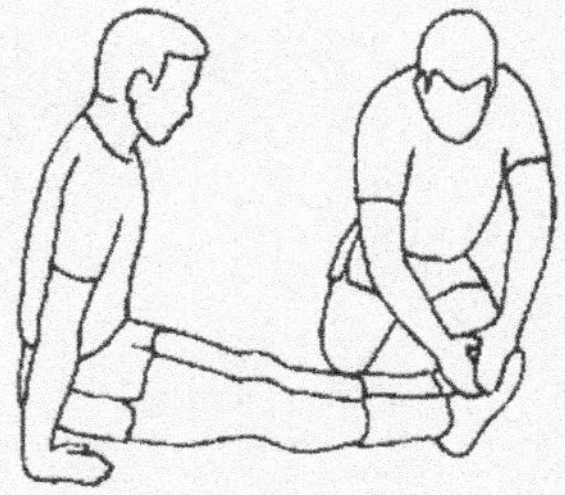

Exerciser

Position: Sit on the floor with your legs together, knees straight, and feet fully extended.

Action: Against the resister's efforts, move your toes toward the knees; then have the resister pull your toes back to the starting position while you resist. Do 8 to 12 repetitions to muscle failure.

Resister

Position: Place your hand(s) on the exerciser's shoelaces near the toes. Press your palms against the exerciser's insteps to resist his foot and ankle movements.

Action: Resist the exerciser's effort to pull his toes toward his knees. Next, pull the exerciser's toes back to the starting position against his resistance.

PUSH-UP

This exercise is for the pectoral and triceps muscles.

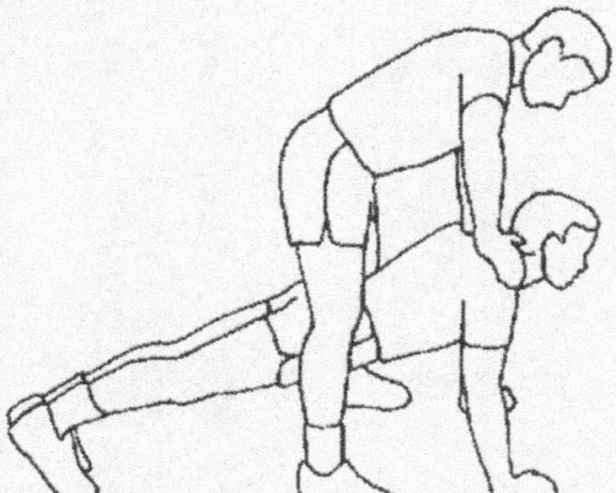

Exerciser

Position: Assume a front-leaning-rest position.

Action: Perform a push-up against your partner's resistance. Do 8 to 12 repetitions to muscle failure.

Resister

Position: Straddle the exerciser's hips. Place your hands on top of his shoulders. Be careful to place your left hand on the upper left part and your right hand on the upper right part of his shoulder.

Action: Apply pressure against the exerciser's push-up movements. As stated earlier, slightly more resistance should be applied during the eccentric phase of contraction (in this case, as the exerciser moves closer to the floor)

SEATED ROW

This exercise is for the biceps, latissimus dorsi, and rhomboid muscles.

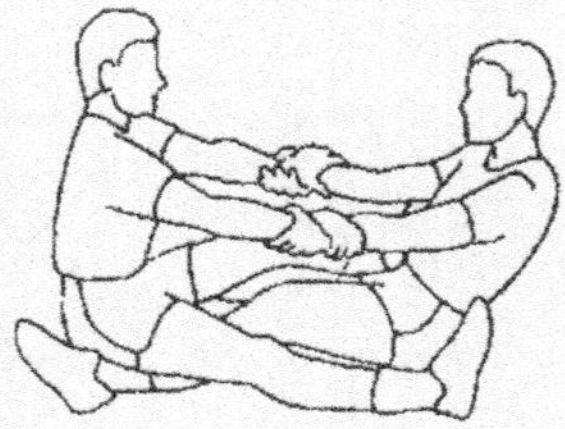

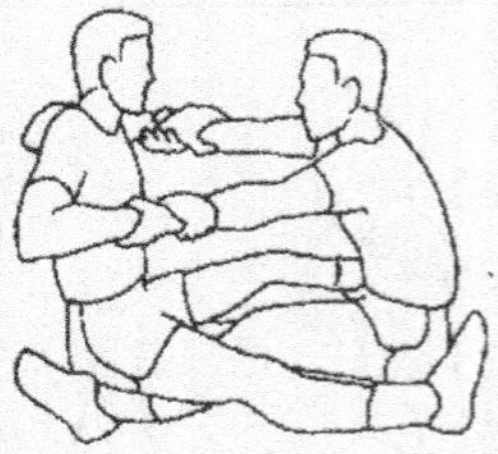

Exerciser

Position: Sit facing the resister with your back straight. Overlap your legs with the resister's, being sure to place your legs on top. Establish a good grip by interlocking your hands with the resister's or by firmly grasping his wrists. The exerciser's palms should be facing downward.

Action: Pull the resister toward you with a rowing motion while keeping your elbows elevated to shoulder height. Be sure to keep your back straight, and move only the arms. Next, slowly return to the starting position as the resister pulls your arms forward. Do 8 to 12 repetitions to muscle failure.

Resister

Position: Face the exerciser and sit with your back straight. Place your legs under the exerciser's legs: establish a good grip by interlocking hands with the resister or by firmly grasping his wrists.

Action: As the exerciser pulls, resist his pulling motion. Next, slowly pull the exerciser back to the starting position by pulling with the muscles of the lower back.

OVERHEAD PRESS

This exercise is for the deltoid and triceps muscles.

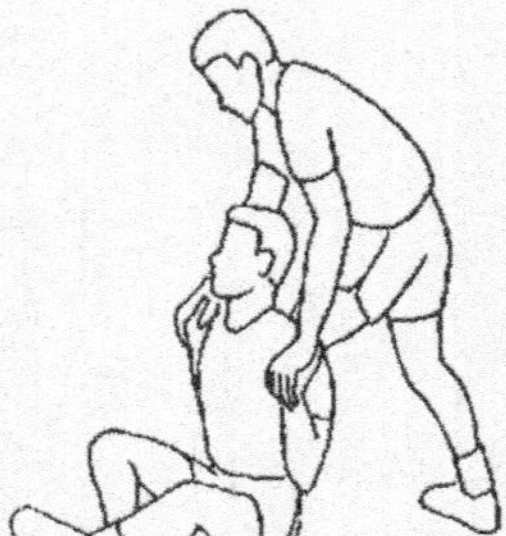

Exerciser

Position: Sit with your legs crossed and your back straight. Raise your hands to shoulder height with your palms flat and facing upward.

Action: Move your arms slowly upward to full extension against your partner's resistance. Next, slowly return to the starting position as the resister applys downward pressure. Do 8 to 12 repetitions to muscle failure.

Resister

Position: Stand behind the exerciser; interlock your thumbs with the exerciser's, and place your hands with the palms down on his hands. Support the exerciser's back with the side of your lower leg.

Action: Resist the exerciser's upward movement: then push his arms back to the starting position. A bar or stick may be used for a better grip and improved leverage.

PULL-DOWN

This exercise is for the latissimus dorsi muscles.

Exerciser

Position: Sit with your legs crossed and back straight. Raise and cross your arms behind your head with your elbows bent.
Action: Pull out and down with your elbows against the partner's resistance until your elbows touch your ribcage. Next, resist as your partner pulls your elbows back to the starting position. Do 8 to 12 repetitions to muscle failure.

Resister

Position: Stand behind the exerciser, and support his back with the side of your lower leg. Place your palms underneath the exerciser's elbows.
Action: Resist the exerciser's downward movements; then pull his elbows back to the up or starting position. VARIATION: A bar or stick may be used for a better grip and leverage and to exercise the biceps and forearm muscles.

SHRUG

This exercise is for the upper trapezius muscle.

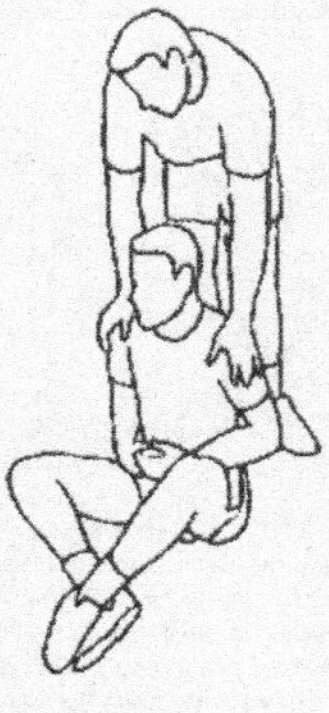

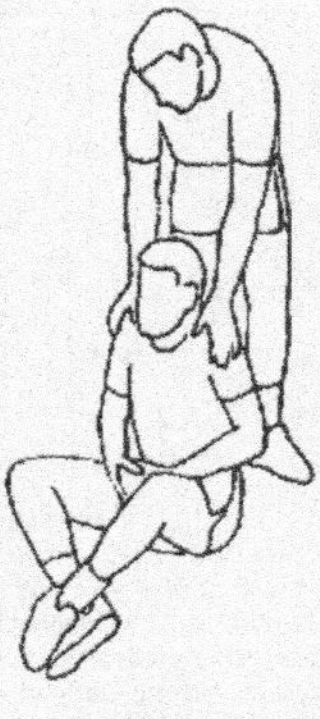

Exerciser

Position: Sit with your legs crossed, back straight, and hands resting in your lap.
Action: Shrug your shoulders as high as possible against your partner's resistance, then resist your partner's pushing motion as you return to the starting position. Do 8 to 12 repetitions to muscle failure.

Resister

Positon: Stand behind the exerciser, and support his back with the side of your lower leg. Place your hands on each of the exerciser's shoulders.
Action: Apply pressure downward with your hands to resist the upward, shrugging movements of the exerciser and, during the second phase of the exercise, push downward as the exerciser resists your pushing movements.

TRICEPS EXTENSION

This exercise is for the triceps muscles.

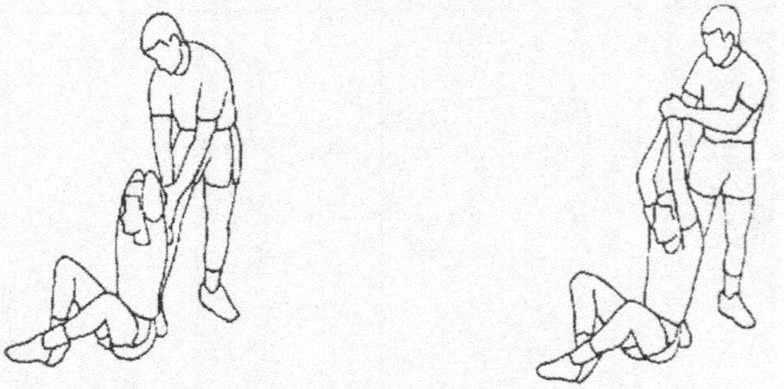

Exerciser

Position: Sit with your legs crossed and back straight. Clasp your hands and place them behind your head while bending your elbows.

Action: Extend your arms upward against the partner's resistance. Next, return to the starting position while resisting your partner's force. Always keep your elbows stationary and pointing straight ahead. Do 8 to 12 repetitions to muscle failure.

Resister

Position: Stand behind the exerciser and support his back with the side of your lower leg. Place your hands, palms down, over the exerciser's hands.

Action: Apply pressure to resist the upward movement of the exerciser, and then push his hands back to the starting position. A bar or a stick may be used for a better grip and/or improved leverage. This exercise may also be done in the prone position with the resister applying a force against the exerciser's movements.

BICEPS CURL

This exercise is for the biceps muscles.

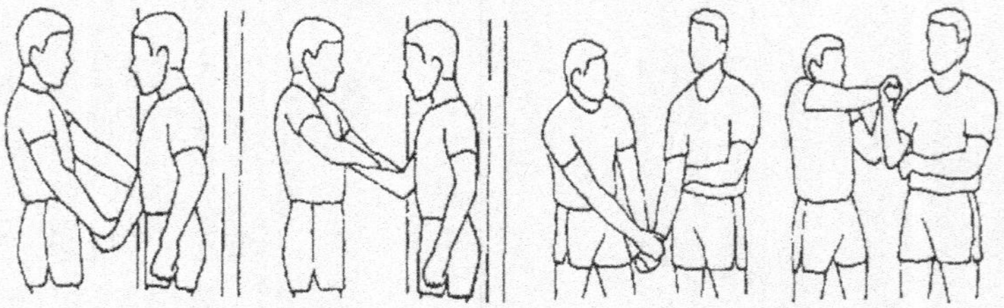

Exerciser

Position: Stand straight with your back supported. Hold the arm to be exercised close to your side.

Action: Bend the elbow, bringing your hand up to your shoulder against your partner's resistance. Return to the starting position by resisting the pushing efforts of your partner. Do 8 to 12 repetitions to muscle failure; repeat with the other arm.

Resister

Position: Face the exerciser with your feet staggered. Use one of your hands to grasp the exerciser's wrist; place the other hand behind his elbow to stabilize it during the exercise movement.

Action: Resist the exerciser's upward movement and provide a downward, pushing force during the lowering movement. A bar may also be used for a better grip and leverage.

VARIATION: A variation may be used if the resister is unable to provide enough resistance to the exerciser with the first exercise. Using this variation, the exerciser places the back of his hand on the non-exercising arm behind the elbow of his exercising arm for support; the resister places both hands on the hand, wrist, or lower part of the exerciser's forearm to apply resistance to the exerciser's movements. The action is the same as before.

ABDOMINAL CURL

This exercise is for the rectus abdominus, iliopsoas, and external and internal oblique muscles.

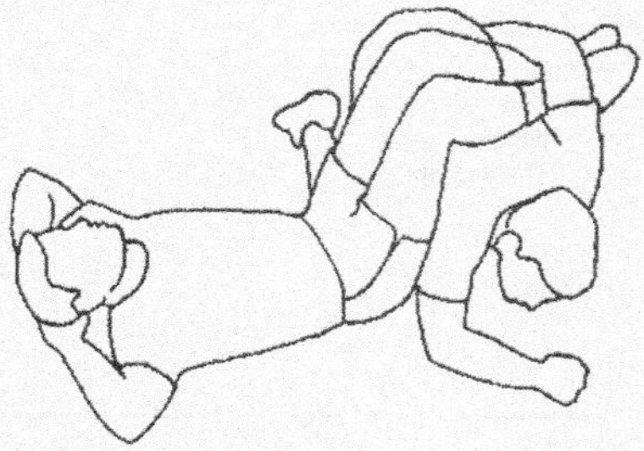

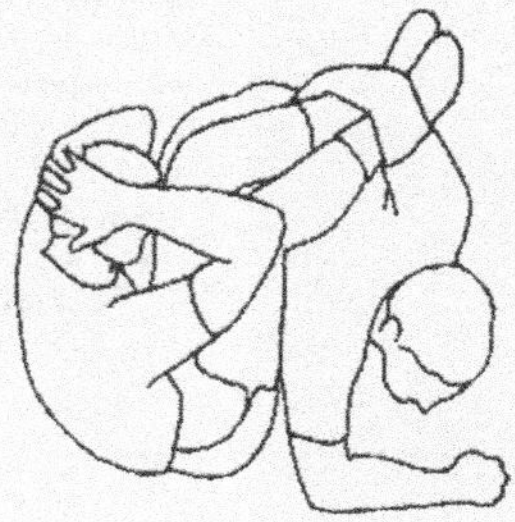

Exerciser

Position: Lie on your back with both legs bent at the knee to about a 90 degree angle. Place your bent legs over the resister's back. Interlace your fingers behind your neck.

Action: Do regular sit-ups, bringing both elbows to your knees.

Do 20 to 50 repetitions to muscle failure.

NOTE: A variation to this exercise is the ROCKY SIT-UP where the exerciser moves the left elbow up to the right knee and then reverses the action, right elbow to left knee.

Resister

Position: Kneel with your inside elbow resting on the ground. With your outside arm, reach back and hold the exerciser's ankles.

Action: Provide a firm foundation upon which the exerciser can place his legs, and keep them tightly anchored during the exercise.

ABDOMINAL CRUNCH

This exercise is for the rectus abdominis and external and internal oblique muscles.

Exerciser

Position: Lie down with your arms crossed over your chest, the backs of your lower legs resting over your partner's back, and your upper leg placed at right angles to the floor.

Action: Curl your neck off the ground, and curl your upper body up toward your knees. (Progressively lift your shoulders, upper back, and finally, lower back off the ground.) Hold this position briefly while forcefully tensing your abdominal muscles. Return slowly to the starting position and repeat. Do 20 to 50 repetitions to muscle failure.

Resister

Position: Kneel with both forearms on the ground.

Action: Allow the exerciser to place the back of his lower legs on your back. DO NOT HOLD HIS LEGS DOWN. (This eliminates the iliopsoas muscle from the exercise and instead isolates the rectus abdominis and external and internal oblique muscles.)

TRAINING WITH EQUIPMENT

Units in garrison usually have access to weight rooms with basic equipment for resistance-training exercises. The exercises described here require free weights and supporting equipment. Although not shown below for the sake of simplicity, all exercises done with free weights require a partner, or spotter, to ensure proper form and the safety of the lifter.

Free-Weight Exercises

SQUAT

This exercise is for the quadriceps and gluteal muscles.

Position: Stand with the feet about shoulder width apart. Hold the weight on your shoulders.
Action: Bend the knees until the tops of your thighs are parallel to the ground. Keep your head and shoulders upright and back straight. In the lowest position, the top of your thighs should not go lower than parallel to the ground. Do 8 to 12 repetitions to muscle failure. A 2" x 4" block may be placed under the heels to increase stability.

HEEL RAISE

This exercise is for the gastrocnemius and soleus muscles.

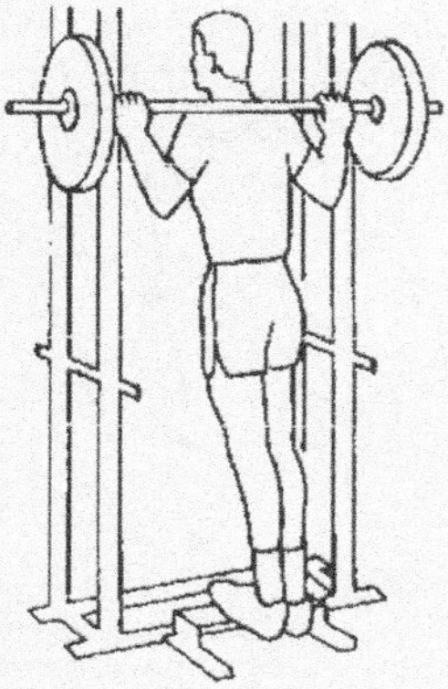

Position: Place a bar on your shoulders behind your neck. Stand with the toes and the balls of the feet on a platform or a 4" x 4 board.

Action: Rise upward on the toes and balls of the feet to full extension, then slowly lower the heels as far as possible. Do not bend the knees or jerk the hips. Do 8 to 12 repetitions to muscle failure.

BENCH PRESS

This exercise is for the pectoralis major, triceps, and deltoid muscles.

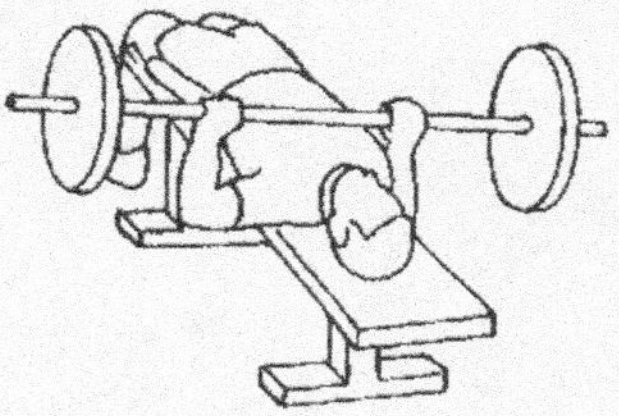

Position: Hold a weight with an overhand grip (palms facing away) slightly wider than shoulder width. Hold the bar directly above your chest at arm's length.

Action: Lower the bar to your chest, keeping the feet flat on the floor. Push the bar up to arm's length. The elbows should be kept wide and away from the body. Keep the buttocks in contact with the bench at all times. Do 8 to 12 repetitions to muscle failure.

BENT-OVER ROW

This exercise is for the latissimus dorsi and biceps muscles.

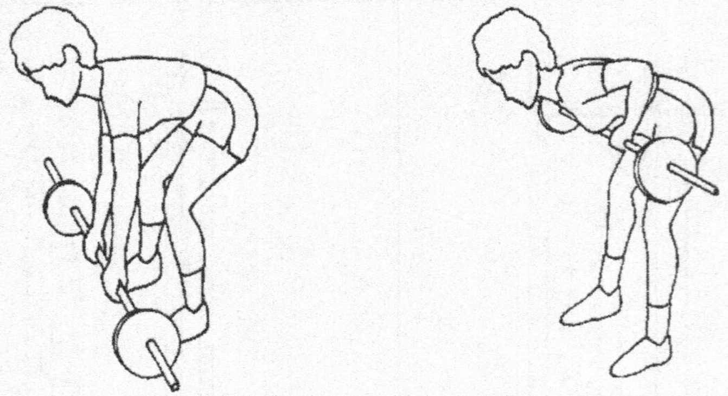

Position: Lean forward at the hips with the back flat; let your arms hang straight down from the shoulders. Keep your knees slightly flexed.

Action: Use an overhand grip with the hands 12 to 24 inches apart. Bend the elbows, bringing the bar up in a straight motion up to the lower portion of the chest. Slowly lower the weight back to the starting position. Do 8 to 12 repetitions to muscle failure.

OVERHEAD PRESS

This exercise is for the deltoids and triceps muscles.

Position: With a barbell, use the overhand grip with the hands spaced slightly greater than shoulder width apart.

Action: Push the bar overhead, moving it upward in a straight line until the elbows are straight. Lower the bar until it touches the chest. Do not bounce the bar off the chest. Dumbbells may also be used. Do 8 to 12 repetitions to muscle failure.

SHRUG

This exercise is for the trapezius muscles.

Position: Using a barbell, start with the bar at thigh-rest position. Use an overhand or reverse grip.

Action: Elevate the bar by contracting the trapezius and raising the shoulders upward toward the ears. In the top position, roll your shoulders backward. Then, slowly lower the shoulders until the bar returns to the starting, thigh-rest position. Keep the arms straight throughout the entire repetition. Dumbbells may also be used. Do 8 to 12 repetitions to muscle failure.

TRICEPS EXTENSION

This exercise is for the triceps muscles.

Position: Using a barbell, hold the bar directly overhead with an overhand grip. Keep the elbows high, close to the head, and stationary.

Action: Lower the bar slowly without bouncing it when it reaches the lower neck area. Extend the bar back to the overhead position while keeping the heels flat and the knees and elbows stationary. A dumbbell may also be used. Do 8 to 12 repetitions to muscle failure.

BICEPS CURL

This exercise is for the biceps muscles.

Position: Start with an underhand grip. Hold the bar at thigh-rest position.

Action: Keep the elbows stationary and close to your sides as you curl the bar to your chest. Do not use your legs or bend your back for assistance. A cambered (bent) bar or dumbbells may also be used. Do 8 to 12 repetitions to muscle failure.

WRIST CURL

This exercise is for the development of the forearm muscles.

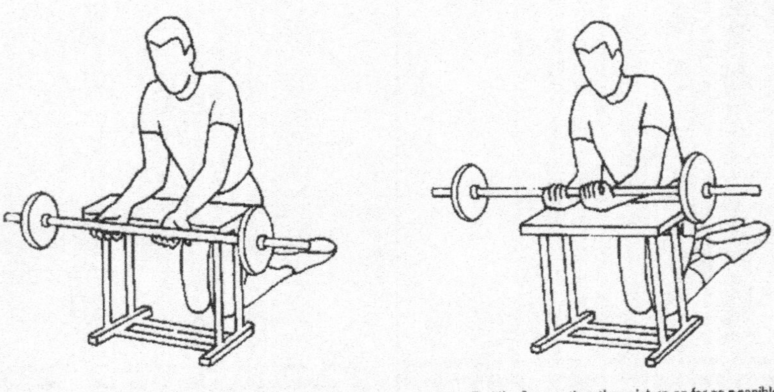

Position: Holding your hand with the palms facing upward, grasp a barbell using only the fingers.

Action: Curl the fingers, then the wrist up as far as possible and then down, keeping the elbows stationary. For the best results, do not grip the barbell; keep it placed on the last few digits of the fingers. Do 8 to 12 repetitions to muscle failure.

BENT-LEG DEAD-LIFTS

This exercise is for the quadriceps, the erector spinae, the gluteals, and the trapezius muscles.

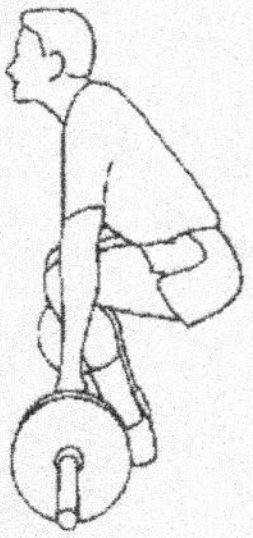

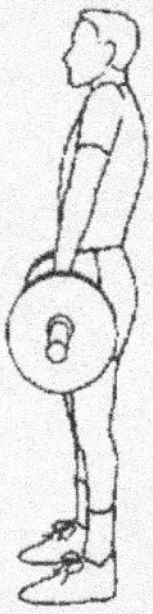

Position: Bend and grasp the bar with the hands shoulder width apart. The legs should be bent, the back flat but inclined forward at a 45 degree angle, the arms straight, and the head up.

Action: Keeping the head erect, gradually straighten the legs and the back together at the same time. Make sure that the back remains flat and the arms remain straight. When the entire body is straight, shrug the shoulders upward as high as possible. In a controlled manner, return to the starting position by first lowering the shoulders. Then, bend at the knees and at the waist simultaneously until the beginning position is attained. Keep the back flat, head up, and the arms straight at all times. Do 8 to 12 repetitions to muscle failure.

Exercises Performed with an Exercise Machine

If exercise machines are available, the exercises described below are also good for strength training. All movements, particularly during the eccentric (negative) phase of contraction, should be done in a deliberate, controlled manner.

LEG PRESS

This **exercise is for the gluteal and quadriceps muscles.**

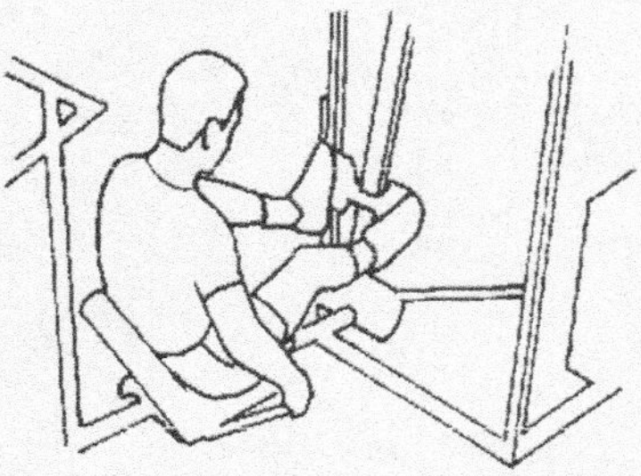

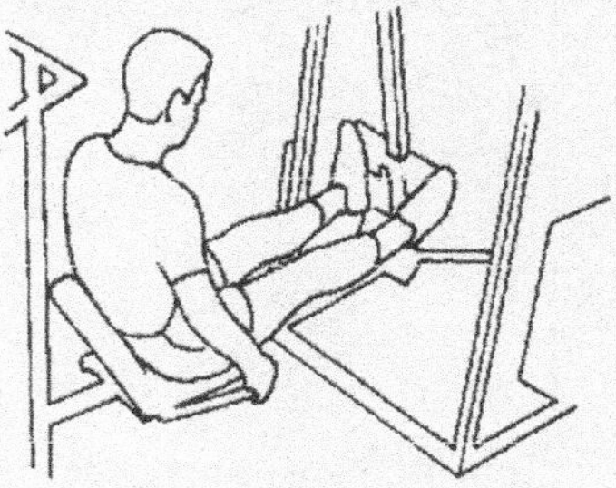

Position: Sit at the leg-press station with the legs bent no more than 90 degrees. Ensure that the balls of both feet are very securely placed on the pedals.

Action: Push the weight with the legs until your knees are straight but not locked. In a controlled manner, return to the starting position. This is one repetition. Do 8 to 12 repetitions to muscle failure.

LEG EXTENSION

This exercise is for the quadriceps muscles.

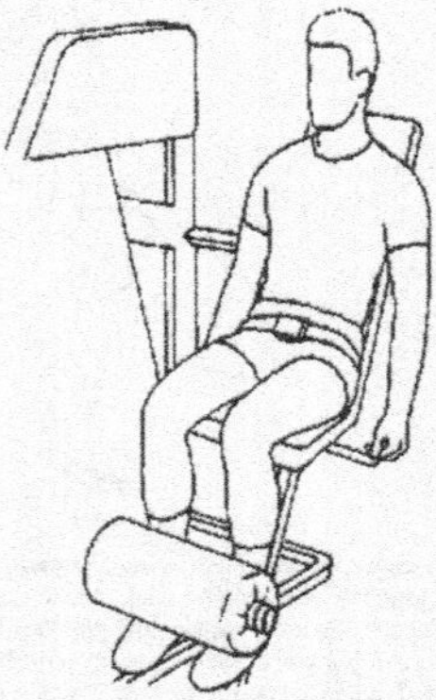

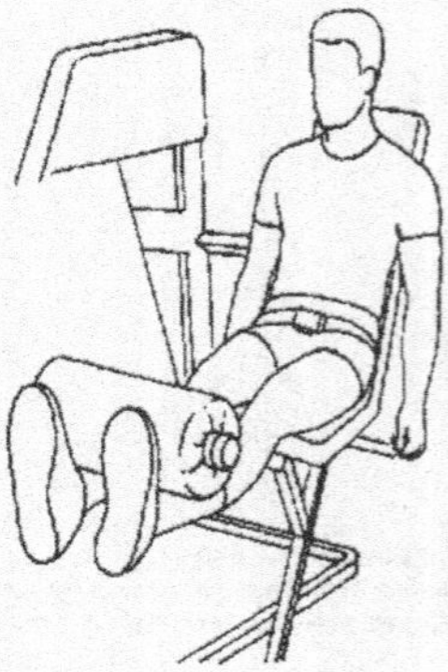

Position: Sit on a bench with your lower legs behind the padded lever. Hold on to the bench or provided handles with your hands to keep the upper body in the correct position.

Action: Straighten the legs as much as possible. In a controlled manner, return to the starting position. This is one repetition. Do 8 to 12 repetitions to muscle failure.

LEG CURL

This exercise is for the hamstring muscles.

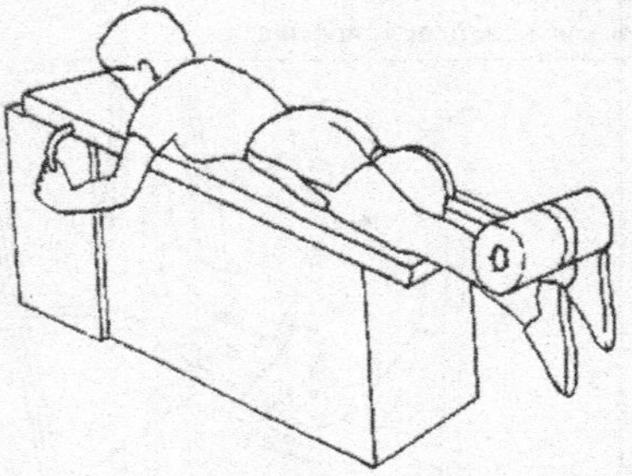

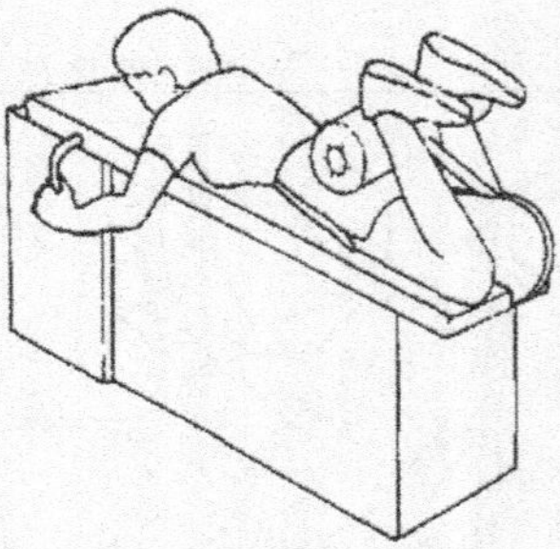

Position: Lie on your stomach with the legs straight and the ankles under the padded lever. Maintain correct upper body position by loosely grasping the sides of the bench or provided handles.

Action: Bend your legs at the knee until the lower legs pass well beyond the perpendicular position and the heels are as close to your buttocks as possible. Return to the starting position. Do 8 to 12 repetitions to muscle failure.

HEEL RAISE

This exercise is for the gastrocnemius and soleus muscles.

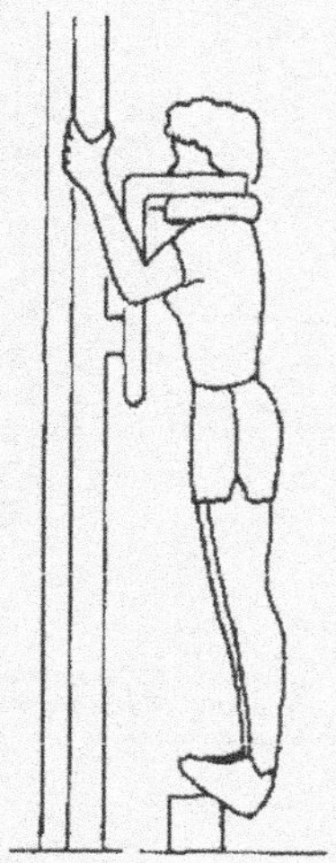

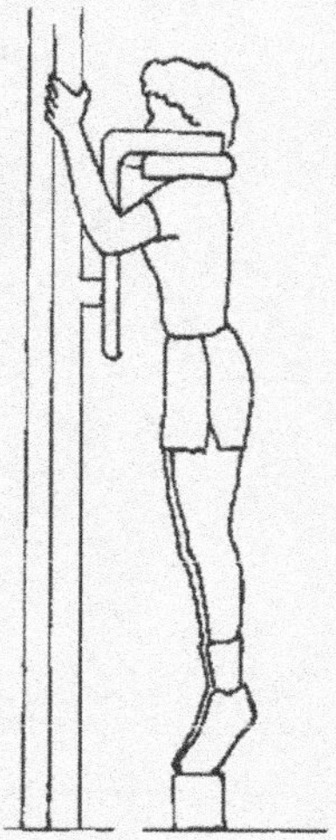

Position: Stand with a weight on your shoulders and the balls of your feet placed firmly on a 4-inch raised surface.

Action: Raise your heels off the floor as far as possible while maintaining your balance. Then, lower them as far as possible. This is one repetition. Do 8 to 12 repetitions to muscle failure.

TOE RAISE

This exercise is for the tibialis anterior muscle.

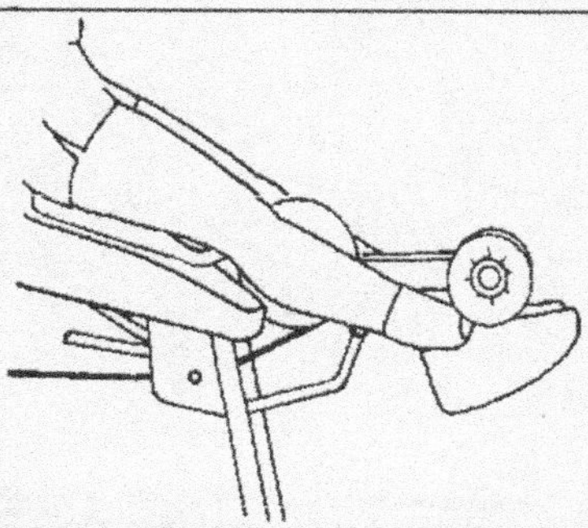

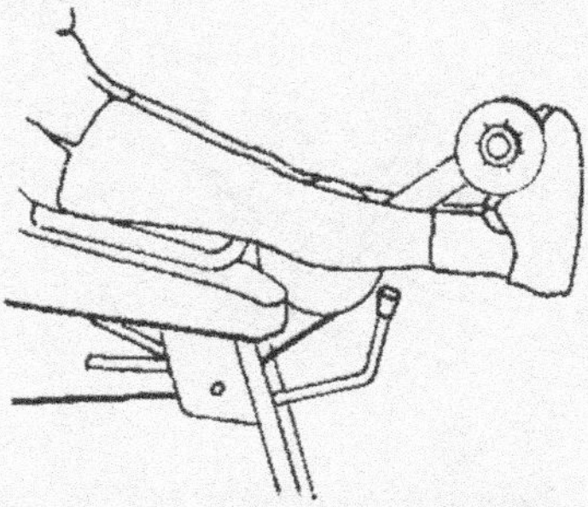

Position: Sit on the leg curl machine with your legs together, knees straight, and toes pointed. Place the top of your feet under the roller pad.

Action: Move your toes toward the knees as far as possible. Then lower the weight to the starting position in a controlled manner. Do 8 to 12 repetitions to muscle failure.

BENCH PRESS

This exercise is for the pectoralis major, triceps, and deltoid muscles.

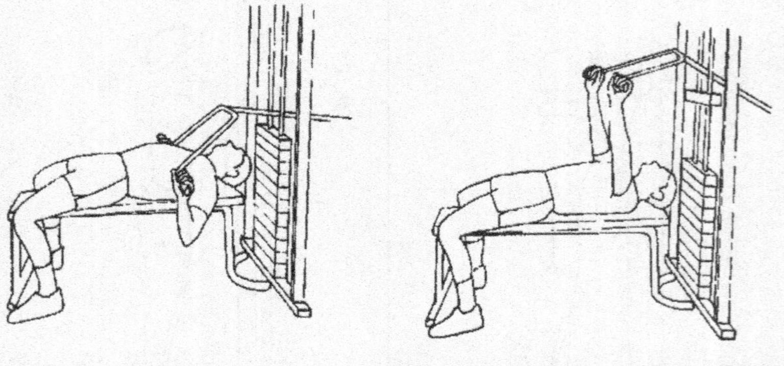

Position: Lie on your back with your hands placed about shoulder width apart on the bar. Generally, the bar or handles should be located at the lower half of the chest.

Action: Push the bar up until your arms are straight. Then, lower the bar to the starting position. This is one repetition. Do 8 to 12 repetitions to muscle failure.

SEATED ROW

This exercise is for the latissimus dorsi and biceps muscles.

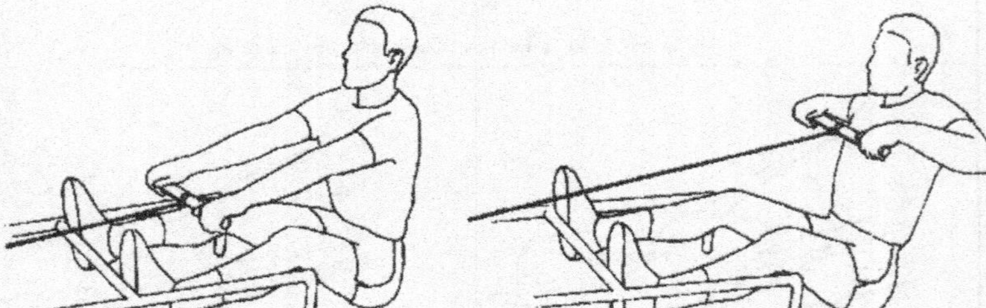

Position: Sit and assume the straight arm position shown above. Use the overhand grip with your hands spaced 6 to 8 inches apart.

Action: Pull the bar to the lower part of your chest, while keeping your elbows elevated to shoulder height, then slowly extend the arms and lower the weight to the beginning position. Be sure to keep the back straight, and move only the arms. Do 8 to 12 repetitions to muscle failure.

LAT PULL-DOWN

This exercise is for the latissimus dorsi and biceps muscles. (Pull-ups or chin-ups may be substituted for this exercise.)

Position: Sit or kneel and grasp the bar with your palms facing away from the body.

Action: Pull the bar down until it touches the back of your neck; return the bar in a controlled manner to that starting position. This is one repetition. Do 8 to 12 repetitions to muscle failure.

SHRUG

This exercise is for the trapezius muscles of the upper back.

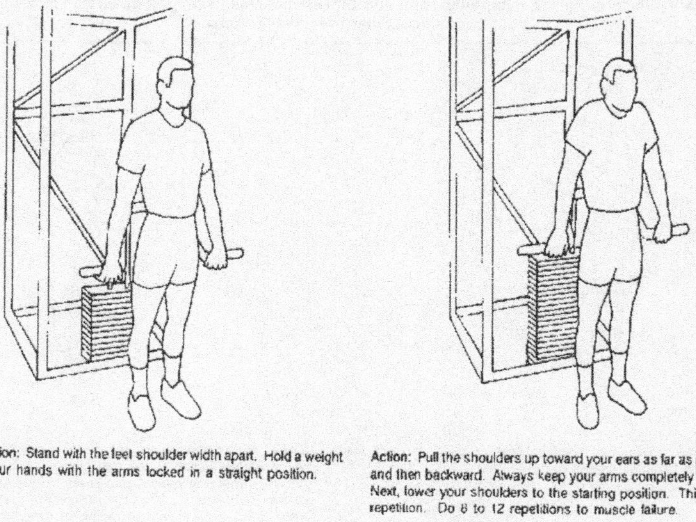

Position: Stand with the feet shoulder width apart. Hold a weight in your hands with the arms locked in a straight position.

Action: Pull the shoulders up toward your ears as far as possible and then backward. Always keep your arms completely straight. Next, lower your shoulders to the starting position. This is one repetition. Do 8 to 12 repetitions to muscle failure.

PARALLEL BAR DIP

This exercise is for the pectoralis major and triceps muscles.

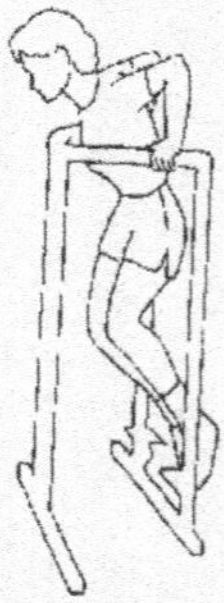

Position: Keep your feet off the floor and support the body's weight on straight arms.

Action: Bend the arms and lower your body until the upper arms are at least parallel to the floor. If necessary, bend your legs at the knees to keep the feet from touching the floor. Straighten your arms to return to the starting position. This is one repetition. Do 8 to 12 repetitions to muscle failure. A weight belt may be worn if additional resistance is needed.

CHIN-UP

This exercise is for the latissimus dorsi and biceps muscles. (Lat pull-downs or pull-ups may be substituted for this exercise.)

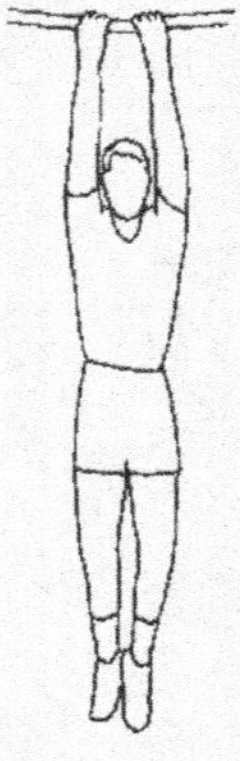

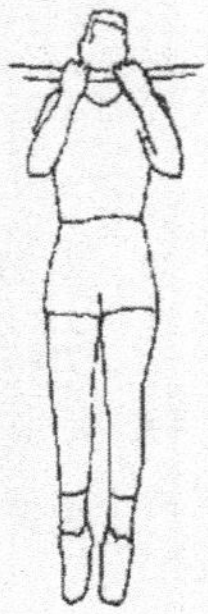

Position: From a standing position, grasp the bar with your palms facing the body.

Action: Bending both arms, pull your body up until your chin clears the bar. Return to the starting position in a controlled manner. If necessary, bend your knees to keep the feet from touching the floor. Do 8 to 12 repetitions to muscle failure. A weight belt may be worn if additional resistance is needed.

TRICEPS EXTENSION

This exercise is for the triceps muscles.

Position: Make seat adjustments to sit with the elbows in line with the axes of the cams and the elbows slightly higher than the shoulders. Place the sides of the hands on the pads with the hands open.

Action: Straighten the arms against the resistance. After doing this, bend the elbows, and return to the starting position in a controlled manner. Do 8 to 12 repetitions to muscle failure.

BICEPS CURL

This exercise is for the biceps muscles.

Position: Stand with the bar in front of your body, arms straight and elbows at the sides. Your hands should be spaced about shoulder width apart and the palms should face away from the body.

Action: Without moving your elbows, bend the arms, bringing the bar to shoulder level. In a controlled manner, lower the weight to the starting position. This is one repetition. Do 8 to 12 repetitions to muscle failure.

The following exercises can be performed to condition the muscles of the mid-section (erector spinae, rectus abdominus and external and internal obliques). As the soldier becomes more conditioned on these exercises, resistance can be added.

BACK EXTENSION

This exercise is for the erector spinae muscle group.

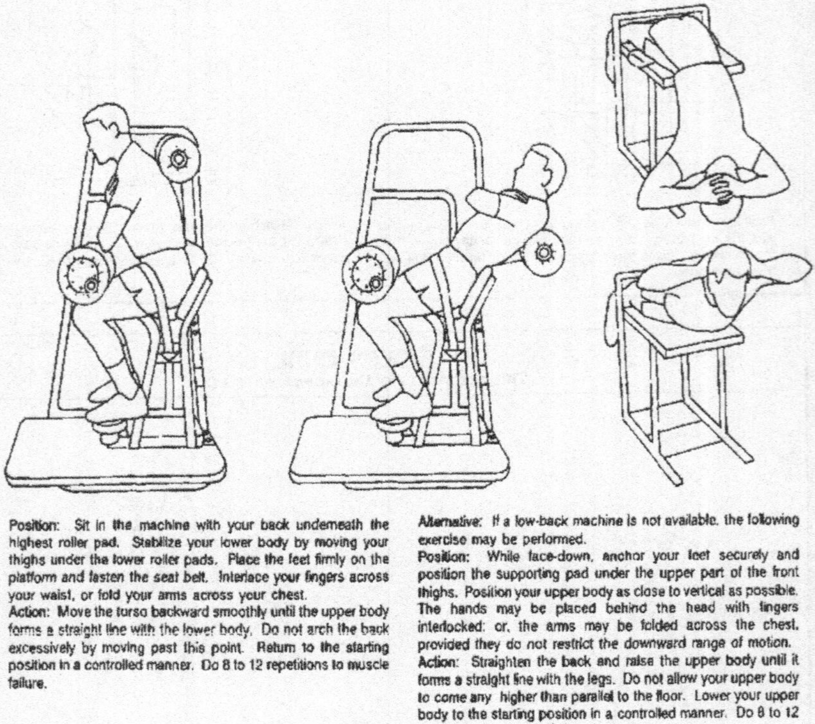

Position: Sit in the machine with your back underneath the highest roller pad. Stabilize your lower body by moving your thighs under the lower roller pads. Place the feet firmly on the platform and fasten the seat belt. Interlace your fingers across your waist, or fold your arms across your chest.

Action: Move the torso backward smoothly until the upper body forms a straight line with the lower body. Do not arch the back excessively by moving past this point. Return to the starting position in a controlled manner. Do 8 to 12 repetitions to muscle failure.

Alternative: If a low-back machine is not available, the following exercise may be performed.

Position: While face-down, anchor your feet securely and position the supporting pad under the upper part of the front thighs. Position your upper body as close to vertical as possible. The hands may be placed behind the head with fingers interlocked; or, the arms may be folded across the chest, provided they do not restrict the downward range of motion.

Action: Straighten the back and raise the upper body until it forms a straight line with the legs. Do not allow your upper body to come any higher than parallel to the floor. Lower your upper body to the starting position in a controlled manner. Do 8 to 12 repetitions to muscle failure.

SIT-UP

This exercise is for the rectus abdominis and iliopsoas (hip flexor) muscles.

Position: Lie on your back with your knees bent at approximately a 90 degree angle and feet anchored. Place your hands behind your head.

Action: Sit up until your trunk is in a vertical position relative to the floor while keeping the knees bent. Lower yourself in a controlled manner to the starting position. The number of repetitions you should do depends on the maximum number of sit-ups you perform in two minutes. Do three sets of 50 percent of your maximum number. For example, if you can do 60 sit-ups in two minutes, do three sets of 30 or more repetitions per set.

INCLINE SIT-UP

This exercise is for the rectus abdominis and iliopsoas muscles.

Position: Lie on an incline board with your knees bent at approximately a 90 degree angle and your feet anchored. The steeper the incline of the board, the more difficult the sit-up will be. Interlace the fingers behind your head.

Action: Curl your torso up as far as comfortably possible. Return to the starting position. This is one repetition. Do 20 to 50 repetitions to muscle failure.

ABDOMINAL CRUNCH

This exercise is for the rectus abdominis muscle.

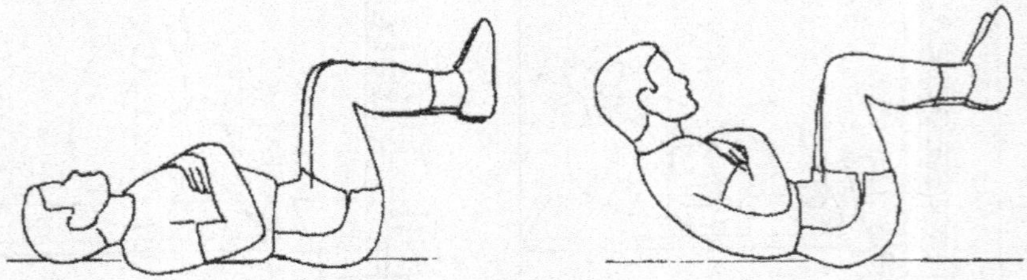

Position: Lie on your back with both legs bent at the knees and the upper legs at right angles to the floor. Your arms should be crossed at chest level with the palms of the hands on their opposite shoulders. Your ankles may be crossed but, in all cases, the feet should not touch the floor.

Action: Roll up your upper body by first lifting your head and tucking the chin. Next, curl your spine by rolling the upper back and then the lower back off the floor. Pause briefly in the up position while tensing the abdominal muscles. Return in a slow, controlled manner to the starting position by "unrolling" the upper body. Do 20 to 50 repetitions to muscle failure.

Exercise Chart

The chart labeled Figure 3-5 will help the soldier select appropriate exercises for use in developing a good muscular endurance and strength workout. For example, if the soldier wants to develop his upper leg muscles, he has several options. He may choose from the following: 1) PREs, concentrating on the split- or single-leg squat; 2) exercises with equipment, doing free weight squats; or, 3) exercises with a machine, doing leg presses, leg curls, and leg extensions.

EXERCISE CHART FOR MUSCULAR STRENGTH AND ENDURANCE

	EXERCISES	LOWER LEGS	UPPER LEGS	WAIST	CHEST	UPPER ARMS	LOWER ARMS	SHOULDERS	BACK
Partner-Resisted Exercises	Split-Squat		X						
	Single-Leg Squat		X						
	Leg Extension		X						
	Leg Curl		X						
	Heel Raise	X							
	Toe Raise	X							
	Push-Up				X	X			
	Seated Row					X			X
	Overhead Press					X		X	
	Pull-Down					X			X
	Shrug							X	
	Triceps Extension					X			
	Biceps Curl					X			
	Abdominal Twist			X					
	Abdominal Curl			X					
	Abdominal Crunch			X					
Exercises with Equipment (Barbell/Dumbbell)	Squat		X						
	Heel Raise	X							
	Bench Press				X	X			
	Bent-Over Row					X			X
	Overhead Press					X		X	
	Shrug							X	
	Triceps Extension					X			
	Biceps Curl					X			
	Wrist Curl						X		
	Bent-Leg Dead Lift		X					X	X
Exercises with an Exercise Machine	Leg Press		X						
	Leg Extension		X						
	Leg Curl		X						
	Heel Raise	X							
	Toe Raise	X							
	Bench Press				X	X			
	Seated Row					X			X
	Lat Pull-Down					X			X
	Shrug							X	
	Parallel Bar Dip				X	X			
	Chin-up					X			X
	Triceps Extension					X			
	Biceps Curl					X			
	Back Extension								X
	Sit-Up			X					
	Incline Sit-Up			X					
	Abdominal Twist			X					
	Abdominal Crunch			X					

Figure 3-5

CHAPTER 4 -

Flexibility

Flexibility refers to the range of movement of a joint.

The four categories of stretching techniques are static,passive, proprioceptive neuromuscular facilitation (PNF), and ballistic.

Flexibility is a component of physical fitness. Developing and maintaining it are important parts of a fitness program. Good flexibility can help a soldier accomplish such physical tasks as lifting, loading, climbing, parachuting, running, and rappelling with greater efficiency and less risk of injury.

Flexibility is the range of movement of a joint or series of joints and their associated muscles. It involves the ability to move a part of the body through the full range of motion allowed by normal, disease-free joints,

No one test can measure total-body flexibility. However, field tests can be used to assess flexibility in the hamstring and low-back areas. These areas are commonly susceptible to injury due, in part, to loss of flexibility. A simple toe-touch test can be used. Soldiers should stand with their legs straight and feet together and bend forward slowly at the waist. A soldier who cannot touch his toes without bouncing or bobbing needs work to improve his flexibility in the muscle groups stretched by this test. The unit's Master Fitness Trainer can help him design a stretching program to improve his flexibility.

Stretching during the warm-up and cool-down helps soldiers maintain overall flexibility. Stretching should not be painful, but it should cause some discomfort because the muscles are being stretched beyond their normal length. Because people differ somewhat anatomically, comparing one person's flexibility with another's should not be done. People with poor flexibility who try to stretch as far as others may injure themselves.

Stretching Techniques

Using good stretching techniques can improve flexibility. There are four commonly recognized categories of stretching techniques: static, passive, proprioceptive neuromuscular facilitation (PNF), and ballistic. These are described here and shown later in this chapter,

STATIC STRETCHING

Static stretching involves the gradual lengthening of muscles and tendons as a body part moves around a joint. It is a safe and effective method for improving flexibility. The soldier assumes each stretching position slowly until he feels tension or tightness. This lengthens the muscles without causing a reflex contraction in the stretched muscles. He should hold each stretch for ten seconds or longer. This lets the lengthened muscles adjust to the stretch without causing injury.

The longer a stretch is held, the easier it is for the muscle to adapt to that length. Static stretching should not be painful. The soldier should feel slight discomfort, but no pain. When pain results from stretching, it is a signal that he is stretching a muscle or tendon too much and may be causing damage.

PASSIVE STRETCHING

Passive stretching involves the soldier's use of a partner or equipment, such as a towel, pole, or rubber tubing, to help him stretch. This produces a safe stretch through a range of motion he could not achieve without help. He should talk with his partner to ensure that each muscle is stretched safely through the entire range of motion,

PNF STRETCHING

PNF stretching uses the neuromuscular patterns of each muscle group to help improve flexibility. The soldier performs a series of intense contractions and relaxations using a partner or equipment to help him stretch. The PNF technique allows for greater muscle relaxation following each contraction and increases the soldier's ability to stretch through a greater range of motion.

BALLISTIC STRETCHING

Ballistic, or dynamic, stretching involves movements such as bouncing or bobbing to attain a greater range of motion and stretch. Although this method may improve flexibility, it often forces a muscle to stretch too far and may result in an injury. Individuals and units should not use ballistic stretching.

FITT Factors

Commanders should include stretching exercises in all physical fitness programs.

The following FITT factors apply when developing a flexibility program.

Frequency: Do flexibility exercises daily. Do them during the warm-up to help prepare the muscles for vigorous activity and to help reduce injury. Do them during the cool-down to help maintain flexibility.

Intensity: Stretch a muscle beyond its normal length to the point of tension or slight discomfort, not pain.

Time: Hold stretches for 10 to 15 seconds for warming up and cooling down and for 30 seconds or longer to improve flexibility.

Type: Use static stretches, assumed slowly and gradually, as well as passive stretching and/or PNF stretching.

Warm-Up and Cool-Down

The warm-up and cool-down are very important parts of a physical training session, and stretching exercises should be a major part of both.

THE WARM-UP

Before beginning any vigorous physical activity, one should prepare the body for exercise. The warm-up increases the flow of blood to the muscles and tendons, thus helping reduce the risk of injury. It also increases the joint's range of motion and positively affects the speed of muscular contraction.

A recommended sequence of warm-up activities follows. Soldiers should do these for five to seven minutes before vigorous exercise.

- Slow joggin-in-place or walking for one to two minutes. This causes a gradual increase in the heart rate, blood pressure, circulation, and increases the temperature of the active muscles.
- Slow joint rotation exercises (for example, arm circles, knee/ankle rotations) to gradually increase the joint's range of motion. Work each major joint for 5 to 10 seconds.
- Slow, static stretching of the muscles to be used during the upcoming activity. This will "loosen up" muscles and tendons so they can achieve greater ranges of motion with less risk of injury. Hold each stretch position for 10 to 15 seconds, and do not bounce or bob.
- Calisthenic exerciese, as described in Chapter 7, to increase the intensity level before the activity or conditioning period.
- Slowly mimic the activities to be performed. For example, lift a lighter weight to warm-up before lifting a heavier one. This helps prepare the neuromuscular pathways.

The warm-up warms the muscles, increasing the flow of blood and reducing the risk of injury.

THE COOL-DOWN

The following information explains the importance of cooling down and how to do it correctly.

- Do not stop suddenly after vigorous exercise, as this can be very dangerous. Gradually bring the body back to its resting state by slowly decreasing the intensity of the activity. After running, for example, one should walk for one to two minutes. Stopping exercise suddenly can cause blood to pool in the muscles, thereby reducing blood flow to the heart and brain. This may cause fainting or abnormal rhythms in the heart which could lead to serious complications.
- Repeat the stretches done in the warm-up to help ease muscle tension and any immediate feeling of muscle soreness. Be careful not to overstretch. The muscles are warm from activity and can possibly be overstretched to the point of injury.
- Hold stretches 30 seconds or more during the cool-down to improve flexiblity. Use partner-assisted or PNF techniques, if possible.

The soldier should not limit flexibility training to just the warm-up and cool-down periods. He should sometimes use an entire PT session on a "recovery" or "easy"training day to work on flexibility improvement. He may also work on it at home. Stretching is one form of exercise that takes very little time relative to the benefits gained.

Rotation Exercises

Rotation exercises are used to gently stretch the tendons, ligments, and muscles associated with a joint and to stimulate lubrication of the joint with synovial fluid. This may provide better movement and less friction in the joint.

The following exercises should be performed slowly.

NECK

Position: Stand with the back straight and feet shoulder width apart. Place the hands on the hips.

Action: Roll the head slowly to the left, making a complete circle with the path of the head. Do this three times in each direction.

ARMS AND SHOULDERS

Position: Stand with the back straight and feet shoulder width apart. Extend the arms outward to shoulder height.

Action: Rotate the shoulders forward, and make a large circular motion with the arms. Repeat the action in the opposite direction. Do this three times in each direction.

HIPS

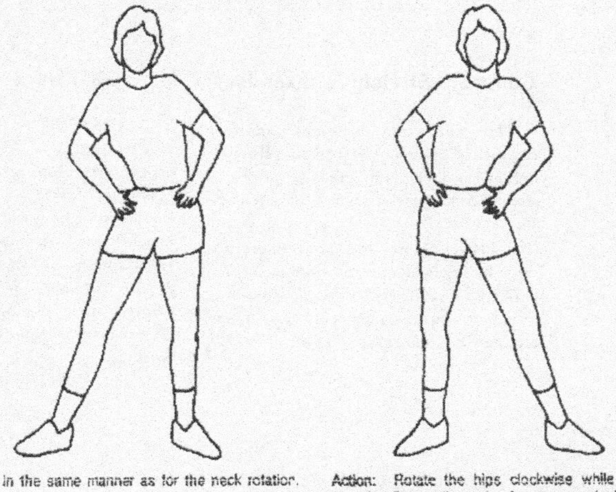

Position: Stand in the same manner as for the neck rotation.

Action: Rotate the hips clockwise while keeping the back straight. Repeat the action in a counterclockwise direction. Do this three times in each direction.

KNEES AND ANKLES

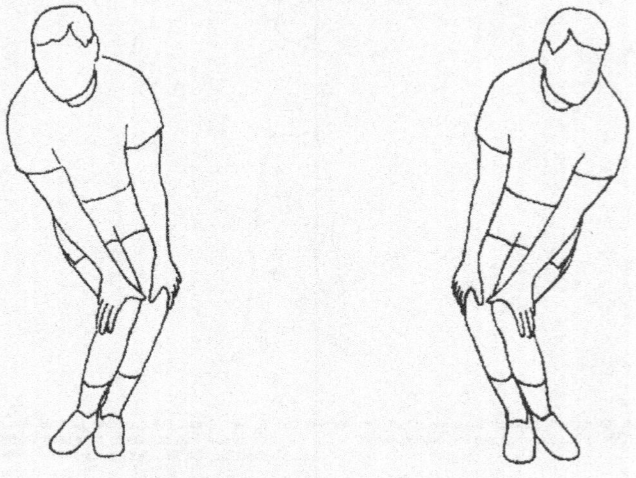

Position: Stand with the feet together, and bend at the waist with the knees slightly bent.

Actio·· Place the hands above the knees, and rotate the legs in a clockwise direction. Repeat the action in a counterclock-wise direction. Do this three times in each direction.

Common Stretching Exercises

The following exercises improve flexibility when performed slowly, regularly, and with gradual progression. Static, passive and PNF stretches are shown.

CAUTION Some of these exercises may be difficult or too strenuous for unfit or medically limited soldiers. Common sense should be used ;n selecting stretching exercises.

STATIC STRETCHES

Assume all stretching positions slowly until you feel tension or slight discomfort. Hold each position for at least 10 to 15 seconds during the warm-up and cool-down. Developmental stretching to improve flexibility requires holding each stretch for 30 seconds or longer.

Choose the appropriate stretch for the muscle groups which you will be working.

NECK AND SHOULDER STRETCH

This stretches the sternocleidomastoid, pectoralis major, and deltoid muscles.

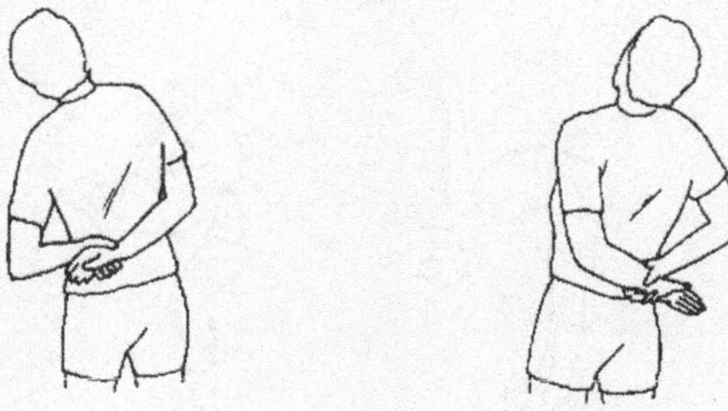

Position: Stand with the feet shoulder width apart and the arms behind the body.

Action: Grasp the left wrist with the right hand. Pull the left arm down and to the right. Tilt the head to the right. Hold this position for 10 to 15 seconds. Repeat the action with the right wrist, pulling the right arm down and to the left. Tilt the head to the left.

ABDOMINAL STRETCH

This stretches the abdominals, obliques, latissimus dorsi, and biceps.

Position: Stand and extend the arms upward and over the head. Interlace the fingers with the palms turned upward.

Action: Stretch the arms up and slightly back. Hold this position for 10 to 15 seconds.

Variation: This stretches the rectus abdominis muscles. Stretch to one side, then the other. Return to the starting position.

CHEST STRETCH

This stretches the pectoralis major, deltoids, and biceps muscle groups.

Position: Stand and interlace the fingers behind the back.

Action: Lift the arms behind the back so that they move outward and away from the body. Lean forward from the waist. Hold this position for 10 to 15 seconds. Bend the knees before moving to the upright position. Return to the starting position.

UPPER-BACK STRETCH

This stretches the lower trapezius and posterior deltoid muscles of the upper back.

Position: Stand with the arms extended to the front at shoulder height with the fingers interlaced and palms facing outward.

Action: Extend the arms and shoulders forward. Hold this position for 10 to 15 seconds. Return to the starting position.

OVERHEAD ARM PULL

This stretches the external and internal obliques, latissimus dorsi, and triceps.

Position: Stand with the feet shoulder width apart. Raise the right arm, bending the right elbow and touching the right hand to the back of the neck.

Action: Grab the right elbow with the left hand, and pull to the left. Hold this position for 10 to 15 seconds. Return to the starting position. Do the same stretch, and pull the left elbow with the right hand for 10 to 15 seconds.

THIGH STRETCH

This stretches the quadriceps and anterior tibialis.

Position: Stand. (For variation, lie on the stomach.)

Action: Bend the left leg up toward the buttocks. Grasp the toes of the left foot with the right hand, and pull the heel to the left buttock. Extend the left arm to the side for balance. Hold this position for 10 to 15 seconds. Return to the starting position. Bend the right leg, grasp the toes of the right foot with the left hand, and pull the heel to the right buttock. Extend the right arm for balance. Hold this position for 10 to 15 seconds. Return to the starting position.

HAMSTRING STRETCH (STANDING)

This stretches the hamstrings, erector spinae, and gluteal muscles.

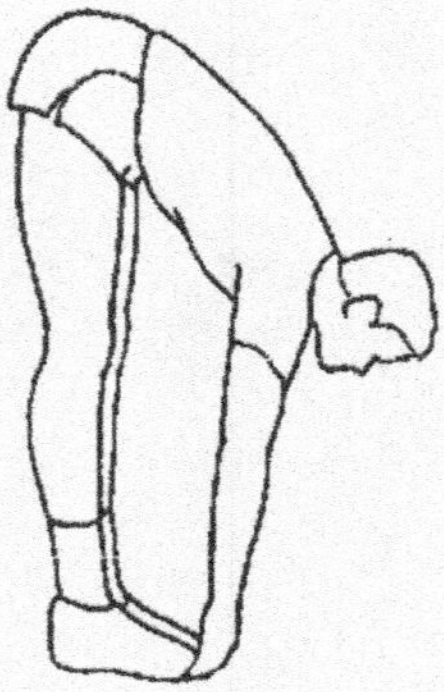

Position: Stand with the knees slightly bent.

Action: Bend forward keeping the head up, and reach toward the toes. Straighten the legs, and hold this position for 10 to 15 seconds.

HAMSTRING STRETCH (SEATED)

In addition to the muscles mentioned in the standing hamstring stretch, this stretches the calf (gastrocnemius and soleus) muscles.

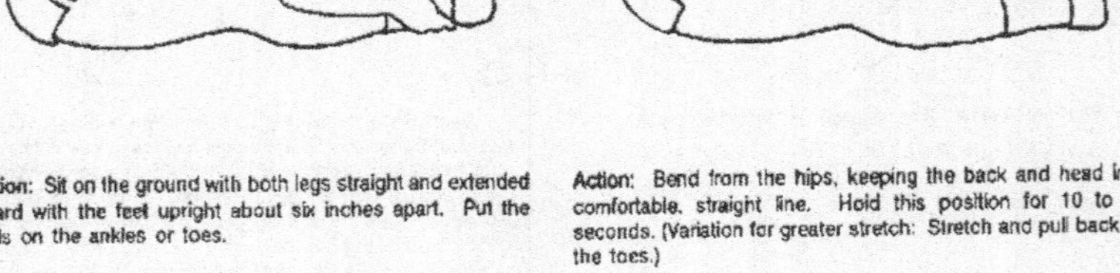

Position: Sit on the ground with both legs straight and extended forward with the feet upright about six inches apart. Put the hands on the ankles or toes.

Action: Bend from the hips, keeping the back and head in a comfortable, straight line. Hold this position for 10 to 15 seconds. (Variation for greater stretch: Stretch and pull back on the toes.)

GROIN STRETCH (STANDING)

This stretches the hip adductor muscles.

Position: Lunge slowly to the left while keeping the right leg straight, the right foot facing straight ahead and entirely on the floor.

Action: Lean over the left leg while stretching the right groin muscles. Hold this position for 10 to 15 seconds. Repeat with the opposite leg.

GROIN STRETCH (SEATED)

This stretches the hip adductor and erector spinae muscles.

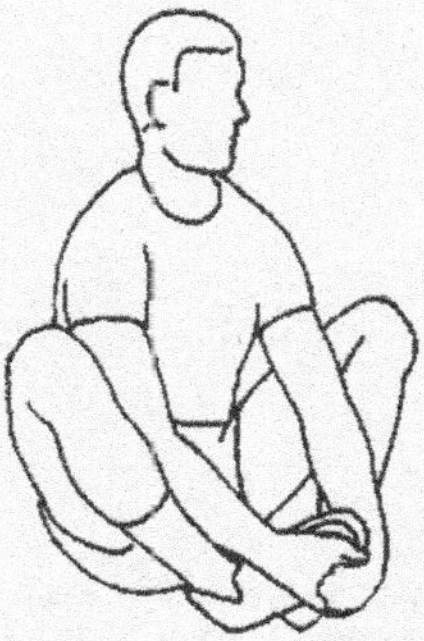

Position: Sit on the ground with the soles together. Place the hands on or near the feet.

Action: Bend forward from the hips, keeping the head up. Hold this position for 10 to 15 seconds.

GROIN STRETCH (SEATED STRADDLE)

This stretches the hip adductor (on the inside of the upper leg), gluteals, erector spinae, and hamstring muscles.

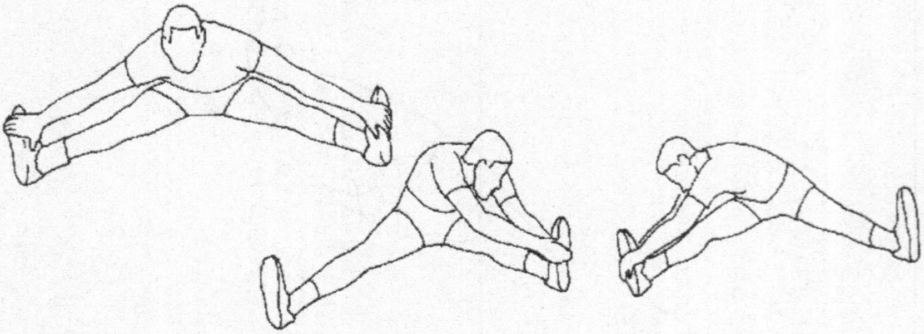

Position: Sit on the ground with the legs straight and spread as far apart as possible.

Action: Bend forward at the hips, keep the head up, and reach toward the feet. Hold this position for 10 to 15 seconds. Variation: Stretch to one side while trying to touch the toes. Next, stretch to the other side.

CALF STRETCH

This stretches the calf (gastrocnemius and soleus) muscles.

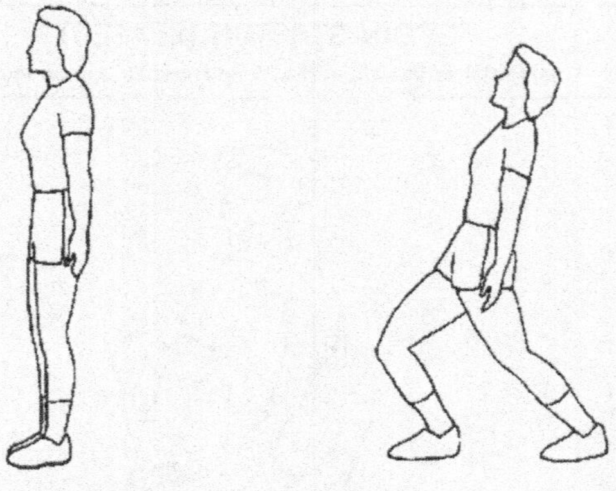

Position: Stand straight with the feet together, arms extended downward, elbows locked, palms facing backward, fingers extended and joined, and head and eyes facing front.

Action: Move the right foot to the rear about two feet, and place the ball of the foot on the ground. Slowly press the right heel to the ground. Slowly bend the left knee while pushing the hips forward and arching the back slightly. Hold this position for 10 to 15 seconds. Return to the starting position. Repeat with the left foot. Return to the starting position.

CALF STRETCH (VARIATION: TOE PULL)

This stretches the calf (gastrocnemius) and to a lesser extent the hamstrings, gluteus maximus, and erector spinae muscles.

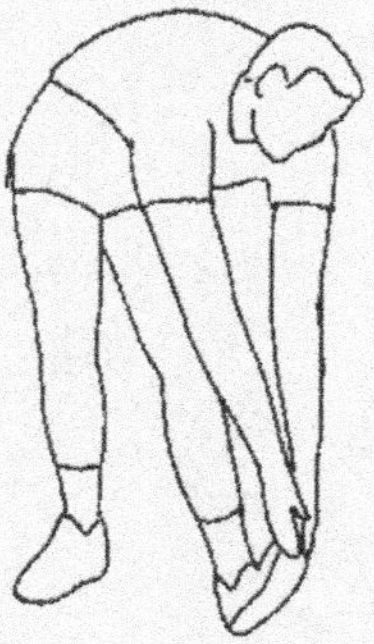

Position: Stand with the feet shoulder width apart and the left foot slightly forward.

Action: Bend forward at the waist. Slightly bend the right knee, and fully extend the left leg. Reach down and pull the toes of the left foot toward the left shin. Hold this position for 10 to 15 seconds. Return to the starting position. In a similar manner, pull the toes of the right foot toward the right shin, and hold for 10 to 15 seconds.

HIP AND BACK STRETCH (SEATED)

This stretches the hip abductors, erector spinae, latissimus dorsi, and oblique muscle groups.

Position: Sit on the ground with the right leg forward and straight. Cross the left leg over the right while sitting erect. Keep the heels of both feet in contact with the ground.

Action: Slowly rotate the upper body to the left and look over the left shoulder. Reach across the left leg with the right arm, and push the left leg to your right. Use the left hand for support by placing it on the ground. Hold this position for 10 to 15 seconds. Repeat this stretch for the other side by crossing and turning in the opposite direction.

HIP AND BACK STRETCH (LYING DOWN)

This stretches the gluteal and erector spinae muscles.

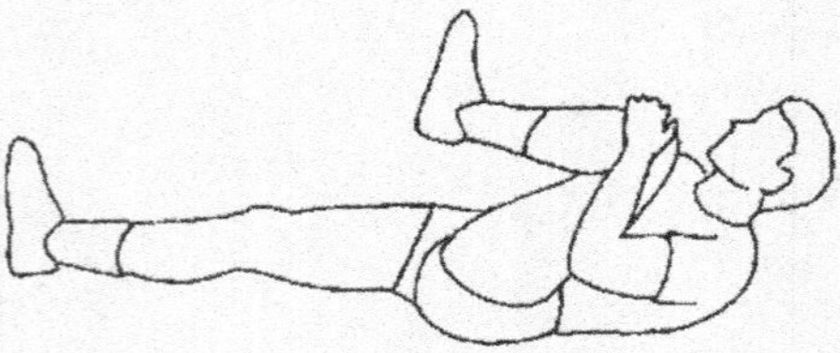

Action 1

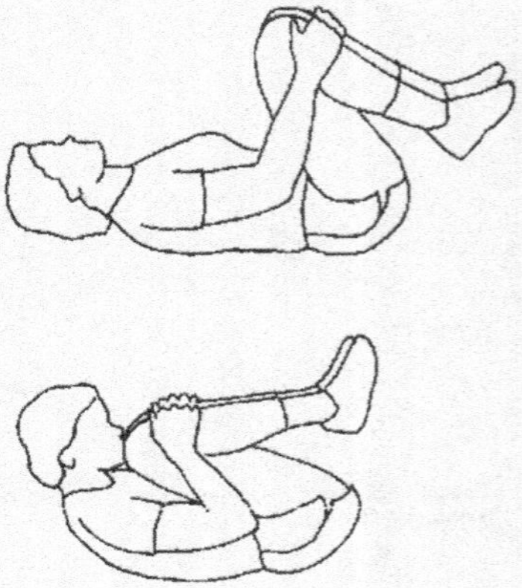

Action 2

Position: Lie on the back with the arms straight beside the body. Keep the legs straight and the knees and feet together.

Action 1: Bring the left leg straight back toward the head, leaving the right leg in the starting position. Bring the head and arms up. Grab the bent left leg below the knee, and pull it gradually to the chest. Hold this position for 10 to 15 seconds. Gradually return to the starting position. Repeat these motions with the opposite leg.

Action 2: Pull both knees to the chest. Pull the head up to the knees. Hold for 10 to 15 seconds. Return to the starting position.

PASSIVE STRETCHES

Passive stretching is done with the help of a partner or equipment. The examples in this chapter show passive stretching done with a towel or with a partner. When stretching alone, using a towel may help the exerciser achieve a greater range of motion.

TOWEL STRETCHES

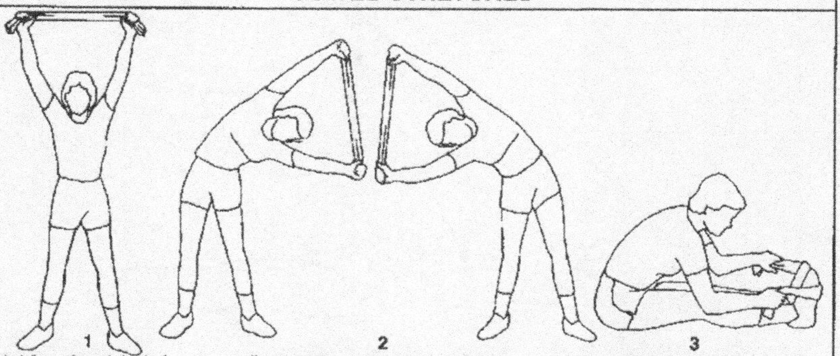

1

This stretches the abdominal and pectoral muscles.
Position: Stand erect with the hands overhead and grasping a towel.
Action: Pull tightly on the towel while reaching up and slightly arching the back. Hold for 10 to 15 seconds.

2

This stretches the abdominals, obliques, and latissimus dorsi.
Position: Stand erect with the hands overhead and grasping a towel.
Action: Slowly bend sideways to the left as far as possible. Hold for 10 to 15 seconds. Repeat for the opposite side. While doing this stretch, pulling on the towel with the bottom arm will enhance the stretch.

3

This stretches the hamstring, calf, and low back muscles.
Position: Sit with the legs straight and together. Grasping each end of a short towel, place the middle of the towel over the balls of the feet.
Action: Pulling on the towel, come forward as far as possible keeping the legs straight and the toes pulled back.

PARTNER-ASSISTED CHEST STRETCH

This exercise stretches the pectoralis major, deltoids, and biceps muscles.

Position: Sit erect with the arms straight, elevated to shoulder height, and the palms facing forward. The partner stands behind the exerciser grasping the arms between the wrists and the elbows.

Action: The partner gradually pulls both of the exerciser's arms toward the rear until the stretch causes the exerciser mild discomfort. Hold this position for 10 to 15 seconds.

PARTNER-ASSISTED HAMSTRING STRETCH

This exercise stretches the hamstrings and erector spinae muscle groups.

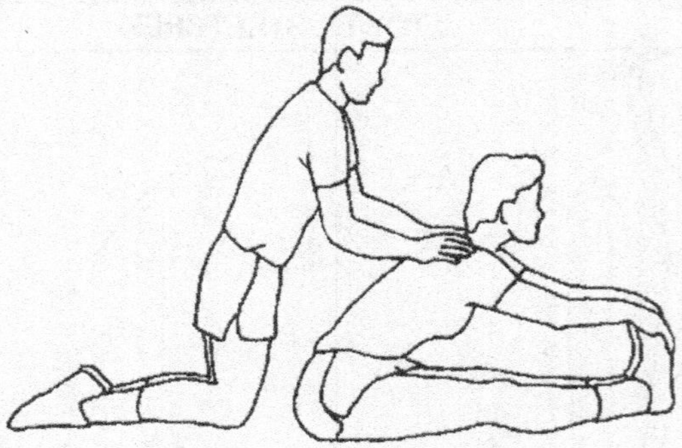

Position: Sit erect on the ground with the legs together. The partner kneels behind the exerciser. If the partner stands, he may apply too much pressure.

Action: The partner places light pressure on the exerciser's upper back until the exerciser's forward motion results in mild discomfort. This position is held for 10 to 15 seconds.

PARTNER-ASSISTED GROIN STRETCH

This exercise stretches the hip adductor and erector spinae muscle groups.

Position: Sit on the ground with knees bent and soles together. The partner kneels behind the exerciser. If the partner stands, he may apply too much pressure.

Action: The partner places light pressure on the exerciser's knees with his hands and leans gently on the exerciser's back with his chest until the stretch causes the exerciser mild discomfort. This position is held for 10 to 15 seconds.

Soldiers can do PNF (Proprioceptive Neuromuscular Facilitation) stretches for most major muscle groups. PNF stretches use a series of contractions, done against a partner's resistance, and relaxations.

Obtaining a safe stretch beyond the muscle's normal length requires a partner's assistance. The following four steps provide general guidance as to how PNF stretches are done. Both the exerciser and partner should follow these instructions:

1. Assume the stretch position slowly with the partner's help.
2. Isometrically contract the muscles to be stretched. Hold the contraction for 5 to 10 seconds against the partner's unyielding resistance.
3. Relax. Next, contract the antagonistic muscles for 5 to 10 seconds while the partner helps the exerciser obtain a greater stretch.
4. Repeat this sequence three times, and try to stretch a little further each time. (Caution: The exerciser should not hold his breath. He should breathe out during each contraction.)

Several examples of PNF stretches are provided below in a stepwise fashion. The numbers given above for each step correspond to the general description listed below.

PNF HAMSTRING AND GLUTEAL STRETCH

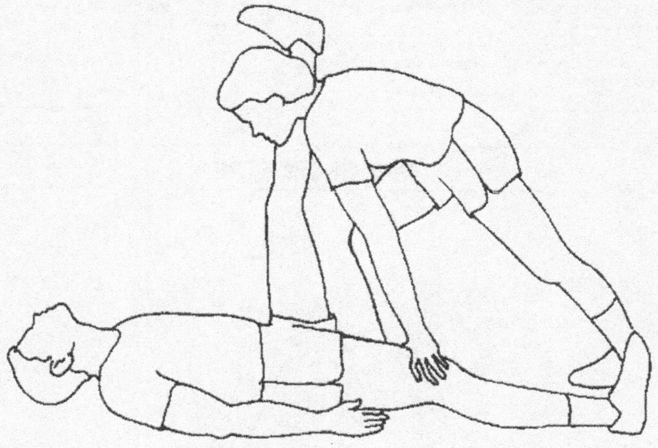

1. The exerciser lies on his back and places the lower part of his left leg on the partner's right shoulder. The exerciser slowly stretches the hamstring and gluteal muscles by gradually bringing the straightened leg toward his head until he feels tension in the stretched muscles. The partner then applies light pressure on the exerciser's lower leg to help maintain or further the stretch.
2. The exerciser isometrically contracts his hamstring and gluteal muscles for 5 to 10 seconds by trying to move his leg downward and away from his head. The partner steadily resists the exerciser's efforts and does not allow any movement to occur.
3. The exerciser relaxes the hamstring and gluteal muscles. He then tries to stretch them farther by using the partner's help and by contracting the antagonistic, hip flexor muscles (the iliopsoas and quadriceps) and the tibialis anterior muscle for 5 to 10 seconds.
4. Perform these movements three times for each leg. Try to stretch a little further each time.

PNF CHEST STRETCH

1. The exerciser and partner assume the positions used in the partner-assisted chest stretch. The exerciser slowly stretches until he feels tension in the stretched muscles. The partner then applies light pressure on the forearm to help maintain or further the stretch.

2. The exerciser contracts the pectorals for 5 to 10 seconds, trying to pull his arms and hands together against the partner's unyielding resistance.

3. The exerciser relaxes his pectorals and attempts to stretch further with the partner's help while contracting the antagonistic muscles of the upper back. He does this for 5 to 10 seconds.

4. Repeat this sequence three times, trying to stretch a little further each time.

PNF GROIN STRETCH

1. The exerciser assumes the position used for the partner-assisted groin stretch. The partner kneels behind him. The exerciser slowly lowers his legs and leans forward until tension is felt in the muscles of the groin (his abductors) and lower back (erector spinae muscles). Next, the partner applies light pressure on the exerciser's thighs and back to help maintain or further increase the stretch.

2. The exerciser then attempts to push upward for 5 to 10 seconds by contracting the groin and lower back muscles while the partner resists and allows no movement to occur.

3. The exerciser relaxes the groin and lower back muscles and tries to stretch further with the partner's help and by contracting the antagonistic muscles (hip abductor and abdominal muscles) for 5 to 10 seconds.

4. Perform these movements three times. Try to stretch a little further each time.

Boay uomposmon

Body composition, which refers to the body's relative amounts of fat and lean body mass (organs, bones, muscles), is one of the five components of physical fitness. Good body composition is best gained through proper diet and exercise. Examples of poor body composition are underdeveloped musculature or excessive body fat. Being overweight (that is, overly fat) is the more common problem.

Poor body composition causes problems for the Army. Soldiers with inadequate muscle development cannot perform as well as soldiers with good body composition. As a soldier gets fat, his ability to perform physically declines, and his risk of developing disease increases. Soldiers with high percentages of body fat often have lower APFT scores than those with lower percentages. Poor body composition, especially obesity, has a negative effect on appearance, self-esteem, and negatively influences attitude and morale.

The Army's weight control program is described in AR 600-9. It addresses body composition standards, programs for the overly fat, and related administrative actions.

The amount of fat on the body, when expressed as a percentage of total body weight, is referred to as the percent body fat. The Army's maximum allowable percentages of body fat, by age and sex, are listed in Figure 5-1.

Evaluation Methods

The Army determines body fat percentage using the girth method. (This is described in AR 600-9, pages 12 to 21.)

Body composition is influenced by age, diet, fitness level, and genetic factors (gender and body type). The Army's screening charts for height and weight (shown in AR 600-9) make allowances for these differences. A soldier whose weight exceeds the standard weight shown on the charts may not necessarily be overfat. For example, some well-muscled athletes have body weights that far exceed the values for weight listed on the charts for their age, gender, and height. Yet, only a small percentage of their total body mass may be fat. In such cases, the lean body mass accounts for a large share of their total body composition, while only a small percentage of the total body mass is composed of fat.

Body composition is influenced by age, fitness level, and genetic factors.

Soldiers who do not meet the weight standards for their height and/or soldiers whose appearance suggests that they have excessive fat are to be evaluated using the circumference (girth measurement) method described in AR 600-9.

BODY FAT STANDARDS

AGES:	17-20	21-27	28-39	40+
MALES 20%	20%	22%	24%	26%
FEMALES	30%	32%	34%	36%

Figure 5-1

A more accurate way to determine body composition is by hydrostatic or underwater weighing. However, this method is very time-consuming and expensive and usually done only at hospitals and universities.

Soldiers who do not meet Army body fat standards are placed on formal, supervised weight (fat) loss programs as stipulated in AR 600-9. Such programs include sensible diet and exercise regimens.

Diet and Exercise

A combination of exercise and diet is the best way to lose excessive body fat. Losing one to two pounds a week is a realistic goal which is best accomplished by reducing caloric intake and increasing energy expenditure. In other words, one should eat less and exercise more. Dieting alone can cause the body to believe it is being starved. In response, it tries to conserve its fat reserves by slowing down its metabolic rate and, as a result, it loses fat at a slower rate.

A combination of exercise and diet is the best way to lose unwanted body fat.

Soldiers must consume a minimum number of calories from all the major food groups, with the calories distributed over all the daily meals including snacks. This ensures an adequate consumption of necessary vitamins and minerals. A male soldier who is not under medical supervision when dieting requires a caloric intake of at least 1 ,500; women require at least 1,200 calories. Soldiers should avoid diets that fail to meet these criteria.

Trying to lose weight with fad diets and devices or by skipping meals does not work for long-term fat loss, since weight lost through these practices is mostly water and lean muscle tissue, not fat. Losing fat safely takes time and patience. There is no quick and easy way to improve body composition.

Aerobic exercise is best for burning fat. examples include jogging, walking, swimming, bicycling, cross-country skiing, and rowing.

The soldier who diets and does not exercise loses not only fat but muscle tissue as well. This can negatively affect his physical readiness. Not only does exercise burn calories, it helps the body maintain its useful muscle mass, and it may also help keep the body's metabolic rate high during dieting.

Fat can only be burned during exercise if oxygen is used. Aerobic exercise, which uses lots of oxygen, is the best type of activity for burning fat. Aerobic exercises include jogging, walking, swimming, bicycling, cross-country skiing, rowing, stair climbing, exercise to music, and jumping rope. Anaerobic activities, such as sprinting or lifting heavy weights, burn little, if any, fat.

Exercise alone is not the best way to lose body fat, especially in large amounts. For an average-sized person, running or walking one mile burns about 100 calories. Because there are 3,500 calories in one pound of fat, he needs to run or walk 35 miles if pure fat were being burned. In reality, fat is seldom the only source of energy used during aerobic exercise. Instead, a mixture of both fats and carbohydrates is used. As a result, most people would need to run or walk over 50 miles to burn one pound of fat.

A combination of proper diet and aerobic exercise is the proven way to lose excessive body fat. Local dietitians and nutritionists can help soldiers who want to lose weight by suggesting safe and sensible diet programs. In addition, the unit's MFT can design tailored exercise programs which will help soldiers increase their caloric expenditure and maintain their lean body mass.

CHAPTER 6

Nutrition and Fitness

In addition to exercise, proper nutrition plays a major role in attaining and maintaining total fitness. Good dietary habits (see Figure 6-1) greatly enhance the ability of soldiers to perform at their maximum potential. A good diet alone, however, will not make up for poor health and exercise habits. This chapter gives basic nutritional guidance for enhancing physical performance. Soldiers must know and follow the basic nutrition principles if they hope to maintain weight control as well as achieve maximum physical fitness, good health, and mental alertness.

Guidelines for Healthy Eating

Eating a variety of foods and maintaining an energy balance are basic guidelines for a healthy diet. Good nutrition is not complicated for those who understand these dietary guidelines.

To be properly nourished, soldiers should regularly eat a wide variety of foods from the major food groups, selecting a variety of foods from within each group. (See Figure 6-2.) A well-balanced diet provides all the nutrients needed to keep one healthy.

Most healthy adults do not need vitamin or mineral supplements if they eat a proper variety of foods. There are no known advantages in consuming excessive amounts of any nutrient, and there may be risks in doing so.

For soldiers to get enough fuel from the food they eat and to obtain the variety of foods needed for nutrient balance they should eat three meals a day. Even snacking between meals can contribute to good nutrition if the right foods are eaten.

Another dietary guideline is to consume enough calories to meet one's energy needs. Weight is maintained as long as the body is in energy balance,

DIETARY GUIDELINES

- Eat a Variety of Foods
- Maintain a Healthy Body Weight
- Choose a Diet Low in Fat, Saturated Fat, and Cholesterol
- Choose a Diet with Plenty of Vegetables, Fruits, and Grain Products
- Use Sugars Only in Moderation
- Use Salt and Sodium Only in Moderation
- If you Drink Alcoholic Beverages, Do So in Moderation

Figure 6-1

DAILY FOOD GUIDE		
Eat a variety of foods from each food group. Most people should have the minimum number of servings; others need more due to their body size and activity level.		
FOOD GROUP	**SUGGESTED NUMBER OF SERVINGS**	**SUGGESTED SIZE OF SERVINGS**
Vegetables (Include dark green, leafy, or deep yellow ones)	3 to 5	1 cup of raw, leafy greens or 1/2 cup of cooked vegetables
Fruits (Include citrus fruits or juices, melons, or berries)	2 to 4	1 medium fruit or 1/2 cup of diced or small fruit or 3/4 cup of juice
Breads, Cereals, Rice, and Pasta (Include whole grain varieties)	6 to 11	1 slice of bread, 1/2 bun or roll, 1/2 cup of cooked cereal, rice or pasta, 1 oz. of ready-to- eat cereal
Milk, Yogurt, and Cheese (Include skim or lowfat varieties)	2 to 3	1 cup of milk or yogurt, 1-1/2 oz. of hard cheese
Meats Pcuftr" Fish Dry Beans or Peas, Eggs, Nuts (Use lean meats and remove skin from poultry)	2 to 3	2 or 3 oz, of cooked meat, fish, or poultry (TOTAL 6 oz/day) 2 eggs, or 1 cup of cooked beans or peas

Figure 6-2

that is, when the number of calories used equals the number of calories consumed.

The most accurate way to control caloric intake is to control the size of food portions and thus the total amount of food ingested. One can use standard household measuring utensils and a small kitchen scale to measure portions of foods and beverages. Keeping a daily record of all foods eaten and physical activity done is also helpful.

Figure 6-3 shows the number of calories burned during exercise periods of different types, intensities, and durations. For example, while participating in archery, a person will burn 0.034 calories per pound per minute. Thus, a 150-pound person would burn 5.1 calories per minute (150 lbs. x 0.034 calories/minute/lb. = 5.1 calories/minute) or about 305 calories/hour, as shown in Figure 6-4. Similarly, a person running at 6 miles per hour (MPH) will burn 0.079 cal./min./lb. and a typical, 150-pound male will burn 11.85 calories/minute (150 lbs. x 0.079 cal./lb./min. = 11.85) or about 710 calories in one hour, as shown in Figure 6-3.

To estimate the number of calories you use in normal daily activity, multiply your body weight by 13 if you are sedentary, 14 if somewhat active, and 15 if moderately active. The result is a rough estimate of the number of calories you need to maintain your present body weight. You will need still more calories if you are more than moderately active. By comparing caloric intake with caloric expenditure, the state of energy balance (positive, balanced, or negative) can be determined.

CALORIC EXPENDITURE CHART

ACTIVITY	CAL/MIN/LB	CAL/HR/150 LB
Archery	.034	305
Badminton:		
Moderate	.039	350
Vigorous	.065	585
Basketball:		
Moderate	.047	420
Vigorous	.066	595
Baseball:		
Infield-outfield	.031	280
Pitching	.039	350
Bicycling:		
Slow (5 MPH)	.025	225
Moderate (10 MPH)	.05	450
Fast (13 MPH)	.072	650
Bowling	.028	255
Calisthenics:		
General	.045	405
Canoeing:		
2.5 MPH	.023	210
4.0 MPH	.047	420
Dancing:		
Slow	.029	260
Moderate	.045	405
Fast	.064	575
Fencing:		
Moderate	.033	300
Vigorous	.057	515
Fishing	.016	145
Football (tag)	.04	360
Gardening	.024	220
Gardening-Weeding	.039	260
Golf	.029	260
Gymnastics:		
Light	.022	200
Heavy	.056	505
Handball	.063	570
Hiking	.042	375
Hill Climbing	.06	540
Hoeing, Raking, Planting	.031	280
Horseback Riding:		
Walk	.019	175
Trot	.046	415
Gallop	.067	600
Jogging:		
4.5 MPH (13:330 mi.)	.063	565

ACTIVITY	CAL/MIN/LB	CAL/HR/150 LB *
Judo, Karate	.087	785
Motor Boating	.016	145
Mountain Climbing	.086	775
Rowing		
(Rec 2.5 MPH)	.036	325
Vigorous	.118	1000
Running:		
6 MPH (10 min/mi)	.079	710
10 MPH (6 min/mi)	.1	900
12 MPH (5 min/mi)	.13	1170
Sailing	.02	180
Skating:		
Moderate (Rec)	.036	325
Vigorous	.064	575
Skiing (Snow):		
Downhill	.059	530
Level (5 MPH)	.078	700
Soccer	.06	570
Squash	.07	630
Stationary Run:		
70-80 cts/min	.078	705
Strength Training (10 rep circuit)		
60% 1RM	.022	198
80% 1RM	.048	432
Swimming (crawl):		
20 yds/min	.032	290
45 yds/min	.058	520
50 yds/min	.071	640
Table Tennis:		
Moderate	.026	235
Vigorous	.06	540
Tennis:		
Moderate	.046	415
Vigorous	.04	540
Volleyball:		
Moderate	.036	325
Vigorous	.065	585
Walking:		
2.0 MPH	.022	200
3.0 MPH	.03	270
4.0 MPH	.039	350
5.0 MPH	.064	575
Water Skiing	.053	480
Wrestling	.091	820

* A 150-pound person will expend the number of calories indicated in one hour for any given activity.

Figure 6-3

Avoiding an excessive intake of fats is an important fundamental of nutrition.

Avoiding an excessive intake of fats is another fundamental dietary guideline. A high intake of fats, especially saturated fats and cholesterol, has been associated with high levels of blood cholesterol.

The blood cholesterol level in most Americans is too high. Blood cholesterol levels can be lowered by reducing both body fat and the amount of fat in the diet. Lowering elevated blood cholesterol levels reduces the risk of developing coronary artery disease (CAD) and of having a heart attack. CAD, a slow, progressive disease, results from the clogging of blood vessels in the heart. Good dietary habits help reduce the likelihood of developing CAD.

Carbohydrates are the primary fuel source for muscles during short-term, high-intensity activities.

It is recommended that all persons over the age of two should reduce their fat intake to 30 percent or less of their total caloric intake. The current national average is 38 percent. In addition, we should reduce our intake of saturated fat to less than 10 percent of the total calories consumed. We should increase our intake of polyunsaturated fat, but to no more than 10 percent of our total calories. Finally, we should reduce our daily cholesterol intake to 300 milligrams or less. Figure 6-4 suggests actions commanders can take to support sound dietary guidelines. Most of these actions concern dining-facility management.

Concerns for Optimal Physical Performance

Carbohydrates, in the form of gly - cogen (a complex sugar), are the primary fuel source for muscles during short-term, high-intensity activities. Repetitive, vigorous activity can use up most of the carbohydrate stores in the exercised muscles.

The body uses fat to help provide energy for extended activities such as a one-hour run. Initially, the chief fuel burned is carbohydrates, 'but as the duration increases, the contribution from fat gradually increases.

The intensity of the exercise also influences whether fats or carbohydrates are used to provide energy. Very intense activities use more carbohydrates. Examples include weight training and the APFT sit-up and push-up events.

Eating foods rich in carbohydrates helps maintain adequate muscle-gly - cogen reserves while sparing amino acids (critical building-blocks needed for building proteins). At least 50 percent of the calories in the diet should come from carbohydrates. Individual caloric requirements vary, depending on body size, sex, age, and training mission. Foods rich in complex carbohydrates (for example, pasta, rice, whole wheat bread, potatoes) are the best sources of energy for active soldiers.

COMMANDER'S CHECKLIST FOR NUTRITION

PRINCIPLES OF NUTRITION	SUPPORTING ACTIONS
1. **Eat a variety of foods.** No single food item provides all essential nutrients.	**In the dining facility:** • Ensure menus provide foods from the basic 4 food groups fruits and vegetables, meats, dairy products, and breads and cereals. • Establish serving lines in the following order, if possible: (1) salads, (2) fruits, (3) entrees, (4) hot vegetables, (5) breads, (6) beverages, (7) desserts.
2. **Maintain a desirable body weight.** Excess body fat detracts from fitness. Weight loss is achieved by increasing physical activity and decreasing total food intake, especially fats, refined sugars, and alcohol.	**In the dining facility, provide:** • Low-calorie menu, including short-order items at each meal. Use the Master Menu (SB 10-260) menu patterns. • Reduced-portion sizes. • No-calorie beverages. • Low-calorie salad dressings. • Posted list of caloric values of menu items, before or on the serving line.
3. **Avoid excess dietary fat.** Too much fat (especially cholesterol and saturated fat) can lead to heart disease and weight problems. Fats contain twice as many calories as equal amounts of carbohydrates or protein.	**In the dining facility, provide:** • Non-fried eggs as an alternative. • Margarine as a butter alternative. • Two percent milk as the primary milk in bulk dispensers. • Skim milk in 1/2-pint cartons. • Sauces, gravies, and margarine separately from the entree or vegetable. • Avoid animal fats, palm oil, and hydrogenated vegetable oil.
4. **Avoid too much sugar.** Sweets are empty calories and may lead to dental cavities and weight problems.	**In the dining facility, provide:** • Fruit as a dessert alternative. • Unsweetened juices. • No-calorie, unsweetened beverages • Non-nutritive, sugar substitute as a granulated sugar alternative. • Unsweetened cereal.
5. **Eat foods with adequate starch and fiber.** Eating complex carbohydrates adds to the diet and reduces symptoms of constipation.	**In the dining facility, provide:** • Whole-grain breads, cereals and legumes. • Fresh fruit. • Salad bars at lunch and dinner.
6. **Avoid too much sodium.** Eating highly-salted foods may lead to excessive sodium intake. This may be a problem for those "at risk" for high blood pressure.	• Reduce salt in recipes by 25 percent.
7. **If you drink alcoholic beverages, do so in moderation.** Alcoholic beverages are high in calories and and low in nutrients. One or two standard-size drinks daily appears to cause no harm in normal, healthy, nonpregnant adults.	• Avoid alcohol; it is detrimental to good health and weight management.
8. **Know the nutrition principles.** Educating soldiers maximizes efforts to improve nutritional fitness.	• Display educational materials on nutrition: (posters, table tents, bulletin boards, and handouts). • Provide food-service personnel with training programs on nutrition standards. • Provide unit-training programs on nutrition for soldiers. (Use installation dietitian).

Reference: AR 30-1, Appendix J.

Figure 6-4

Because foods eaten one to three days before an activity provide part of the fuel for that activity, it is important to eat foods every day that are rich in complex carbohydrates. It is also important to avoid simple sugars, such as candy, up to 60 minutes before exercising, because they can lead to low blood sugar levels during exercise.

Soldiers often fail to drink enough water, especially when training in the heat. Water is an essential nutrient that is critical to optimal physical performance. It plays an important role in maintaining normal body temperature. The evaporation of sweat helps cool the body during exercise. As a result, water lost through sweating must be replaced or poor performance, and possibly injury, can result. Sweat consists primarily of water with small quantities of minerals like sodium. Cool, plain water is the best drink to use to replace the fluid lost as sweat. Soldiers should drink water before, during, and after exercise to prevent dehydration and help enhance performance. Figure 6-5 shows recommendations for fluid intake when exercising.

Sports drinks, which are usually simple carbohydrates (sugars) and electrolytes dissolved in water, are helpful under certain circumstances. There is evidence that solutions containing up to 10 percent carbohydrate will enter the blood fast enough to deliver additional glucose to the active muscles. This can improve endurance.

During prolonged periods of exercise (1.5+ hours) at intensities over **50** percent of heart rate reserve, one can benefit from periodically drinking sports drinks with a concentration of 5 to 10 percent carbohydrate. Soldiers on extended road marches can also benefit from drinking these types of glucose-containing beverages. During intense training, these beverages can provide a source of carbohydrate for working muscles. On the other hand, drinks that exceed levels of 10 percent carbohydrate, as do regular soda pops and most fruit juices, can lead to abdominal cramps, nausea, and diarrhea. Therefore, these drinks should be used with caution during intense endurance training and other similar activities.

Many people believe that body builders need large quantities of

RECOMMENDATIONS FOR FLUID INTAKE

- Drink cool (40 degrees F) water. This is the best drink to sustain performance. Fluid also comes from juice, milk, soup, and other beverages.
- Do not drink coffee, tea, and soft drinks even though they provide fluids. The caffeine in them acts as a diuretic which can increase urine production and fluid loss. Avoid alcohol for the same reason.
- Drink large quantities (20 oz.) of water one or two hours before exercise to promote hyperhydration. This allows time for adequate hydration and urination.
- Drink three to six ounces of fluid every 15 to 30 minutes during exercise.
- Replace fluid sweat losses by monitoring pre-and post-exercise body weights. Drink two cups of fluid for every pound of weight lost.

Figure 6-5

protein to promote better muscle growth. The primary functions of protein are to build and repair body tissue and to form enzymes. Protein is believed to contribute little, if any, to the total energy requirement of heavy-resistance exercises. The recommended dietary allowance of protein for adults is 0.8 grams per kilogram of body weight. Most people meet this level when about 15 percent of their daily caloric intake comes from protein. During periods of intense aerobic training, one's need for protein might be somewhat higher (for example, 1.0 to 1.5 grams per kilogram of body weight per day). Weight lifters, who have a high proportion of lean body mass, can easily meet their protein requirement with a well-balanced diet which has 15 to 20 percent of its calories provided by protein. Recent research suggests that weight trainers may need no more protein per kilogram of body weight than average, nonathletic people. Most Americans routinely consume these levels of protein, or more. The body converts protein consumed in excess of caloric needs to fat and stores it in the body.

Nutrition in the Field

Soldiers in the field must eat enough food to provide them with the energy they need. They must also drink plenty of water or other non-alcoholic beverages. The "meal, ready to eat" (MRE) supplies the needed amount of carbohydrates, protein, fat, vitamins, and minerals. It is a nutritionally adequate ration when all of its components are eaten and adequate amounts of water are consumed. Because the foods are enriched and fortified with vitamins and minerals, each component is a major source of nutrients. Soldiers must eat all the components in order to get the daily military recommended dietary allowances (MRDA) and have an adequate diet in the field. Soldiers who are in weight control programs or who are trying to lose weight can eat part of each MRE item, as recommended by dietitians.

CHAPTER 7

Circuit Training and Exercise Drills

This chapter gives commanders and trainers guidance in designing and using exercise circuits. It describes calisthenic exercises for developing strength, endurance, coordination, and flexibility. It also describes grass drills and guerilla exercises which are closely related to soldiering skills and should be regularly included in the unit's physical fitness program.

Circuit training is a term associated with specific training routines. Commanders with a good understanding of the principles of circuit training may apply them to a wide variety of training situations and environments.

Circuits

A circuit is a group of stations or areas where specific tasks or exercises are performed.

A circuit is a group of stations or areas where specific tasks or exercises are performed. The task or exercise selected for each station and the arrangement of the stations is determined by the objective of the circuit.

Circuits are designed to provide exercise to groups of soldiers at intensities which suit each person's fitness level. Circuits can promote fitness in a broad range of physical and motor fitness areas. These include CR endurance, muscular endurance, strength, flexibility, and speed. Circuits can also be designed to concentrate on sports skills, soldiers' common tasks, or any combination of these. In addition, circuits can be organized to exercise all **the** fitness components in a short period of time. A little imagination can make circuit training an excellent addition to a unit's total physical fitness program. At the same time, it can provide both fun and a challenge to soldiers' physical and mental abilities. Almost any area can be used, and any number of soldiers can exercise for various lengths of time.

TYPES OF CIRCUITS

The two basic types of circuits are the free circuit and the fixed circuit. Each has distinct advantages.

Free Circuit

In a free circuit, there is no set time for staying at each station, and no signal is given to move from one station to the next. Soldiers work at their own pace, doing a fixed number of repetitions at each station. Progress is measured by the time needed to complete a circuit. Because soldiers may do incomplete or fewer repetitions than called for to reduce this time, the quality and number of the repetitions done should be monitored. Aside from this, the free circuit requires little supervision.

Fixed Circuit

In a fixed circuit, a specific length of time is set for each station. The time is monitored with a stopwatch, and soldiers rotate through the stations on command.

There are three basic ways to increase the intensity or difficulty of a fixed circuit:

- Keep the time for completion the same, but increase the number of repetitions.
- Increase the time per station along with the number of repetitions.
- Increase the number of times soldiers go through the circuit.

VARIABLES IN CIRCUIT TRAINING

Several variables in circuit training must be considered. These include the time, number of stations, number of

time, number of stations, number of soldiers, number of times the circuit is completed, and sequence of stations. These are discussed below.

Time

One of the first things to consider is how long it should take to complete the circuit. When a fixed circuit is run, the time at each station should always be the same to avoid confusion and help maintain control. Consider also the time it takes to move from one station to the next. Further, allow from five to seven minutes both before and after running a circuit for warming up and cooling down, respectively.

Number of Stations

The objective of the circuit and time and equipment available strongly influence the number of stations. A circuit geared for a limited objective (for example, developing lower-body strength) needs as few as six to eight stations. On the other hand, circuits to develop both strength and CR fitness may have as many as 20 stations.

Number of Soldiers

If there are 10 stations and 40 soldiers to be trained, the soldiers should be divided into 10 groups of four each. Each station must then be equipped to handle four soldiers. For example, in this instance a rope jumping station must have at least four jump ropes. It is vital in a free circuit that no soldier stand around waiting for equipment. Having enough equipment reduces bottlenecks, slowdowns, and poor results.

Number of Times a Circuit is Completed

To achieve the desired training effect, soldiers may have to repeat the same circuit several times. For example, a circuit may have ten stations. Soldiers may run through the circuit three times, exercising for 30 seconds at each station, and taking 15 seconds to move between stations. The exercise time at each station may be reduced to 20 seconds the second and third time through. The whole workout takes less than 45 minutes including warm-up and cool-down. As soldiers become better conditioned, exercise periods may be increased to 30 seconds or longer for all three rotations. Another option is to have four rotations of the circuit.

Sequence of Stations

Stations should be arranged in a sequence that allows soldiers some recovery time after exercising at strenuous stations. Difficult exercises can be alternated with less difficult ones. After the warm-up, soldiers can start a circuit at any station and still achieve the objective by completing the full circuit.

DESIGNING A CIRCUIT

The designer of a circuit must consider many factors. The six steps below cover the most important aspects of circuit development.

Determine Objectives

The designer must consider the specific parts of the body and the components of fitness on which soldiers need to concentrate. For example, increasing muscular strength may be the primary objective, while muscular endurance work may be secondary. On the other hand, improving cardiorespiratory endurance may be the top priority. The designer must first identify the training objective in order to choose the appropriate exercises.

The designer must consider the specific parts of the body and the components of fitness on which soldiers need to concentrate.

Select the Activities

The circuit designer should list all the exercises or activities that can help meet the objectives. Then he should look at each item on the list and ask the following questions:

- Will equipment be needed? Is it available?
- Will supervision be needed? Is it available?
- Are there safety factors to consider?

Answering these questions helps the designer decide which exercises to use. He can choose from the exercises, calisthenics, conditioning drills, grass drills, and guerrilla drills described in this chapter. However, he should not limit the circuit to only these activities. Imagination and field expediency are important elements in developing circuits that hold the interest of soldiers. (See Figures 7-1 through 7-3.)

Arrange the Stations

A circuit usually has 8 to 12 stations, but it may have as many as 20. After deciding how many stations to include, the designer must decide how to arrange them. For example, in a circuit for strength training, the same muscle group should not be exercised at consecutive stations.

One approach is to alternate "pushing" exercises with "pulling" exercises which involve movement at the same joint(s). For example, in a strength training circuit, exercisers may follow the pushing motion of a bench press with the pulling motion of the seated row. This could be followed by the pushing motion of the overhead press which could be followed by the pulling motion of the lat pull-down. Another approach might be to alternate between upper and lower body exercises.

The choice of exercises for circuit training depends on the objectives of the circuit.

By not exercising the same muscle group twice in a row, each muscle has a chance to recover before it is used in another exercise. If some exercises are harder than others, soldiers can alternate hard exercises with easier ones. The choice of exercises depends on the objectives of the circuit.

Select the Training Sites

Circuits may be conducted outdoors or indoors. If the designer wants to include running or jogging a certain distance between stations, he may do this in several ways. In the gymnasium, soldiers may run five laps or for 20 to 40 seconds between stations. Outdoors, they may run laps or run between spread-out stations if space is available. However, spreading the stations too far apart may cause problems with control and supervision.

Prepare a Sketch

The designer should draw a simple sketch that shows the location of each station in the training area. The sketch should include the activity and length of time at each station, the number of stations, and all other useful information.

Lay Out the Stations

The final step is to lay out the stations which should be numbered and clearly marked by signs or cards. In some cases, instructions for the stations are written on the signs. The necessary equipment is placed at each station.

Sample Conditioning Circuits

Figures 7-1, 7-2, and 7-3 show different types of conditioning circuits. Soldiers should work at each station 45 seconds and have 15 seconds to rotate to the next station.

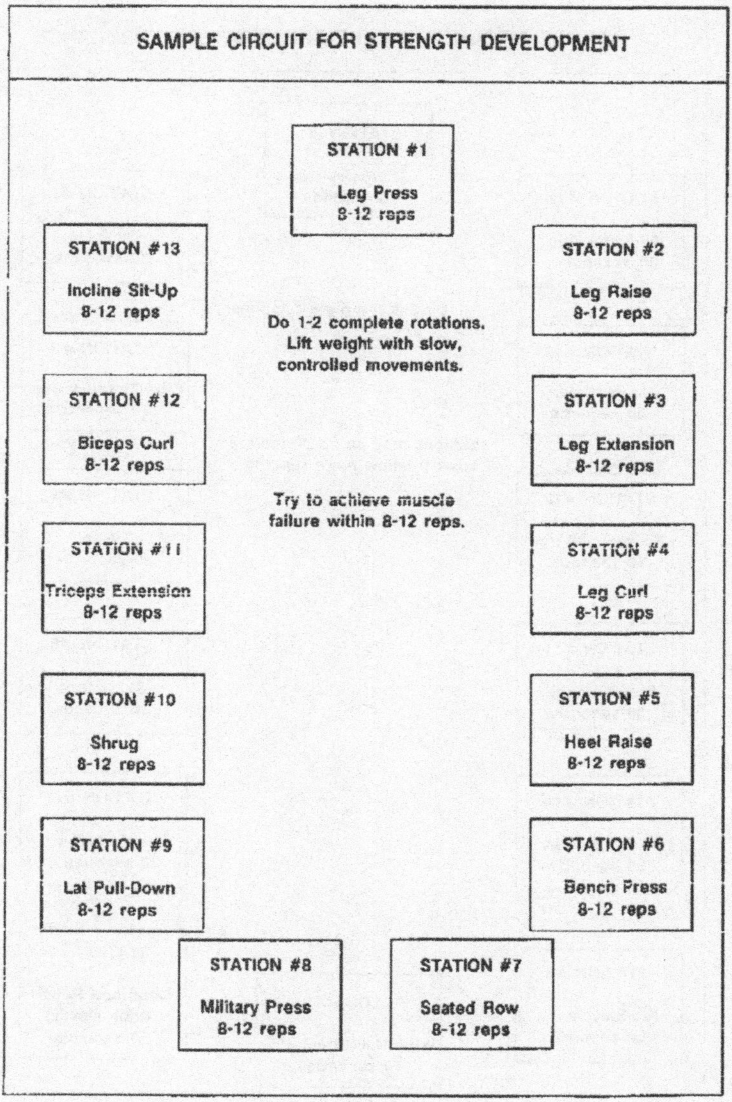
SAMPLE CIRCUIT FOR STRENGTH DEVELOPMENT
STATION #1
Leg Press
8-12 reps
STATION #2
Leg Raise
8-12 reps
STATION #3
Leg Extension
8-12 reps
STATION #4
Leg Curl
8-12 reps
STATION #5
Heel Raise
8-12 reps
STATION #6
Bench Press
8-12 reps
STATION #7
Seated Row
8-12 reps
STATION #8
Military Press
8-12 reps
STATION #9
Lat Pull-Down
8-12 reps
STATION #10
Shrug
8-12 reps
STATION #11
Triceps Extension
8-12 reps
STATION #12
Biceps Curl
8-12 reps
STATION #13
Incline Sit-Up
8-12 reps
Do 1-2 complete rotations.
Lift weight with slow,
controlled movements.
Try to achieve muscle
failure within 8-12 reps.

Figure 7-1

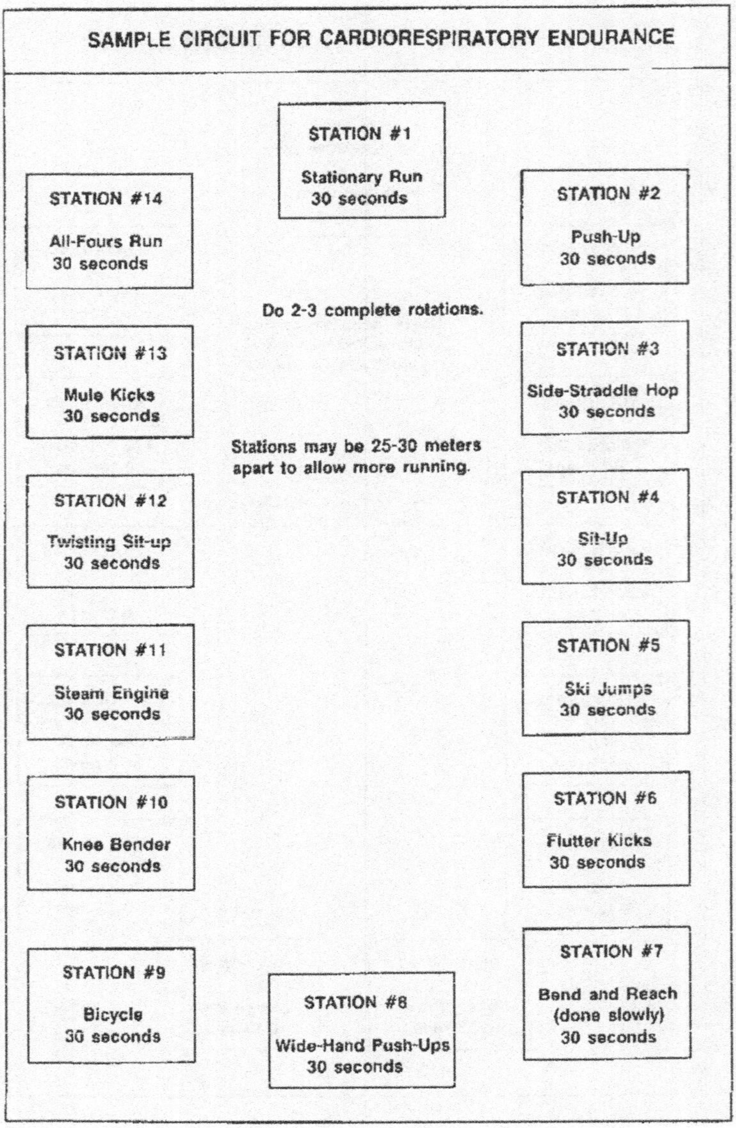
SAMPLE CIRCUIT FOR CARDIORESPIRATORY ENDURANCE
STATION #1
Stationary Run
30 seconds
STATION #2
Push-Up
30 seconds
STATION #3
Side-Straddle Hop
30 seconds
STATION #4
Sit-Up
30 seconds
STATION #5
Ski Jumps
30 seconds
STATION #6
Flutter Kicks
30 seconds
STATION #7
Bend and Reach
(done slowly)
30 seconds
STATION #8
Wide-Hand Push-Ups
30 seconds
STATION #9
Bicycle
30 seconds
STATION #10
Knee Bender
30 seconds
STATION #11
Steam Engine
30 seconds
STATION #12
Twisting Sit-up
30 seconds
STATION #13
Mule Kicks
30 seconds
STATION #14
All-Fours Run
30 seconds
Do 2-3 complete rotations.
Stations may be 25-30 meters
apart to allow more running.

Figure 7-2

SAMPLE CIRCUIT FOR PUSH-UP AND SIT-UP IMPROVEMENT

STATION #1
Elevated Push-Up
30 seconds

STATION #2
Twisting Sit-Up
30 seconds

STATION #3
Parallel Dips
30 seconds

STATION #4
Sit-Up
30 seconds

STATION #5
Wide-Hand Push-Up
30 seconds

STATION #6
Flutter Kick
30 seconds

STATION #7
Close-Hand Push-Up
30 seconds

STATION #8
Bicycle
30 seconds

Do 1-2 complete rotations.

Time may decrease to 20 sec on the second rotation.

Move immediately from station to station. If too fatigued, push-ups may be done on the knees.

Figure 7-3

Calisthenics

Calisthenics can be used to help develop coordination. CR and muscular encurance, flexibility, and strength.

Calisthenics can be used to exercise most of the major muscle groups of the body. They can help develop coordination. CR and muscular endurance, flexibility, and strength. Poorly-coordinated soldiers, however, will derive the greatest benefit from many of these exercises

Although calisthenics have some value when included in a CR circuit or when exercising to music, for the average soldier, calisthenics such as the bend and reach, squat bender, lunger, knee bender, and side-straddle hop can best be used in the warm-up and cool-down periods. Exercises such as the push-up, sit-up, parallel bar dip, and chin-up/pull-up, on the other hand, can effectively be used in the conditioning period to develop muscular endurance or muscular strength.

Please note that exercises such as the bend and reach, lunger, and leg spreader, which were once deleted from FM 21-20 because of their potential risk to the exerciser, have been modified and reintroduced in this edition. All modifications should be strictly adhered to.

Few exercises are inherently unsafe. Nonetheless, some people, because of predisposing conditions or injuries, may find certain exercises less safe than others. Leaders must consider each of their soldier's physical limitations and use good judgment before letting a soldier perform these exercises. However, for the average soldier who is of sound body, following the directions written below will produce satisfactory results with a minimum risk of injury.

Finally, some of the calisthenics listed below may be done in cadence. These calisthenics are noted, and directions are provided below with respect to the actions and cadence. When doing exercises at a moderate cadence, use 80 counts per minute. With a slow cadence, use 50 counts per minute unless otherwise directed.

SAFETY FACTORS

While injury is always possible in any vigorous physical activity, few calisthenic exercises are really unsafe or dangerous. The keys to avoiding injury while gaining training benefits are using correct form and intensity. Also, soldiers with low fitness levels, such as trainees, should not do the advanced exercises highly fit soldiers can do. For example, with the lower back properly supported, flutter kicks are an excellent way to condition the hip flexor muscles. However, without support, the possibility of straining the lower back increases. It is not sensible to have recruits do multiple sets of flutter kicks because they probably are not conditioned for them. On the other hand, a conditioned Ranger company may use multiple sets of flutter' kicks with good results.

The key to doing calisthenic exercises safely is to use common sense. Also, ballistic (that is, quick-moving) exercises that combine rotation and bending of the spine increase the risk of back injury and should be avoided. This is especially true if someone has had a previous injury to the back. If this type of action is performed, slow stretching exercises, not conditioning drills done to cadence, should be used.

Some soldiers complain of shoulder problems resulting from rope climbing, horizontal ladder, wheelbarrow, and crab-walk exercises. These exercises are beneficial when the soldier is fit and he does them in a regular, progressive manner. However, a certain level of muscular strength is needed to do them safely. Therefore, soldiers should progressively train to build up to these exercises. Using such exercises for unconditioned soldiers increases the risk of injury and accident.

Progression and Recovery

Other important principles for avoiding injury are progression and recovery. Programs that try to do too much too soon invite problems. The day after a "hard" training day, if soldiers are working the same muscle groups and/or fitness components, they should work them at a reduced intensity to minimize stress and permit recovery.

The best technique is to train alternate muscle groups and/or fitness components on different days. For example, if the Monday-Wednesday-Friday (M-W-F) training objective is CR fitness, soldiers can do ability group running at THR with some light calisthenics and stretching. If the Tuesday-Thursday (T-Th) objective is muscular endurance and strength, soldiers can benefit from doing partner-resisted exercises followed by a slow run. To ensure balance and regularity in the program, the next week should have muscle endurance and strength development on M-W-F and training for CR endurance on T-Th. Such a program has variety, develops all the fitness components, and follows the seven principles of exercise while, at the same time, it minimizes injuries caused by overuse.

Leaders should plan PT sessions to get a positive training effect, not to conduct "gut checks." They should know how to correctly do all the exercises in their program and teach their soldiers to train using good form to help avoid injuries.

Key Points for Safety

Doing safe exercises correctly improves a soldier's fitness with a minimum risk of injury.

The following are key points for ensuring safety during stretching and calisthenic exercises:

- Stretch slowly and without pain and unnatural stress to a joint. Use static (slow and sustained) stretching for warming up, cooling down, ballistic (bouncy or jerky) stretching movements.
- Do not allow the angle formed by the upper and lower legs to become less than 90 degrees when the legs are bearing weight.
- A combination of spinal rotation and bending should generally be avoided. However, if done, use only slow, controlled movements with little or no extra weight.

Leaders must be aware of the variety of methods they may use to attain their physical training goals. The unit's Master Fitness Trainer is schooled to provide safe, effective training methods and answer questions about training techniques.

CALISTHENIC EXERCISES

The following are some common calisthenic exercises.

SIDE-STRADDLE HOP

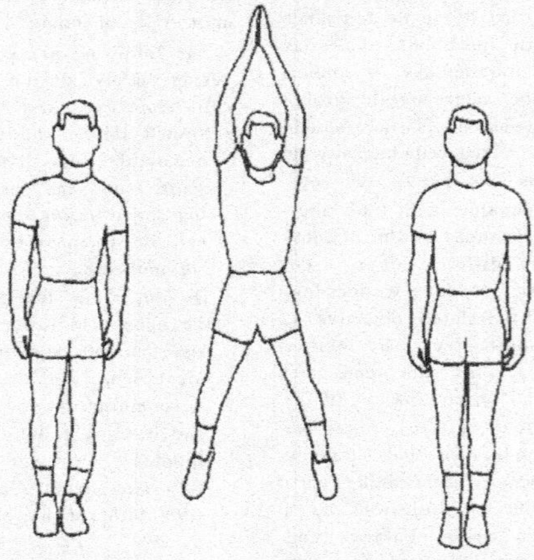

Position: Assume the position of attention.
Action: (1) Jump slightly into the air while moving the legs more than shoulder-width apart, swinging the arms overhead, and clapping the palms together. (2) Jump slightly into the air while swinging the arms sideward and downward and returning to the position of attention. (3) Repeat action 1. (4) Repeat action 2. Use a moderate cadence.

Variation: (1) Jump slightly into the air while moving the left leg forward and the right leg backward, swinging the arms overhead, and clapping the palms together. (2) Jump slightly into the air while swinging the arms sideward and downward and returning to the position of attention. (3) Repeat the jumping and arm movements of action 1 while moving the right leg forward and the left leg backward. (4) Repeat action 2. Use a moderate cadence.

MULE KICK

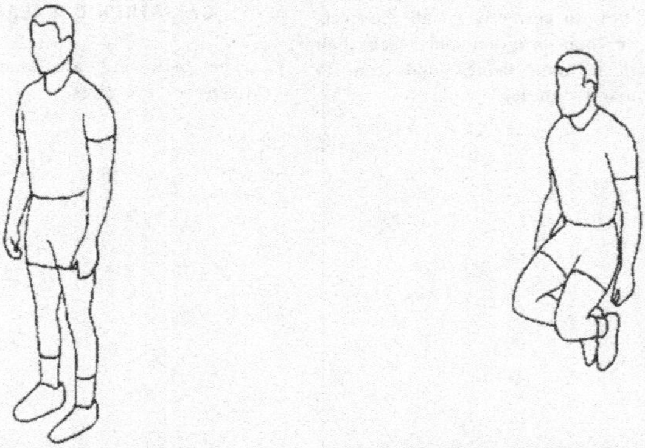

Position: Stand with the feet shoulder-width apart.

Action: Jump up repeatedly while kicking the heels to the buttocks. To do the Mule Kick to cadence, do one repetition per count. Use a moderate cadence.

SKI JUMP

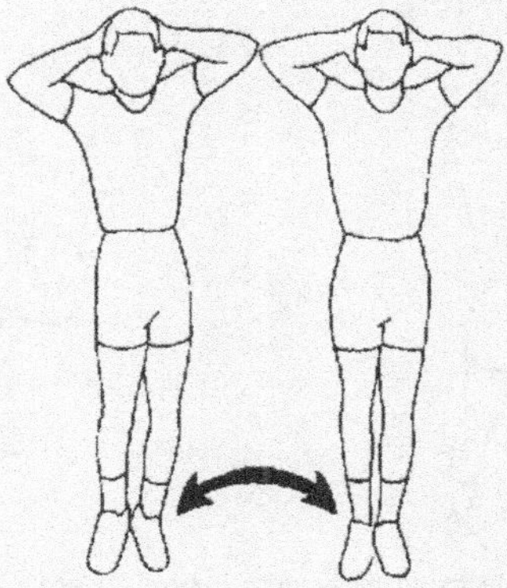

Position: Stand with the feet together, the hands placed behind the head with the fingers interlaced.

Action: (1) Keeping the feet together, jump sideways to the left. (2) Keeping the feet together, jump sideways to the right. (3) Repeat action 1. (4) Repeat action 2. Use a moderate cadence.

FLUTTER KICK

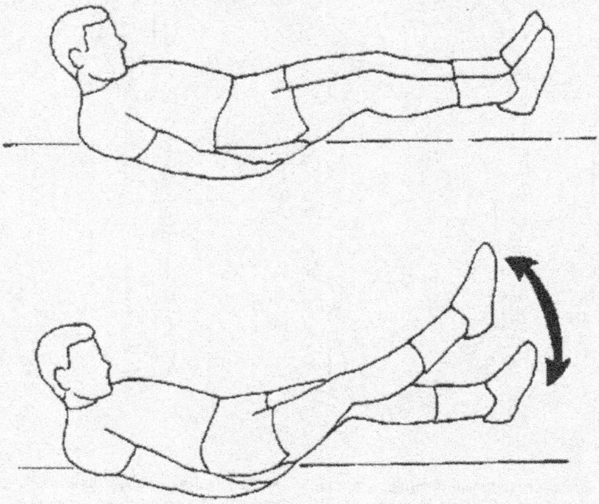

Position: Lie on your back with the hands beneath the buttocks, the head raised, and the knees slightly bent.

Action: Alternately raise and lower the legs, keeping the knees slightly bent and the feet elevated 6 to 18 inches above the floor. To do the flutter kick to cadence, do one repetition per count. Use a moderate cadence.

BEND AND REACH

Position: Stand in a wide, side-straddle position with the palms facing each other and the arms overhead and straight.

Action: (1) Bend at the knees and waist. Slowly bring the arms down, and reach between the legs as far as possible. Make sure the angle formed by the upper and lower leg is never less than 90 degrees. (2) Recover slowly to the start position. (3) Repeat action 1. (4) Repeat action 2. Use a slow cadence.

HIGH JUMPER

Position: Place the feet about shoulder-width apart with the knees flexed. Bend forward at the waist, aligning the arms with the trunk and hips. Keep the arms straight at all times during the exercise. Keep the palms facing each other with the head and eyes initially to the front.

Action: (1) Take a slight jump into the air while swinging the arms forward and up to shoulder level. (2) Take a slight jump while swinging the arms backward, returning to the start position. (3) Jump strongly upward while swinging the arms forward and up to the overhead position; at the same time, briefly look skyward. While descending, return the head and eyes to the front, and flex the knees. (4) Repeat action 2. Use a moderate cadence.

SQUAT BENDER

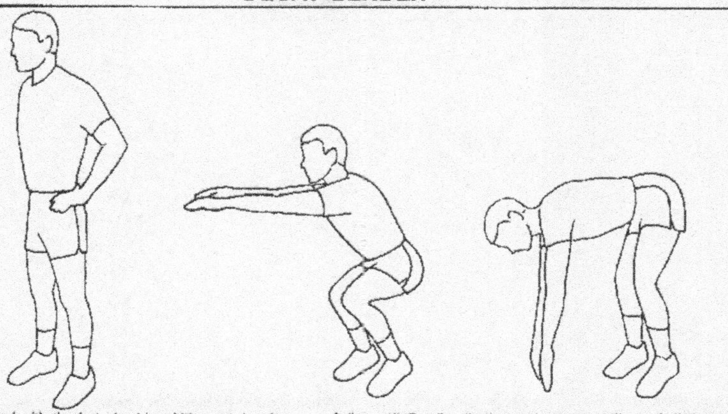

Position: Stand with the feet shoulder-width apart, hands on hips, thumbs in the small of the back, and the elbows back.

Action: (1) Bending the knees, lower yourself to a half-squat position while maintaining balance on the balls of the feet. With the trunk inclined slightly forward, thrust the arms forward to shoulder level with the elbows locked and the palms down. (2) Recover to the start position. (3) Keeping the knees slightly bent, bend forward at the waist, touching the ground in front of the toes. (4) Recover to the start position. Use a moderate cadence.

LUNGER

Position: Start from the position of attention.

Action: (1) Lunge diagonally forward to the left by stepping in that direction with the left foot, placing the left knee over the left foot. At the same time, place the arms sideward at shoulder level, the palms up, and the head and shoulders squarely to the front. (2) Bend slowly forward and downward over the left thigh, and wrap the arms around the thigh, hands grasping the opposite arms above the elbows. (3) Recover slowly to the second position by releasing the arms, straightening the trunk, and extending the arms sideward, palms up. (4) Resume the position of attention by dropping the arms and returning the left foot to the side of the right. Repeat the exercise to the right side. Use a moderate cadence.

KNEE BENDER

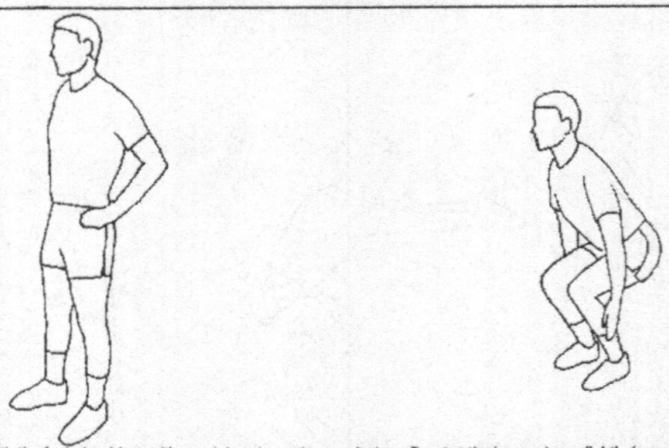

Position: Stand with the feet shoulder-width apart, hands on the hips, the thumbs in the small of the back, and the elbows back.

Action: Bend at the knees, lean slightly forward at the waist with the head up, and slide the hands along the outside of the legs until the extended fingers reach the top of the boots or the middle of the lower leg. (2) Recover to the start position. (3) Repeat action 1. (4) Repeat action 2. Use a moderate cadence.

THE SWIMMER

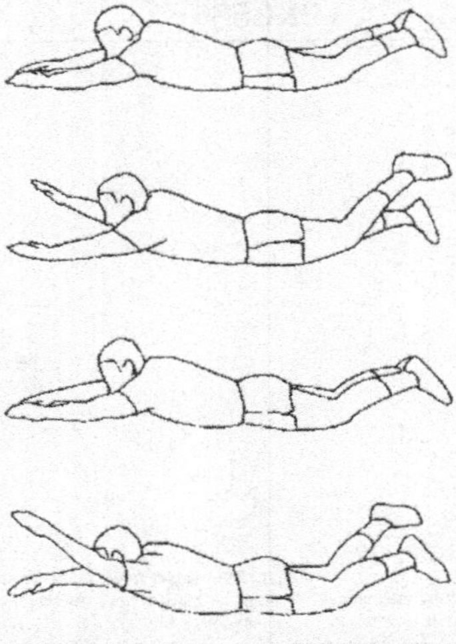

Position: Lie prone with the feet together and with the arms together and extended forward in front of the body. Keep the arms and legs straight at all times during this exercise.

Action: (1) Move the right arm and left leg up. (2) Return to the start position. (3) Move the left arm and right leg up. (4) Return to the start position. Continue in an alternating manner. Use a moderate cadence.

SUPINE BICYCLE

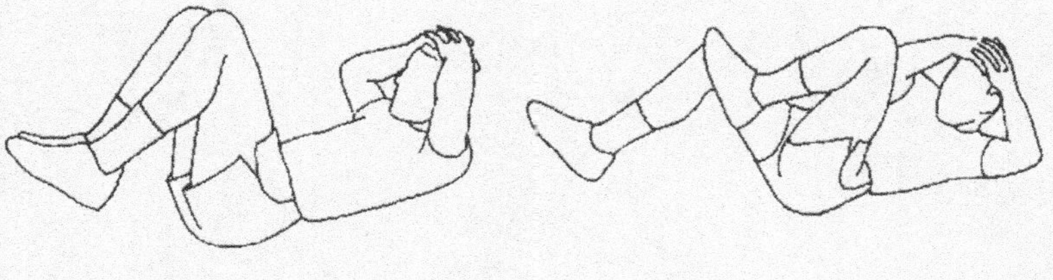

Position: Assume a supine position with the hips and knees flexed. Place the palms directly on top of the head with the fingers interlaced.

Action: (1) Bring the left knee upward while curling the trunk upward, and touch the right elbow to the left knee. (2) Repeat action 1 with the other leg and elbow. (3) Repeat action 1. (4) Repeat action 2. Use a slow cadence.

THE ENGINE

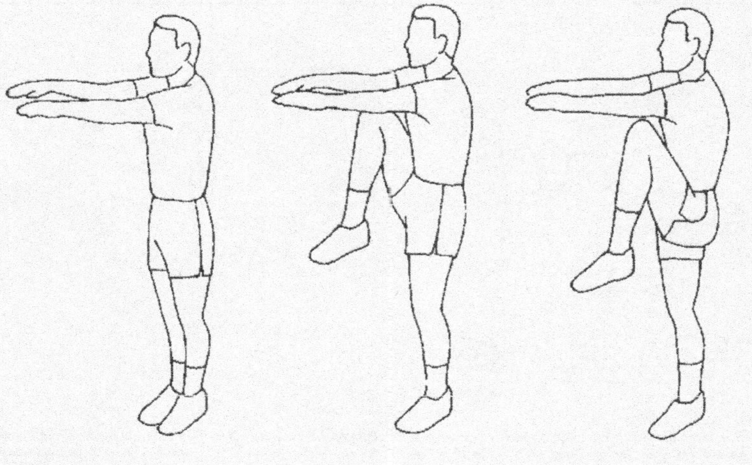

Position: Stand with the arms straight and in front of the body. The arms should be parallel to the ground with the palms facing downward.

Action: (1) Bring the left knee upward to the left elbow. (2) Return to the start position. (3) Touch the right knee to the right elbow. (4) Recover to the start position. Be sure to keep the arms parallel to the ground throughout the entire exercise. Use a moderate cadence.

CROSS-COUNTRY SKIER

Position: Assume a position of attention.
Action: Jump slightly into the air, and move the left foot forward and the right foot backward, landing with both knees slightly bent. At the same time, move the right arm upward and forward to shoulder height and the left arm back as far as possible, always keeping the arms straight and the palms facing each other.
(2) Jump slightly into the air, and move the right foot forward and the left foot backward. At the same time, move the left arm upward and forward to shoulder height and the right arm back as far as possible. (3) Repeat action 1. (4) Repeat action 2. Use a moderate cadence.

PUSH-UP

Position: Assume the front-leaning rest position with the hands placed comfortably apart, the feet together or up to 12 inches apart, and the body forming a generally straight line from the shoulders to the ankles.
Action: Keeping the body straight throughout the exercise, lower the body until the upper arms are at least parallel to the ground. Then, push yourself up to the initial position by completely straightening the arms.

Push-Up Variations: To train the muscles more completely, place the hands at varying widths. They may be wider apart or closer together than shoulder width. Elevating the feet to different heights makes push-ups more difficult. The higher the feet, the more difficult the exercise. Push-ups are also more difficult when the hands and feet are placed on boxes or chairs. This helps the soldier exercise through a fuller range of motion. To do extra repetitions when fatigued, drop to the knees while keeping the knees, hips, and shoulders in a straight line.

SIT-UP

Position: Lie on the back with the feet together or up to 12 inches apart, the knees bent so that an angle of 90 degrees is formed by the upper and lower legs, and the fingers interlocked behind the head.

Action: Raise your upper body forward to the vertical position so that the base of the neck is above the base of the spine, then lower yourself in a controlled manner until the bottom of the shoulder blades touch the ground.

Sit-Up Variations: Variations include keeping the feet elevated and crossing the hands on the chest.

CHIN-UP (PULL-UP)

Position: Hang from a horizontal bar with the arms fully extended.

Action: Bend your elbows and pull yourself upward until your chin is above the bar; do not swing or kick your legs. Return to the starting position in a controlled manner.

Variations: Use overhand (pull-up), underhand (chin-up), or alternating grips, with the hands close together, far apart, or at shoulder-width. If unable to complete a chin-up using proper form, elevate yourself to the up position with help and hang there, or slowly lower yourself to the starting position. Repeat this several times, gradually adding more repetitions from workout to workout.

PARALLEL BAR DIP

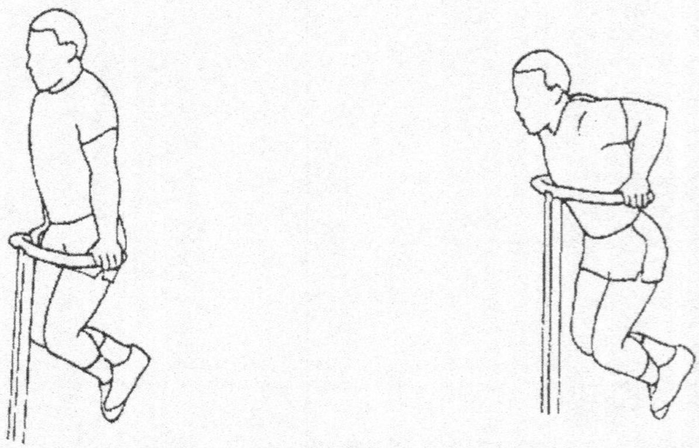

Position: Keep the feet off the floor and support the body's weight on straight arms.

Action: Bend the arms and lower the body in a controlled manner until the upper arms are at least parallel to the floor. If necessary, bend the legs at the knees to keep the feet from touching the floor. Straighten the arms to return to the starting position.

CONDITIONING DRILLS

Conditioning drills are intended to supplement muscular strength and endurance training sessions.

Some large units prefer to use sets of calisthenic exercises as part of their PT sessions. Figure 7-4 shows three calisthenic conditioning drills for both the poorly conditioned and physically fit soldiers. The drills are designed to be done progressively and are intended to supplement muscular strength and endurance training sessions.

Leaders can mix the exercises to provide greater intensity, based on the fitness level of the soldiers being trained. However, they should choose and sequence them to alternate the muscle groups being worked. Soldiers should do each exercise progressively from 15 to 40 or more repetitions (20 to 60 seconds for timed sets) based on their level of conditioning. They may also do each exercise in cadence unless timed sets are specified. For timed sets, soldiers do as many repetitions of an exercise as possible in the allowed time. Using timed sets, both the well-conditioned and less-fit soldiers can work themselves to their limits.

The following conditioning drills (Figure 7-4) are arranged according to the phase of training.

Grass Drills

Grass drills are exercise movements that feature rapid changes in body position. These are vigorous drills which, when properly done, exercise all the major muscle groups. Soldiers should respond to commands as fast as possible and do all movements at top speed. They continue to do multiple repetitions of each exercise until the next command is given. No cadence is counted.

TRAINING-PHASE CONDITIONING DRILLS

#1 PREPARATORY TRAINING

High Jumper
Push-Up (TS* 20-45** seconds)
Sit-Up (TS 20-45** seconds)
Side-Straddle Hop
Side Bender
Knee Bender
Stationary Run

#2 CONDITIONING TRAINING

Push-UP (varied hand positions)
(TS 30-60 seconds)
Supine Bicycle
High Jumper
Sit-Up (all types)
(TS 30-60 seconds)
The Engine or Cross-Country Skiier
All-Fours Run (stationary)

#3 MAINTENANCE TRAINING

Ski Jump
Sit-Ups (all types) (TS 30-60 seconds)
Push-Up (varied hand positions) (TS 30-60 seconds)
Mule Kick
Flutter Kick
The Engine
The Swimmer

*TS = timed set

** Because of a lower level of fitness, 45 seconds will usually be the upper limit.

Figure 7-4

Performing grass drills can improve CR endurance, help develop muscular endurance and strength, and speed up reaction time. Since these drills are extremely strenuous, they should last for short periods (30 to 45 seconds per exercise). The two drills described here each have four exercises. Leaders can develop additional drills locally.

The soldiers should do a warm-up before performing the drills and do a cool-down afterward. The instructor does all the activities so that he can gauge the intensity of the session. The commands for grass drills are given in rapid succession without the usual preparatory commands. To prevent confusion, commands are given sharply to distinguish them from comments or words of encouragement.

As soon as the soldiers are familiar with the drill, they do all the exercises as vigorously and rapidly as possible, and they do each exercise until the

Grass drills are exercise movements that feature rapid changes in body position.

Soldiers should do a warm-up before performing grass drills and do a cool-down afterward.

Progression with grass drills is made by a gradual increase in the time devoted to the drills.

next command is given. Anything less than a top-speed performance decreases the effectiveness of the drills.

Once the drills start, soldiers do not have to resume the position of attention. The instructor uses the command "Up" to halt the drill for instructions or rest. At this command, soldiers assume a relaxed, standing position.

Grass drills can be done in a short time. For example, they may be used when only a few minutes are available for exercise or when combined with another activity. Sometimes, if time is limited, they are a good substitute for running.

Most movements are done in place. The extended-rectangular formation is best for a platoon- or company-sized unit. The circle formation is more suitable for squad- or section-sized groups.

When soldiers are starting an exercise program, a 10- to 15-minute workout may be appropriate. Progression is made by a gradual increase in the time devoted to the drills. As the fitness of the soldiers improves, the times should be gradually lengthened to 20 minutes. The second drill is harder than the first. Therefore, as soldiers progress in the first drill, the instructor should introduce the second. If he sees that the drill needs to be longer, he can repeat the exercises or combine the two drills.

STARTING POSITIONS

After the warm-up, bring the soldiers to a position of ATTENTION. The drills begin with the command *GO*. Other basic commands are FRONT, BACK, and STOP. (See Figure 7-5 for the positions and actions associated with these commands.)

- ATTENTION: The position of attention is described in FM 22-5, Drill and Ceremonies.
- GO This involves running in place at top speed on the balls of the feet. The soldier raises his knees high, pumps his arms, and bends forward slightly at the waist.
- FRONT The soldier lies prone with elbows bent and palms directly under the shoulders as in the down position of the push up. The legs are straight and together with the head toward the instructor.
- BACK: The soldier lies flat on his back with his arms extended along his sides and his palms facing down ward. His legs are straight and together; his feet face the instructor.
- STOP The soldier assumes the stance of a football lineman with feet spread and staggered. His left arm is across his left thigh; his right arm is straight. His knuckles are on the ground; his head is up, and his back is roughly parallel to the ground.

To assume the FRONT or BACK position from the standing GO or STOP positions, the soldier changes positions vigorously and rapidly. (See Figure 7-5.)

To change from the FRONT to the BACK position (Figure 7-5), the soldier does the following:

- Takes several short steps to the right or left.
- Lifts his arm on the side toward which his feet move,
- Thrusts his legs vigorously to the front.

To change from the BACK to the FRONT position, the soldier sits up quickly. He places both hands on the ground to the right or left of his legs. He takes several short steps to the rear on the side opposite his hands. When his feet are opposite his hands, he thrusts his legs vigorously to the rear and lowers his body to the ground. (See Figure 7-5.)

STARTING POSITIONS FOR GRASS DRILLS
GO
FRONT
BACK
STOP
CHANGING FROM FRONT TO BACK
CHANGING FROM BACK TO FRONT

Figure 7-5

GRASS DRILL ONE

Exercises for grass drill one are described below and shown in Figure 7-6.

Bouncing Ball

From the FRONT position, push up and support the body on the hands (shoulder-width apart) and feet. Keep the back and legs generally in line and the knees straight. Bounce up and down in a series of short, simultaneous, upward springs from the hands, hips, and feet.

Supine Bicycle

From the BACK position, flex the hips and knees. Place the palms directly on top of the head, and interlace the fingers. Bring the knee of one leg upward toward the chest. At the same time, curl the trunk and head upward while touching the opposite elbow to the elevated knee. Repeat with the other leg and elbow. Continue these movements as opposite legs and arms take turns.

Knee Bender

From the position of ATTENTION, do half-knee bends with the feet in line and the hands at the sides. Make sure the knees do not bend to an angle less than 90 degrees.

Roll Left and Right

From the FRONT position, continue to roll in the direction commanded until another command is given. Then, return to the FRONT position.

GRASS DRILL TWO

Exercises for grass drill two are described below and shown in Figure 7-6.

The Swimmer

From the FRONT position, extend the arms forward. Move the right arm and left leg up and down; then, move the left arm and right leg up and down. Continue in an alternating manner.

Bounce and Clap Hands

The procedure is almost the same as for the bouncing ball in grass drill one. However, while in the air, clap the hands. This action requires a more vigorous bounce or spring. The push-up may be substituted for this exercise.

Leg Spreader

From the BACK position, raise the legs until the heels are no higher than six inches off the ground. Spread the legs apart as far as possible, then put them back together. Keep the head off the ground. Throughout, place the hands under the upper part of the buttocks, and slightly bend the knees to ease pressure on the lower back. Open and close the legs as fast as possible. The curl-up may be substituted for this exercise.

Forward Roll

From the STOP position, place both hands on the ground, tuck the head, and roll forward. Keep the head tucked while rolling.

Stationary Run

From the position of ATTENTION, start running in place at the GO command by lifting the left foot first. Follow the instructor as he counts two repetitions of cadence. For example, "One, two, three, four; one, two, three, four." The instructor then gives informal commands such as the following: "Follow me," "Run on the toes and balls of your feet," "Speed it up," "Increase to a sprint, raise your knees high, lean forward at your waist, and pump your arms vigorously," and "Slow it down."

To halt the exercise, the instructor counts two repetitions of cadence as the left foot strikes the ground: "One, two, three, four, one, two, three, HALT."

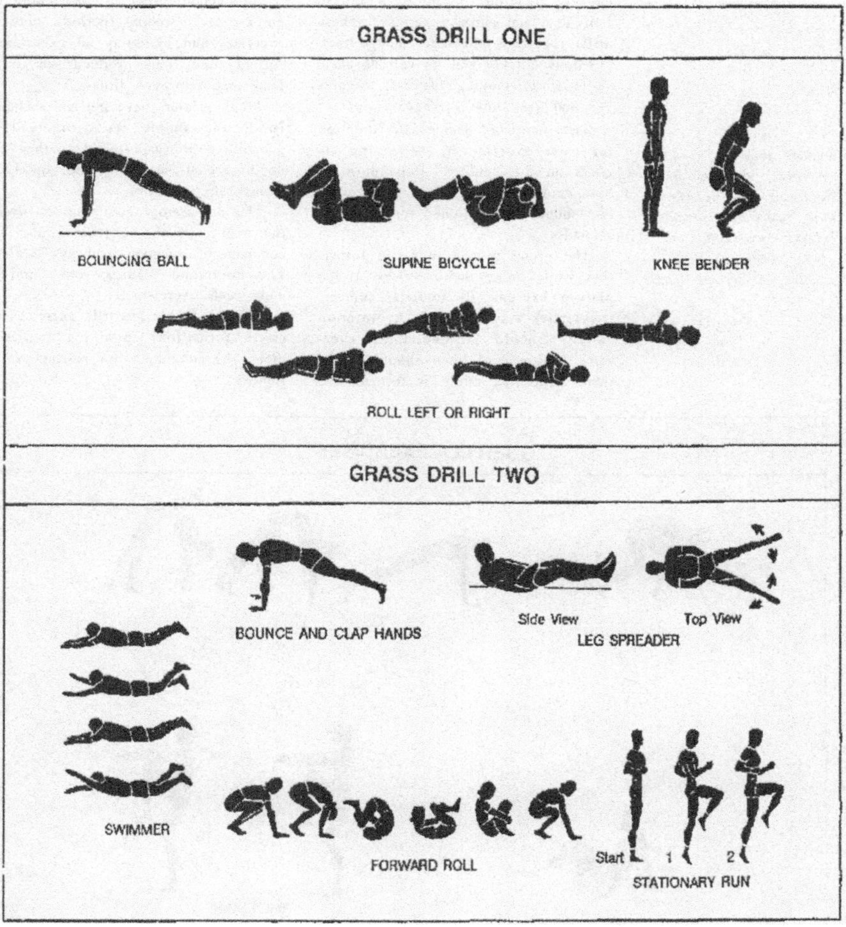

Figure 7-6

Soldiers progress with guerilla exercises by shortening the quick-time marching periods between exercises and by doing all the exercises a second time.

Guerilla Exercises

Guerrilla exercises, which can be used to improve agility, CR endurance, muscular endurance, and to some degree muscular strength, combine individual and partner exercises. These drills require soldiers to change their positions quickly and do various basic skills while moving forward. Figures 7-7 and 7-8 show these exercises.

The instructor decides the duration for each exercise by observing its effect on the soldiers. Depending on how vigorously it is done, each exercise should be continued for 20 to 40 seconds.

The group moves in circle formation while doing the exercises. If the platoon exceeds 30 soldiers, concentric circles may be used. A warm-up activity should precede these exercises, and a cool-down should follow them. After the circle is formed, the instructor steps into the center and issues commands.

EXERCISE AND PROGRESSION

Soldiers progress by shortening the quick-time marching periods between exercises and by doing all exercises a second time. This produces an overload that improves fitness.

Many soldiers have not had a chance to do the simple skills involved in guerrilla exercises. However, they can do these exercises easily and quickly in almost any situation.

The preparatory command is always the name of the exercise, and the command of execution is always "March." The command "Quick time, march" ends each exercise.

For the double guerrilla exercises (in circle formation) involving two soldiers, the commands for pairing are as follows:

Figure 7-7

- "Platoon halt."
- "From (soldier is designated), by twos, count off." (For example: 1-2, 1-2, 1-2.)
- "Even numbers, move up behind odd numbers." (Pairs are adjusted according to height and weight.)
- "You are now paired up for double guerrillas." The command "Change" is given to change the soldiers' positions.

After the exercises are completed, the instructor halts the soldiers and positions the base soldier or platoon guide by commanding, "Base man (or platoon guide), post." He then commands "Fall out and fall in on the base man (or platoon guide)."

EXERCISE DESCRIPTIONS

Brief explanations of guerrilla exercises follow.

All-Fours Run

Face downward, supporting the body on the hands and feet. Advance forward as fast as possible by moving the arms and legs forward in a coordinated way.

Bottoms-Up Walk

Take the front-leaning rest position, and move the feet toward the hands in short steps while keeping the knees locked. When the feet are as close to the hands as possible, walk forward on the hands to the front-leaning-rest position.

Crab Walk

Assume a sitting position with the hips off the ground and hands and feet supporting the body's weight. Walk forward, feet first.

The Engine

Stand with the arms straight and in front of the body. The arms should be parallel to the ground with the palms facing downward. While walking forward, bring the left knee upward to the left elbow. Return to the start position. Continuing to walk forward, touch the right knee to the right elbow. Recover to the start position. Be sure to keep the arms parallel to the ground throughout the entire exercise.

Double Time

Do a double-time run while maintaining the circle formation.

Broad Jump

Jump forward on both feet in a series of broad jumps. Swing the arms vigorously to help with the jumps.

Straddle Run

Run forward, leaping to the right with the left foot and to the left with the right foot.

Hobble Hopping

Hold one foot behind the back with the opposite hand and hop forward. On the command "Change," grasp the opposite foot with the opposite hand and hop forward.

Two-Man Carry

For two-man carries, soldiers are designated as number one (odd-numbered) and number two (even-numbered). A number-one and number-two soldier work as partners.

Fireman's Carry

Two soldiers do the carry. On command, number-two soldier bends at the waist, with feet apart in a balanced stance. Number-one soldier moves toward his partner. He places himself by his partner's left shoulder and bends himself over his partner's shoulders and back. When in position, number-two soldier, with his left hand, reaches between his partner's legs and grasps his left wrist. On command, they move forward until the command for changeover. They then change positions. The fireman's carry can also be done from the other side,

Single-Shoulder Carry

Two soldiers do the carry. On command, number-two soldier bends at the waist with feet apart in a balanced stance. At the same time, number-one soldier moves toward his partner. He places his abdominal area onto his partner's right or left shoulder and leans over. Number-two soldier puts his arms around the back of his partner's knees and stands up. On command, they move forward until the command for changeover. They then change positions.

Cross Carry

On command, number-two soldier bends over at the waist. He twists slightly to the left with feet spread apart in a balanced position. At the same time, number-one soldier moves toward his partner's left side and leans over his partner's back. Number two soldier, with his left arm, reaches around his partner's legs. At the same time, he reaches around his partner's back with his right arm, being careful not to grab his partner's neck or head. He then stands up straight, holding his partner on his back. On command, they move forward until the command for changeover. They then change positions.

Saddle-Back (Piggyback) Carry

On command, number-two soldier bends at the waist and knees with his hand on his knees and his head up. To assume the piggyback position, number-one soldier moves behind his partner, places his hands on his partner's shoulders, and climbs carefully onto his partner's hips. As number-one soldier climbs on, number-two soldier grasps his partner's legs to help support him. Number-one soldier places his arms over his partner's shoulders and crosses his hands over his partner's upper chest. They move forward until the command for changeover is given. They then change positions.

ADDITIONAL GUERILLA EXERCISES
DOUBLE TIME
BROAD JUMP
STRADDLE RUN
HOBBLE HOPPING
FIREMAN'S CARRY
SINGLE-SHOULDER CARRY
CROSS CARRY
SADDLE-BACK CARRY

Figure 7-8

CHAPTER 8

Obstacle Courses and Additional Drills

There are two types of obstacle courses-conditioning and confidence.

This chapter describes obstacle courses as well as rifle drills, log drills, and aquatic exercises. These are not designed to develop specific components of physical fitness. Commanders should use them to add variety to their PT programs and to help soldiers develop motor fitness including speed, agility, coordination, and related skills and abilities. Many of these activities also give soldiers the chance to plan strategy, make split-second decisions, learn teamwork, and demonstrate leadership.

Obstacle Courses

Physical performance and success in combat may depend on a soldier's ability to perform skills like those required on the obstacle course. For this reason, and because they help develop and test basic motor skills, obstacle courses are valuable for physical training.

There are two types of obstacle courses-conditioning and confidence. The conditioning course has low obstacles that must be negotiated quickly. Running the course can be a test of the soldier's basic motor skills and physical condition. After soldiers *receive* instruction and practice the skills, they run the course against time.

A confidence course has higher, more difficult obstacles than a conditioning course. It gives soldiers confidence in their mental and physical abilities and cultivates their spirit of daring. Soldiers are encouraged, but not forced, to go through it. Unlike conditioning courses, confidence courses are not run against time.

NONSTANDARD COURSES AND OBSTACLES

Commanders may build obstacles and courses that are nonstandard (that is, not covered in this manual) in order to create training situations based on their unit's METL. When planning and building such facilities, designers should, at a minimum, consider the following guidance:

- Secure approval from the local installation's commander.
- Prepare a safety and health-risk assessment to support construction of each obstacle.
- Coordinate approval for each obstacle with the local or supporting safety office. Keep a copy of the approval in the permanent records.
- Monitor and analyze all injuries.
- Inspect all existing safety precautions on-site to verify their effectiveness.
- Review each obstacle to determine the need for renewing its approval.

SAFETY PRECAUTIONS

Instructors must always be alert to safety. They must take every precaution to minimize injuries as soldiers go through obstacle courses. Soldiers must do warm-up exercises before they begin. This prepares them for the physically demanding tasks ahead and helps minimize the chance of injury. A cool-down after the obstacle course is also necessary, as it helps the body recover from strenuous exercise.

Commanders should use ingenuity in building courses, making good use of streams, hills, trees, rocks, and other natural obstacles. They must inspect courses for badly built obstacles, protruding nails, rotten logs, unsafe landing pits, and other safety hazards.

There are steps which designers can take to reduce injuries. For example, at the approach to each obstacle, they should post an instruction board or sign with text and pictures showing how to negotiate it. Landing pits for jumps or vaults, and areas under or around obstacles where soldiers may fall from a height, should be filled with loose sand or sawdust. All landing areas should be raked and refilled before each use. Puddles of water under obstacles can cause a false sense of security. These could result in improper landing techniques and serious injuries. Leaders should postpone training on obstacle courses when wet weather makes them slippery.

Units should prepare their soldiers to negotiate obstacle courses by doing conditioning exercises beforehand. Soldiers should attain an adequate level of conditioning before they run the confidence course. Soldiers who have not practiced the basic skills or run the conditioning course should not be allowed to use the confidence course.

Instructors must explain and demonstrate the correct ways to negotiate all obstacles before allowing soldiers to run them. Assistant instructors should supervise the negotiation of higher, more dangerous obstacles. The emphasis is on avoiding injury. Soldiers should practice each obstacle until they are able to negotiate it. Before they run the course against time, they should make several slow runs while the instructor watches and makes needed corrections. Soldiers should never be allowed to run the course against time until they have practiced on all the obstacles.

CONDITIONING OBSTACLE COURSES

If possible, an obstacle course should be shaped like a horseshoe or figure eight so that the finish is close to the start. Also, signs should be placed to show the route.

A course usually ranges from 300 to 450 yards and has 15 to 25 obstacles that are 20 to 30 yards apart. The obstacles are arranged so that those which exercise the same groups of muscles are separated from one another.

The obstacles must be solidly built. Peeled logs that are six to eight inches wide are ideal for most of them. Sharp points and corners should be eliminated, and landing pits for jumps or vaults must be filled with sand or sawdust. Courses should be built and marked so that soldiers cannot sidestep obstacles or detour around them. Sometimes, however, courses can provide alternate obstacles that vary in difficulty.

Each course should be wide enough for six to eight soldiers to use at the same time, thus encouraging competition. The lanes for the first few obstacles should be wider and the obstacles easier than those that follow. In this way, congestion is avoided and soldiers can spread out on the course. To minimize the possibility of falls and injuries due to fatigue, the last two or three obstacles should not be too difficult or involve high climbing.

Trainers must always be aware that falls from the high obstacles could cause serious injury. Soldiers must be in proper physical condition, closely supervised, and adequately instructed.

Instructors must explain and demonstrate the correct ways to negotiate all obstacles before allowing soldiers to run them.

The best way for the timer to time the runners is to stand at the finish and call out the minutes and seconds as each soldier finishes. If several watches are available, each wave of soldiers is timed separately. If only one watch is available, the waves are started at regular intervals such as every 30 seconds. If a soldier fails to negotiate an obstacle, a previously determined penalty is imposed.

When the course is run against time, stopwatches, pens, and a unit roster are needed. Soldiers may run the course with or without individual equipment.

Obstacles for Jumping

These obstacles are ditches to clear with one leap, trenches to jump into, heights to jump from, or hurdles. (See Figure 8-1.)

Obstacles for Dodging

These obstacles are usually mazes of posts set in the ground at irregular intervals. (See Figure 8-2.) The spaces between the posts are narrow so that soldiers must pick their way carefully through and around them. Lane guides are [illegible] to guide soldiers in dodging and changing direction.

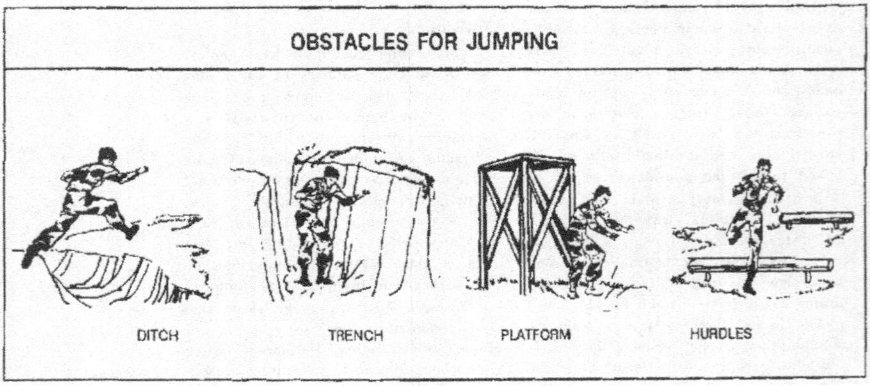

Figure 8-1

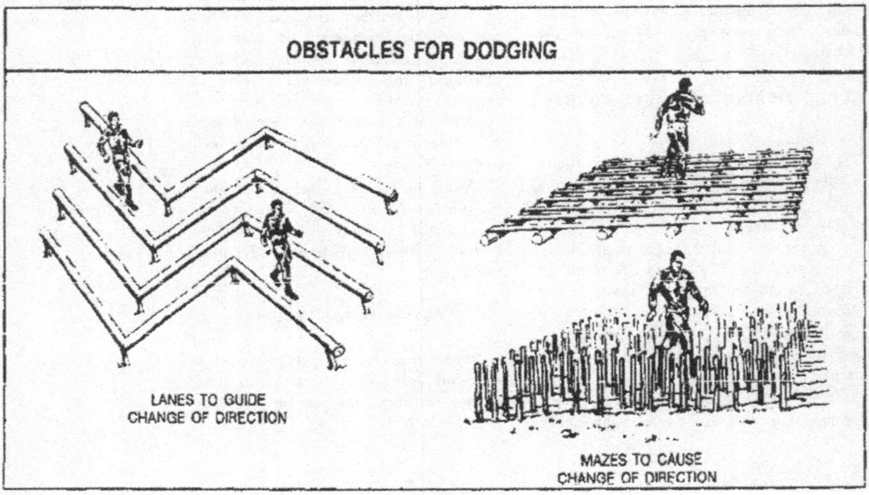

Figure 8-2

Obstacles for Vertical Climbing and Surmounting

These obstacles are shown at Figure 8-3 and include the following:

- Climbing ropes that are 1 1/2 inches wide and either straight or knotted.
- Cargo nets.
- Walls 7 or 8 feet high.
- Vertical poles 15 feet high and 6 to **8** inches wide.

Obstacles for Horizontal Traversing

Horizontal obstacles may be ropes, ,i,,s, or beams. (See Figure 8-4.)

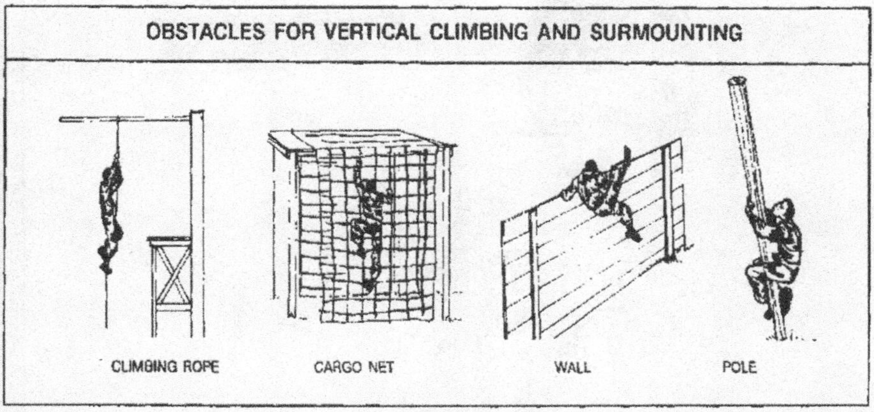

Figure 8-3

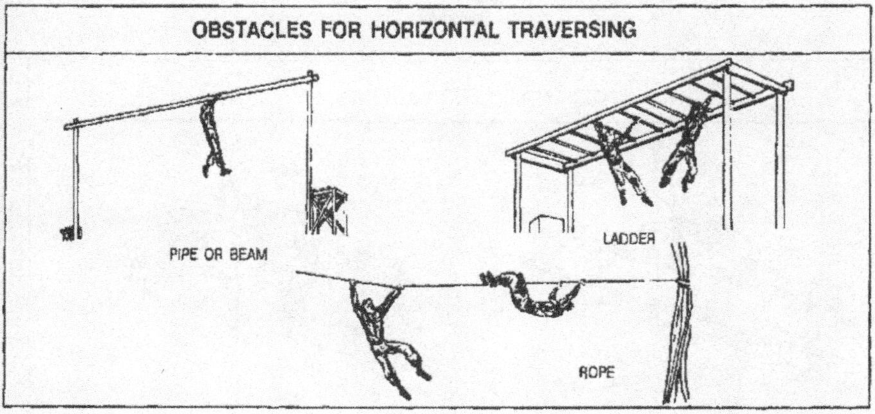

Figure 8-4

Obstacles for Crawling

These obstacles may be built of large pipe sections, low rails, or wire. (See Figure 8-5.)

Obstacles for Vaulting

These obstacles should be 3 to 3 1/2 feet high. Examples are fences and low walls. (See Figure 8-6.)

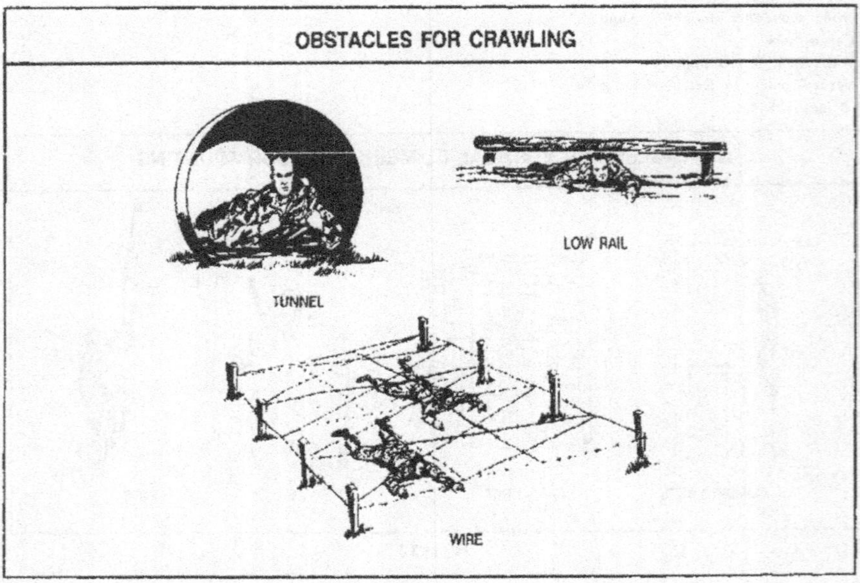

Figure 8-5

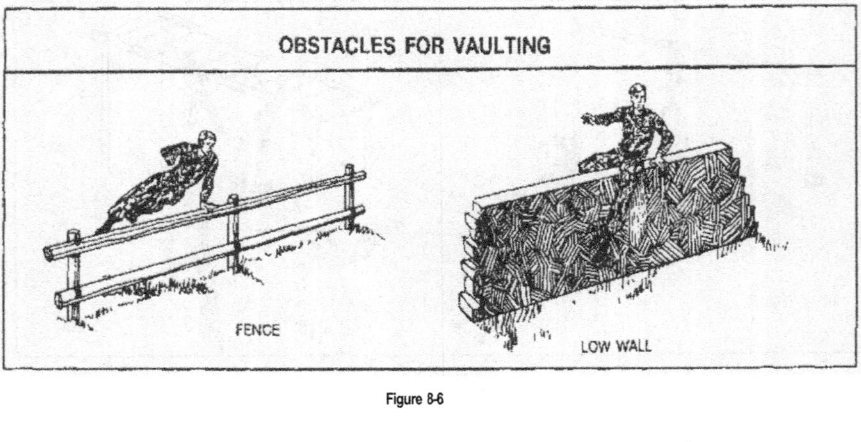

Figure 8-6

Obstacles for Balancing

Beams, logs, and planks may be used. These may span water obstacles and dry ditches, or they may be raised off the ground to simulate natural depressions. (See Figure 8-7.)

OBSTACLES FOR BALANCING

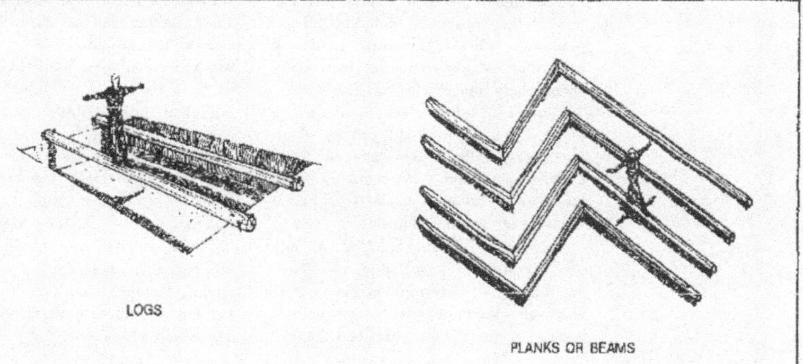

Figure 8-7

CONFIDENCE OBSTACLE COURSES

Confidence obstacle courses must be built in accordance with Folio No. 1, "Training Facilities," Corps of Engineers Drawing Number 28-13-95, You can obtain this publication from the Directorate of Facilities Engineering at most Army installations.

Confidence courses can develop confidence and strength by using obstacles that train and test balance and muscular strength. Soldiers do not negotiate these obstacles at high speed or against time. The obstacles vary from fairly easy to difficult, and some are high. For these, safety nets are provided. Soldiers progress through the course without individual equipment. Only one soldier at a time negotiates an obstacle unless it is designed for use by more than one.

Confidence courses should accommodate four platoons, one at each group of six obstacles. Each platoon begins at a different starting point. In the example below, colors are used to group the obstacles. Any similar method may be used to spread a group over the course. Soldiers are separated into groups of 8 to 12 at each obstacle. At the starting signal, they proceed through the course.

Soldiers may skip any obstacle they are unwilling to try. Instructors should encourage fearful soldiers to try the easier obstacles first. Gradually, as their confidence improves, they can take their places in the normal rotation. Soldiers proceed from one obstacle to the next until time is called. They then assemble and move to the next group of obstacles.

Rules for the Course

Supervisors should encourage, but not force, soldiers to try every obstacle. Soldiers who have not run the course before should receive a brief orientation at each obstacle, including an explanation and demonstration of the best way to negotiate it. Instructors should help those who have problems. Trainers and soldiers should not try to make obstacles more difficult by shaking ropes, rolling logs, and so forth. Close supervision and common sense must be constantly used to enhance safety and prevent injuries.

Soldiers need not conform to any one method of negotiating obstacles, but there is a uniformity in the general approach. Recommended ways to negotiate obstacles are described below.

Red Group

This group contains the first six obstacles. These are described below and numbered 1 through 6 in Figure 8-8, Belly Buster. Soldiers vault, jump, or climb over the log. They must be warned that it is not stationary. Therefore, they should not roll or rock the log while others are negotiating it.

Reverse Climb. Soldiers climb the reverse incline and go down the other side to the ground.

Weaver. Soldiers move from one end of the obstacle to the other by weaving their bodies under one bar and over the next.

Hip-Hip. Soldiers step over each bar; they either alternate legs or use the same lead leg each time.

Balancing Logs. Soldiers step up on a log and walk or run along it while keeping their balance.

Island Hopper. Soldiers jump from one log to another until the obstacle is negotiated.

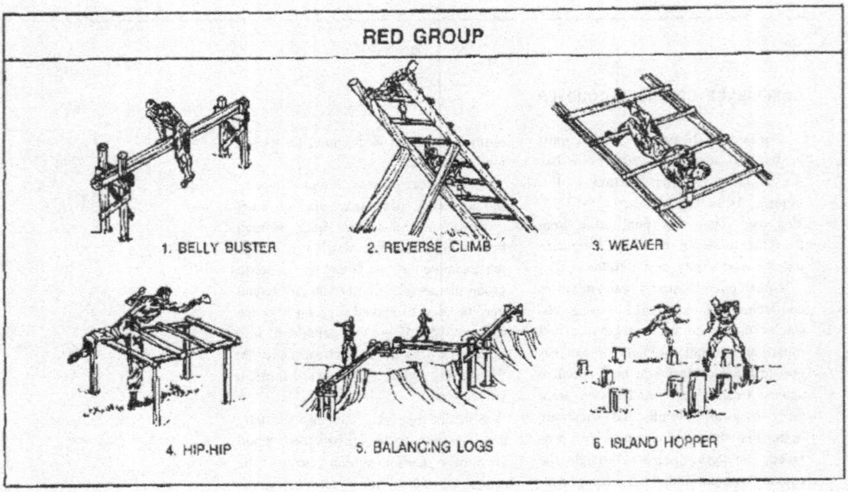

Figure 8-8

White Group

This group contains the second six obstacles. These are described below and numbered 7 through 12 in Figure 8-9.

Tough Nut. Soldiers step over each X in the lane.

Inverted Rope Descent. Soldiers climb the tower, grasp the rope firmly, and swing their legs upward. They hold the rope with their legs to distribute the weight between their legs and arms. Braking the slide with their feet and legs, they proceed down the rope. Soldiers must be warned that they may get rope burns on their hands. This obstacle can be dangerous when the rope is slippery. Soldiers leave the rope at a clearly marked point of release. Only one soldier at a time is allowed on the rope. Soldiers should not shake or bounce the ropes. This obstacle requires two instructors—one on the platform and the other at the base.

Low Belly-Over. Soldiers mount the low log and jump onto the high log. They grasp over the top of the log with both arms, keeping the belly area in contact with it. They swing their legs over the log and lower themselves to the ground.

Belly Crawl. Soldiers move forward under the wire on their bellies to the end of the obstacle. To reduce the tendency to push the crawling surface, it is filled with sand or sawdust to the far end of the obstacle. The direction of negotiating the crawl is reversed from time to time.

Easy Balancer. Soldiers walk up one inclined log and down the one on the other side to the ground.

Tarzan. Soldiers mount the lowest log, walk the length of it, then each higher log until they reach the horizontal ladder. They grasp two rungs of the ladder and swing themselves into the air. They negotiate the length of the ladder by releasing one hand at a time and swinging forward, grasping a more distant rung each time.

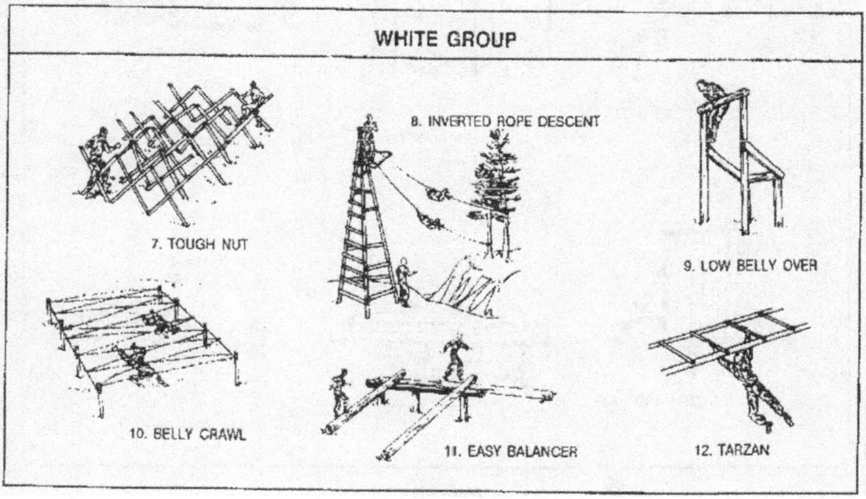

Figure 8-9

Blue Group

This group contains the third six obstacles. These are described below and numbered 13 through 18 in Figure 8-10.

High **Step-over. Soldiers step over** each log while alternating their lead foot or using the same one.

Swinger. Soldiers climb over the swing log to the ground on the opposite side.

Low Wire. Soldiers move under the wire on their backs while raising the wire with their hands to clear their bodies. To reduce the tendency to push the crawling surface, it is filled with sand or sawdust to the far end of the obstacle. The direction of negotiating the obstacle is alternated.

Swing, Stop, and Jump. Soldiers gain momentum with a short run, grasp the rope, and swing their bodies forward to the top of the wall. They release the rope while standing on the wall and jump to the ground.

Six Vaults. Soldiers vault over the logs using one or both hands.

Wall Hanger. Soldiers walk up the wall using the rope. From the top of the wall, they grasp the bar and go hand-over-hand to the rope on the opposite end. They use the rope to descend.

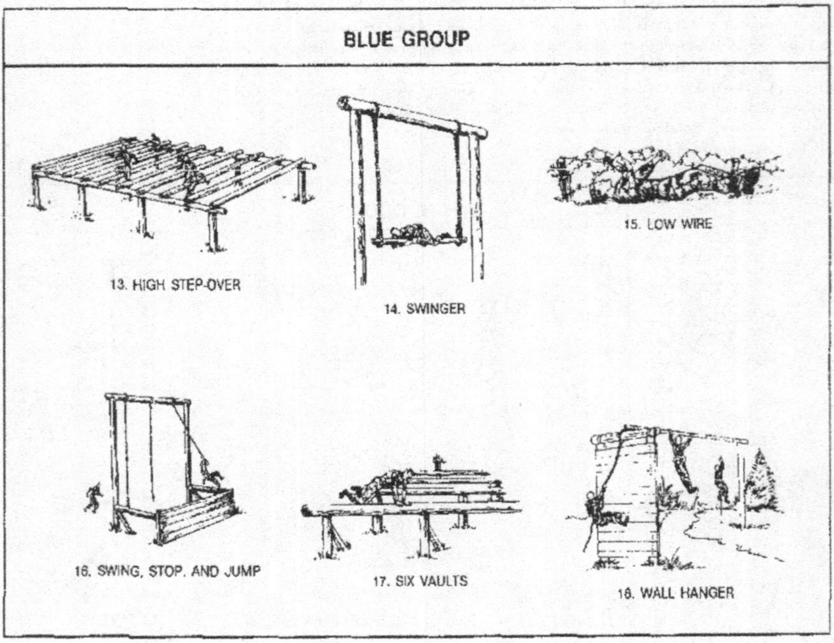

Figure 8-10

Black Group

This group contains the last six obstacles. These are described below and numbered 19 through 24 in Figure 8-II.

Inclining Wall. Soldiers approach the underside of the wall, jump up and grasp the top, and pull themselves up and over. They slide or jump down the incline to the ground.

Skyscraper. Soldiers jump or climb to the first floor and either climb the corner posts or help one another to the higher floors. They descend to the ground individually or help one another down. The top level or roof is off limits, and the obstacle should not be overloaded. A floor must not become so crowded that soldiers are bumped off. Soldiers should not jump to the ground from above the first level.

Jump and Land. Soldiers climb the ladder to the platform and jump to the ground.

Confidence Climb. Soldiers climb the inclined ladder to the vertical ladder, they go to the top of the vertical ladder, then down the other side to the ground.

Belly Robber. Soldiers step on the lower log and take a prone position on the horizontal logs. They crawl over the logs to the opposite end of the obstacle. Rope gaskets must be tied to the ends of each log to keep the hands from being pinched and the logs from falling.

The Tough One. Soldiers climb the rope or pole on the lowest end of the obstacle. They go over or between the logs at the top of the rope. They move across the log walkway, climb the ladder to the high end, then climb down the cargo net to the ground.

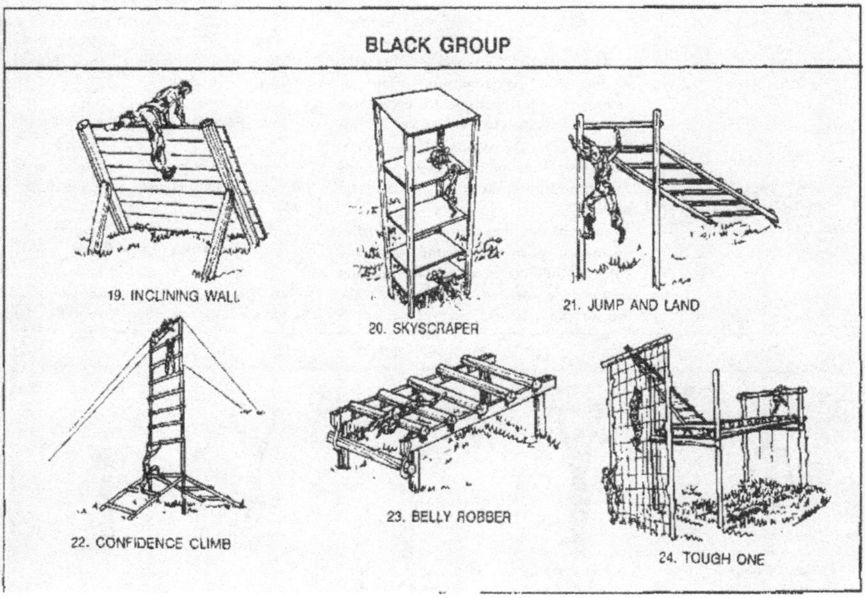

Figure 8-11

Rifle Drills

Rifle drills are suitable activities for fitness training while bivouacking or during extended time in the field. In most situations, the time consumed in drawing weapons makes this activity cumbersome for garrison use. However, it is a good conditioning activity, and the use of individual weapons in training fosters a warrior's spirit.

There are four rifle-drill exercises that develop the upper body. They are numbered in a set pattern. The main muscle groups strengthened by rifle drills are those of the arms, shoulders, and back.

Rifle drill is a fast-moving method of exercising that soldiers can do in as little as 15 minutes. With imagination, the number of steps and/or rifle exercises can be expanded beyond those described here.

EXERCISE PROGRESSION

The rifle-drill exercise normally begins with six repetitions and increases by one repetition for each three periods of exercise. This rate continues until soldiers can do 12 repetitions. However, the number of repetitions can be adjusted as the soldiers improve.

In exercises that start from the rifle-downward position, on the command "Move," soldiers execute port arms and assume the starting position. At the end of the exercise, the command to return soldiers to attention is "Position of attention, move."

In exercises that end in other than the rifle-downward position, soldiers assume that position before executing port arms and order arms.

These movements are done without command and need not be precise. Effective rifle exercises are strenuous enough to tire the arms. When the arms are tired, moving them with precision is difficult.

RIFLE DRILL EXERCISES

The following exercises are for use in rifle drills.

Up and Forward

This is a four-count exercise done at a fast cadence. (See Figure 8-12.)

Fore-Up, Squat

This is a four-count exercise done at a moderate cadence. (See Figure 8-13.)

Fore-Up, Behind Back

This is a four-count exercise done at a moderate cadence. (See Figure 8-14.)

Fore-Up, Back Bend

This is a four-count exercise done at moderate cadence. (See Figure 8- 15.)

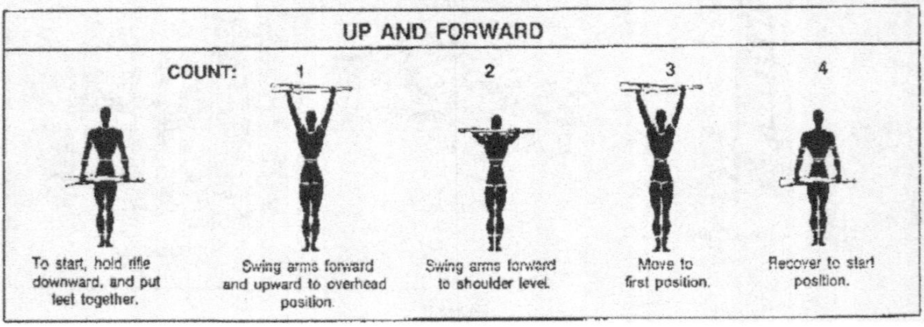

Figure 8-12

Figure 8-13

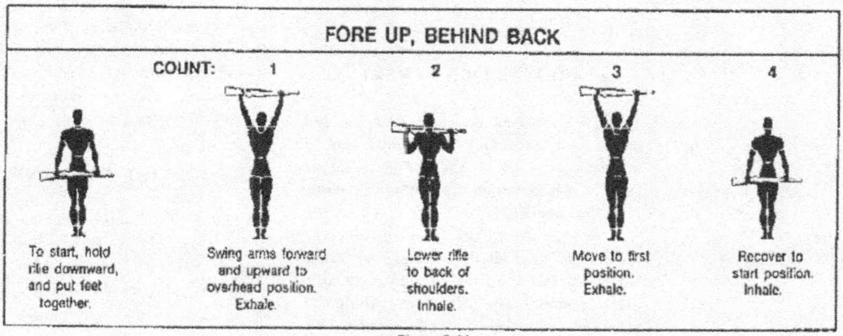

Figure 8-14

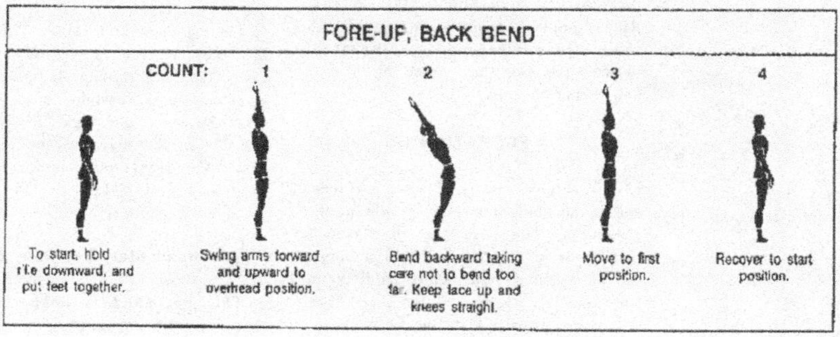

Figure 8-15

Log Drills

Log drills are excellent for developing strength and muscular endurance, because they require the muscles to contract under heavy loads.

Log drills are team-conditioning exercises. They are excellent for developing strength and muscular endurance because they require the muscles to contract under heavy loads. They also develop teamwork and add variety to the PT program.

Log drills consist of six different exercises numbered in a set pattern. The drills are intense, and teams should complete them in 15 minutes. The teams have six to eight soldiers per team. A principal instructor is required to teach, demonstrate, and lead the drill. He must be familiar with leadership techniques for conditioning exercises and techniques peculiar to log drills.

AREA AND EQUIPMENT

Any level area is good for doing log drills. All exercises are done from a standing position. If the group is larger than a platoon, an instructor's stand may be needed.

The logs should be from six to eight inches thick, and they may vary from 14 to 18 feet long for six and eight soldiers, respectively. The logs should be stripped, smoothed, and dried. The 14-foot logs weigh about 300 pounds, the 18-foot logs about 400 pounds. Rings should be painted on the logs to show each soldier's position. When not in use, the logs are stored on a rack above the ground.

FORMATION

All soldiers assigned to a log team should be about the same height at the shoulders. The best way to divide a platoon is to have them form a single file or column with short soldiers in front and tall soldiers at the rear. They take their positions in the column according to shoulder height, not head height. When they are in position, they are divided into teams of six or eight. The command is "Count off by sixes (or eights), count off." Each team, in turn, goes to the log rack, shoulders a log, and carries it to the exercise area.

The teams form columns in front of the instructor. Holding the logs in chest position, they face the instructor and ground the log. Ten yards should separate log teams within the columns. If more than one column is used, 10 yards should separate columns.

STARTING DOSAGE AND PROGRESSION

The starting session is six repetitions of each exercise. The progression rate is an increase of one repetition for each three periods of exercise. Soldiers continue this rate until they do 12 repetitions with no rest between exercises. This level is maintained until another drill is used.

START POSITIONS

The soldiers fall in facing their log, with toes about four inches away. Figure 8-16 shows the basic starting positions and commands.

Right-Hand Start Position, Move

On the command "Move," move the left foot 12 inches to the left, and lower the body into a flatfooted squat. Keep the back straight, head up, and arms between the legs. Encircle the far side of the log with the left hand. Place the right hand under the log. (See I, Figure 8-16.)

Left-Hand Start Position, Move

This command is done the same way as the preceding command. However, the left hand is under the log, and the right hand encircles its far side. (See 2, Figure 8-16.)

Right-Shoulder Position, Move

This command is given from the right-hand-start position. On the command "Move," pull the log upward in one continuous motion to the right shoulder. At the same time, move the left foot to the rear and stand up, facing left. Balance the log on the right shoulder with both hands. (See 3, Figure 8-16.) This movement cannot be done from the left-hand-start position because of the position of the hands.

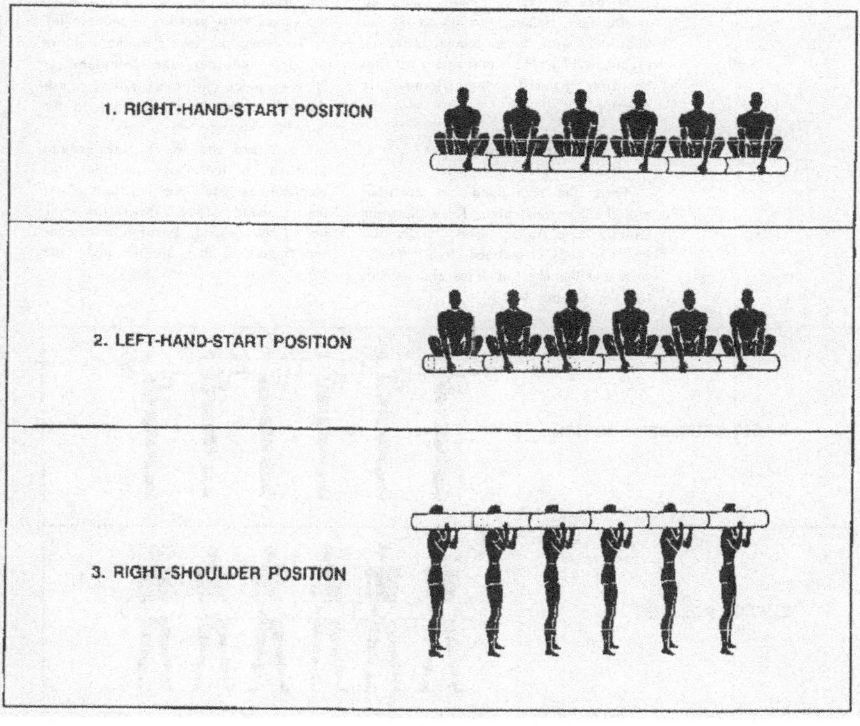

Figure 8-16

Left-Shoulder Position, Move

This command is given from the left-hand-start position. On the command "Move, " pull the log upward to the left shoulder in one continuous motion. At the same time, move the right foot to the rear, and stand up facing right. Balance the log on the left shoulder with both hands. (See 4, Figure 8-17.) This movement cannot be done from the right-hand-start position.

Waist Position, Move

From the right-hand-start position, pull the log waist high. Keep the arms straight and fingers laced under the log. The body is inclined slightly to the rear, and the chest is lifted and arched. (See 5, Figure 8-17.)

Chest Position, Move

This command is given after taking the waist position. On the command "Move," shift the log to a position high on the chest, bring the left arm under the log, and hold the log in the bend of the arms. (See 6, figure 8-17.) Keep the upper arms parallel to the ground.

To move the log from the right to the left shoulder, the command is "Left-shoulder position, move." Push the log overhead, and lower it to the opposite shoulder.

To return the log to the ground from any of the above positions, the command is "Start position, move." At the command "Move," slowly lower the log to the ground. Position the hands and fingers so they are not under the log.

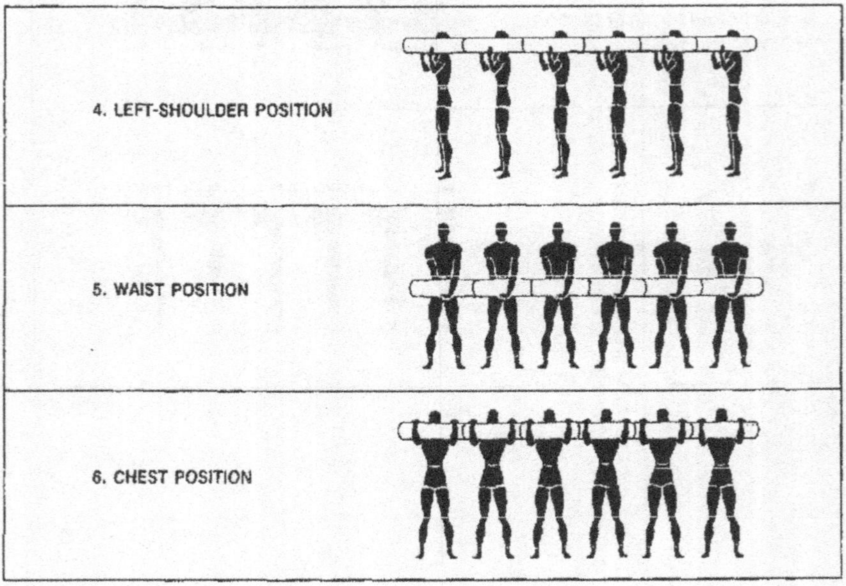

Figure 8-17

LOG-DRILL EXERCISES

The following are log-drill exercises.

Exercise 1. Two-Arm Push-Up

Start Position: Right- or left-shoulder position, with feet about shoulder-width apart. (See 1, Figure 8-18.)

Cadence: Moderate.

Movement: A four-count exercise; at the count of —

"One"-Push the log overhead until the elbows lock.

"Two"-Lower the log to the opposite shoulder.

"Three"-Repeat the action of count one.

"Four"-Recover to the start position.

Exercise 2. Forward Bender

Start Position: Chest position, with feet about shoulder-width apart. (See 2, Figure 8-18.)

Cadence: Moderate.

Movement A four-count exercise; at the count of -

"One"-Bend forward at the waist while keeping the back straight and the knees slightly bent.

"Two"-Recover to the start position.

Three"-Repeat the action of count one.

"Four"-Recover to the start position.

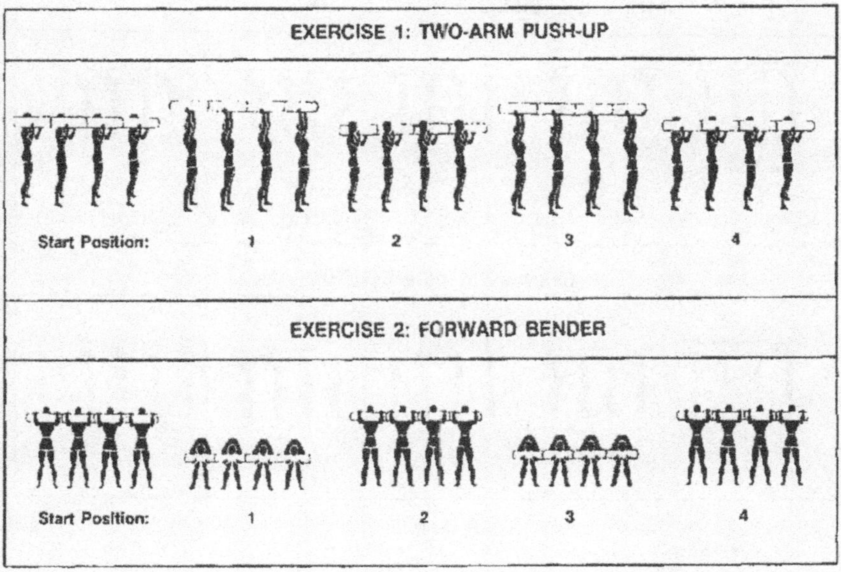

Figure 8-18

Exercise 3. Straddle Jump

Start Position Right- or left-shoulder position, with feet together, and fingers locked on top of the log. Pull the log down with both hands to keep it from bouncing on the shoulder. (See 3, Figure 8-19.)

Cadence: Moderate.

Movement A four-count exercise; at the count of—

"One"-Jump to a side straddle.

"Two"-Recover to the start position.

Threes-Repeat the action of count one.

"Four"-Recover to the start position.

Exercise 4. Side Bender

Start Position: Right-shoulder position with the feet about shoulder-width apart. (See 4, Figure 8- 19.)

Cadence Moderate.

Movement: A four-count exercise; at the count of—

"One"-Bend sideward to the left as far as possible, bending the left knee.

"Two"-Recover to the start position.

"Three"-Repeat the action of count one.

"Four"-Recover to the start position.

NOTE: After doing the required number of repetitions, change shoulders and do an equal number to the right side.

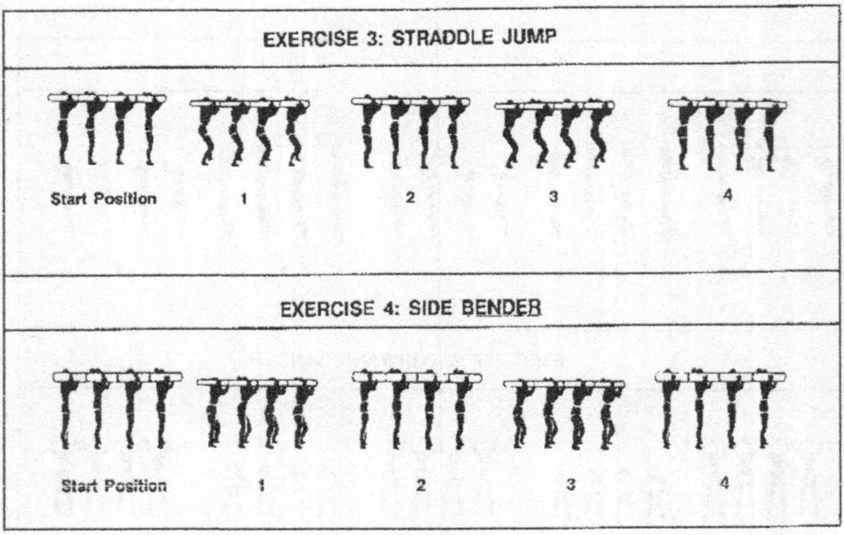

Figure 8-19

Exercise 5. Half-Knee Bend

Start Position: Right- or left-shoulder position, with feet about shoulder-width apart, and fingers locked on top of the log. (See 5, Figure 8-20.)

Cadence: Slow.

Movement: A four-count exercise; at the count of -

"One"-Flex the knees to a half-knee bend.

"Two"-Recover to the start position.

"Three"-Repeat the action of count one.

"Four"-Recover to the start position.

(NOTE: Pull forward and downward on the log throughout the exercise.)

Exercise 6. Overhead Toss (NOTE; Introduce this exercise only after soldiers have gained experience and strength by doing the other exercises for several sessions.)

Start Position: Right-shoulder position with the feet about shoulder-width part. The knees are at a quarter bend. (See 6, Figure 8-20.)

Cadence: Moderate.

Movement: A four-count exercise; at the count of -

"One"-Straighten the knees and toss the log about 12 inches overhead. Catch the log with both hands, and lower it toward the opposite shoulder. As the log is caught, lower the body into a quarter bend.

"Two"-Again, toss the log into the air and, when caught, return it to the original shoulder.

"Three"-Repeat the action of count one.

"Four"-Recover to the start position.

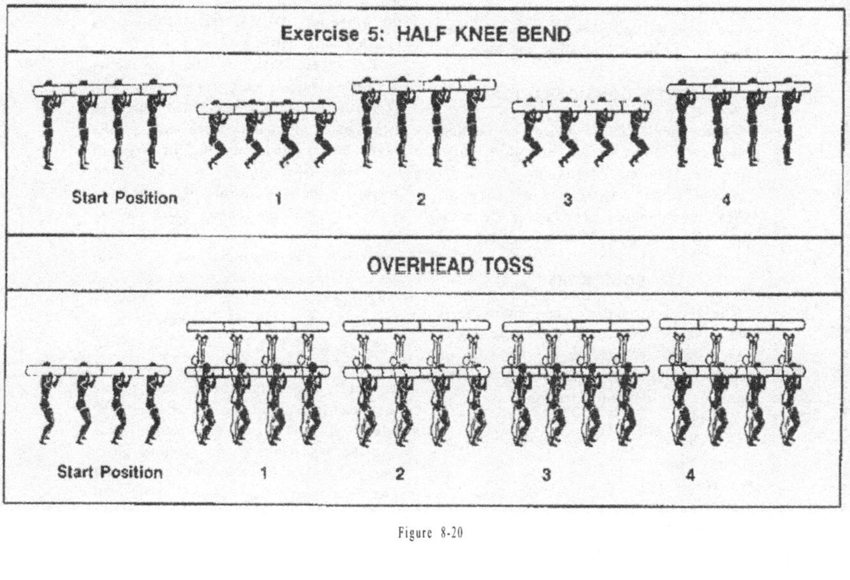

Figure 8-20

Aquatic Exercise

Aquatics is a mode of physical training which helps one attain and maintain physical fitness through exercises in the water. It is sometimes called slimnastics. Aquatic training can improve muscular endurance, CR endurance, flexibility, coordination, and muscular strength.

Because of its very low impact to the body, an aquatic exercise program is ideal for soldiers who are overweight and those who are limited due to painful joints, weak muscles, or profiles. The body's buoyancy helps minimize injuries to the joints of the lower legs and feet. It exercises the whole body without jarring the bones and muscles. Leaders can tailor the variety and intensity of the exercises to the needs of all the soldiers in the unit.

Aquatic training is a good supplement to a unit's PT program. Not only is it fun, it exposes soldiers to water and can make them more comfortable around it. Most Army installations have swimming pools for conducting aquatic, physical training sessions.

SAFETY CONSIDERATIONS

One qualified lifeguard is needed for every 40 soldiers at all aquatic training sessions. Nonswimmers must remain in the shallow end of the pool. They should never exercise in the deep end with or without flotation devices.

EQUIPMENT

Soldiers normally wear swim suits for aquatics, but they can wear boots and fatigues to increase the intensity of the activities. The following equipment is optional for training:

- Goggles.
- Kickboard.
- Pull buoy.
- Ear/nose plugs.
- Fins.
- Hand paddles.

SAMPLE TRAINING PROGRAM

'Warm-Up

As in any PT session, a warm-up is required. It can be done in the water or on the deck. Allow five to seven minutes for the warm-up.

Conditioning Phase

Soldiers should exercise vigorously to get a training effect. Energetic music may be used to keep up the tempo of the workout. The following are some exercises that can be used in an aquatic workout. (See Figure 8-21.)

Side Leg-Raises. Stand in chest to shoulder-deep water with either side of the body at arm's length to the wall of the pool, and grasp the edge with the nearest hand. Raise the outside leg sideward and upward from the hip. Next, pull the leg down to the starting position. Repeat these actions. Then, turn the other side of the body to the wall, and perform the exercise with the other leg. DURATION: 30 seconds (15 seconds per leg).

Leg-Over. Stand in chest-to shoulder-deep water, back facing the wall of the pool. Reach backward with the arms extended, and grasp the pool's edge. Next, raise one leg in front of the body away from the wall, and move it sideward toward the other leg as far as it can go. Then, return the leg to the front-extended position, and lower it to the starting position. Repeat these actions with the other leg, and continue to alternate legs. DURATION: 30 seconds (15 seconds per leg).

Rear Leg Lift. Stand in chest-to shoulder-deep water with hands on the pool's edge, chest to the wall. Raise one leg back and up from the hip, extend it, and point the foot. Then, pull the leg back to the starting position. Alternate these actions back and forth with each leg. DURATION: 20 seconds (10 seconds each leg).

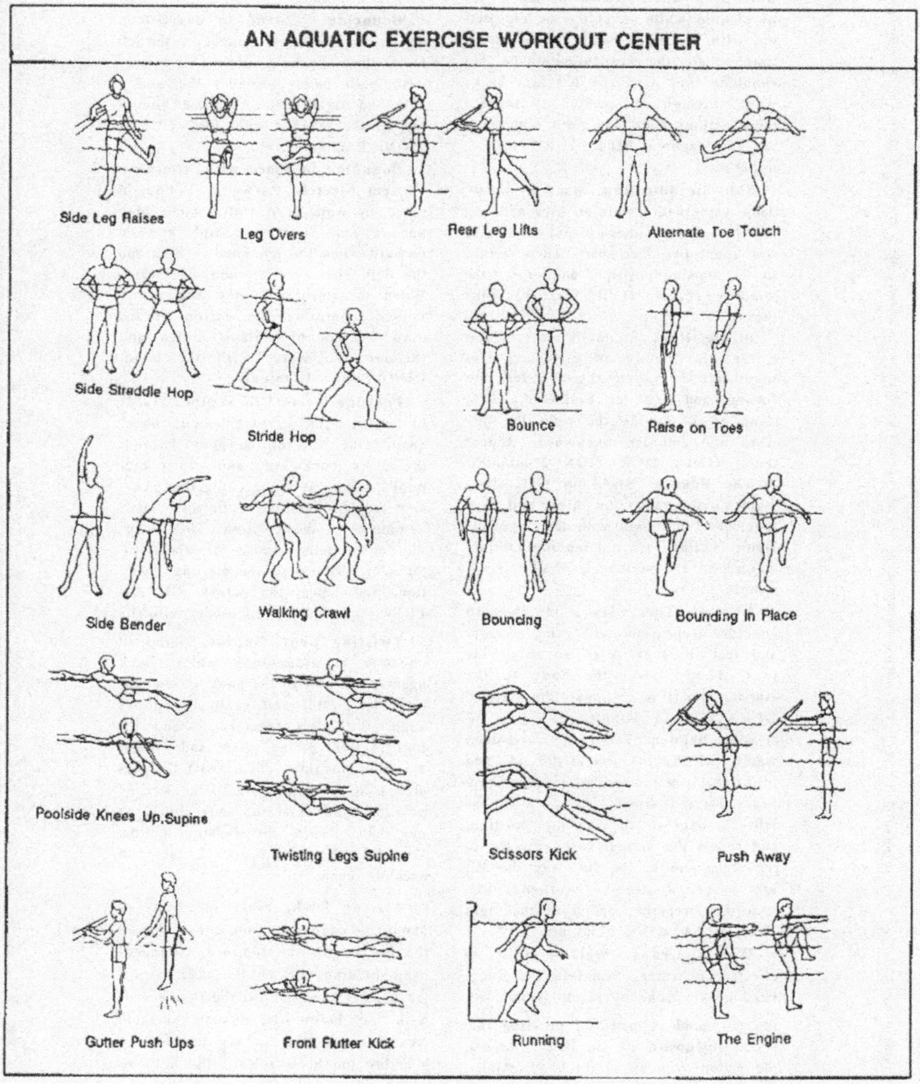
AN AQUATIC EXERCISE WORKOUT CENTER
Side Leg Raises
Leg Overs
Rear Leg Lifts
Alternate Toe Touch
Side Straddle Hop
Stride Hop
Bounce
Raise on Toes
Side Bender
Walking Crawl
Bouncing
Bounding In Place
Poolside Knees Up.Supine
Twisting Legs Supine
Scissors Kick
Push Away
Gutter Push Ups
Front Flutter Kick
Running
The Engine

Figure 8-21

Alternate Toe Touch. Stand in waist-deep water. Raise the left leg as in kicking while touching the elevated toe with the right hand. At the same time, rotate the head toward the left shoulder, and push the left arm backward through the water. Alternate these actions back and forth with each leg and opposite hand. DURATION 2 minutes.

Side Straddle Hop. Stand in waist-deep water with hands on hips and feet together. Jump sideward and land with feet about two feet apart. Then, return to the starting position, and repeat the jumping action. DURATION 2 minutes.

Stride Hop. Stand in waist-deep water with hands on hips and feet together. Jump, moving the left leg forward and right leg backward. Then, jump again moving the right leg forward and left leg backward. Repeat these actions. DURATION 2 minutes.

The Bounce. Stand in waist-deep water with hands on hips and feet together. Jump high with feet together. Upon landing, use a bouncing motion, and repeat the action. DURATION: 1 minute.

Rise on Toes. Stand in chest-to shoulder-deep water with arms at sides and feet together. Rise up using the toes. Then, lower the body to the starting position. Repeat the action, DURATION: **I** minute.

Side Bender. Stand in waist-deep water with the left arm at the side and the right arm extended straight overhead. Stretch slowly, bending to the left. Recover to the starting position, and repeat the action. Next, reverse to the right arm at the side and the left arm extended straight overhead. Repeat the stretching action to the right side. DURATION: 1 minute.

Walking Crawl. Walk in waist- to chest-deep water. Simulate the overhand crawl stroke by reaching out with the left hand cupped and pressing the water downward to the thigh. Repeat the action with the right hand. Alternate left and right arm action. DURATION: 2 minutes.

Bouncing. Stand in chest-deep water, arms at sides. Bounce on the left foot while pushing down vigorously with both hands. Repeat the action with the right foot. Alternate bouncing on the left and right foot. DURATION: 2 minutes.

Bounding in Place with Alternate Arm Stretch, Forward. Bound in place in waist-deep water using high knee action. Stretch the right arm far forward when the left knee is high and the left arm is stretched backward. When the position of the arm is reversed, simulate the action of the crawl stroke by pulling down and through the water with the hand. DURATION **1** minute.

Poolside Knees Up, Supine. Stand in chest-to shoulder-deep water, back against the wall of the pool. Extend the arms backward, and grasp the pool's edge. With feet together, extend the legs in front of the torso, and assume a supine position. Then with the legs together, raise the knees to the chin. Return to the starting position, and repeat the action. DURATION: 2 minutes (maximum effort).

Twisting Legs, Supine. Stand in chest-to shoulder-deep water, back against the wall of the pool. Extend the arms backward, and grasp the pool's edge. With feet together, extend the legs in front of the torso, and assume a supine position. Then, twist the legs slowly to the left, return to the starting position, and twist the legs slowly to the right. Repeat this twisting action. DURATION: 1 minute (2 sets, 30 seconds each).

Scissor Kick. Float in chest- to shoulder- deep water on either side of the body with the top arm extended, hand holding the pool's edge. Brace the bottom hand against the pool's wall with feet below the water's surface. Next, assume a crouching position by gringing the heels toward the hips by bending the knees. Then, straighten and spread the legs with the top leg extending backward. When the legs are extended and spread, squeeze them back together (scissoring). Pull with the top hand, and push with the bottom hand. The propulsive force of the kick will tend to cause the body to rise to the water's surface. DURATION 1 minute (2 sets, 30 seconds each, maximum effort).

Push Away. Stand in chest-to shoulder-deep water facing the pool's wall and at arm's length from it. Grasp the pool's edge, and bend the arms so that the body is leaning toward the wall of the pool. Vigorously push the chest back from the wall by straightening the arms. Then, with equal vigor, pull the upper body back to the wall. Repeat these actions. DURATION: 2 minutes (maximum effort).

Gutter Push-Ups. Stand in chest-to shoulder- deep water facing the pool's wall. Place the hands on the edge or gutter of the pool. Then, raise the body up and out of the water while extending the arms, repeat this action. DURATION: 2 minutes (4 sets, 30 seconds each with 5-second rests between sets).

Front Flutter Kick. Stand in chest-to shoulder-deep water facing the pool's wall. Grasp the pool's edge or gutter and assume a prone position with legs extended just below the water's surface. Then, kick flutter style, toes pointed, ankles flexible, knee joint loose but straight. The legs should simulate a whip's action. DURATION 1 minute (2 sets, 30 seconds each).

Running. Move in a running gait in chest-to shoulder-deep water with arms and hands under the water's surface. This activity can be stationary, or the exerciser may run from poolside to poolside. Runners must concentrate on high knee action and good arm movement. DURATION 10 to 20 minutes.

The Engine. Stand in chest-to shoulder-deep water, arms straight and in front of the body and parallel to the water with the palms facing downward. While walking forward, raise the left knee to the left elbow, then return to the starting position. Continuing to walk forward, touch the right knee to the right elbow, and return to the starting position. Be sure to keep the arms parallel to the water throughout the exercise. DURATION 1 to 2 minutes (2 sets).

Cool-Down

This is required to gradually bring the body back to its pre-exercise state. It should last from five to seven minutes.

CHAPTER 9

Competitive Fitness Activities

Competitive fitness activities help in the development of assets that are vital to combat effectiveness.

Physical fitness is one of the foundations of combat readiness, and maintaining it must be an integral part of every soldier's life. This chapter discusses competitive fitness activities and athletic events that commanders can use to add variety to a unit's physical fitness program. There is also a section on developing a unit intramural program. Athletic and competitive fitness activities are sports events which should only be used to supplement the unit's PT program. They should never replace physical training and conditioning sessions but, rather, should exist to give soldiers a chance for healthy competition. Only through consistent, systematic physical conditioning can the fitness components be developed and maintained.

Crucial to the success of any program is the presence and enthusiasm of the leaders who direct and participate in it. The creativity of the physical training planners also plays a large role. Competitive fitness and athletic activities must be challenging. They must be presented in the spirit of fair play and good competition.

It is generally accepted that competitive sports have a tremendous positive influence on the physical and emotional development of the participants. Sports competition can enhance a soldier's combat readiness by promoting the development of coordination, agility, balance, and speed. Competitive fitness activities also help develop assets that are vital to combat effectiveness. These include team spirit, the will to win, confidence, toughness, aggressiveness, and teamwork.

Intramural

The Army's sports mission is to give all soldiers a chance to participate in sports activities. A unit-level intramural program can help achieve this important goal. DA Pam 28-6 describes how to organize various unit-level intramural programs.

Factors that affect the content of the sports program differ at every Army installation and unit. Initiative and ingenuity in planning are the most vital assets. They are encouraged in the conduct of every program.

OBJECTIVES

A well-organized and executed intramural program yields the following:

- Team spirit, the will to win, confidence, aggressiveness, and teamwork. All are vital to combat effectiveness.
- A change from the routine PT program.
- The chance for **all** soldiers to take part in organized athletics.

ORGANIZATION

The command level best suited to organize and administer a broad intramural program varies according to a unit's situation. If the objective of maximum participation is to be achieved, organization should start at company level and then provide competition up through higher unit levels. Each command level should have its own program and support the next higher program level.

To successfully organize and conduct an intramural program, developers should consider the following factors and elements.

Authority

The unit commander should publish and endorse a directive giving authorization and guidance for a sports program. A detailed SOP should also be published.

Personnel

Leaders at all levels of the intramural program should plan, organize, and supervise it. Appointments at all echelons should be made for at least one year to provide continuity. The commander must appoint a qualified person to be the director, regardless of the local situation, type, and size of the unit. The director must be a good organizer and administrator and must have time to do the job correctly. He should also have a sense of impartiality and some athletic experience.

Commanders should form an intramural sports council in units of battalion size or larger and should appoint members or require designated unit representatives. The council should meet at least once a month or as often as the situation requires. The council serves as an advisory body to the unit commander and intramural director. It gives guidance about the organization and conduct of the program.

Facilities and Equipment

Adequate facilities and equipment must be available. When facilities are limited, leaders must plan activities to ensure their maximum use. In all cases, activities must be planned to ensure the safety of participants and spectators.

Funds and Budget

Adequate funds are essential to successfully organize and operate a sports program. Therefore, beforehand, organizers must determine how much money is available to support it. To justify requests for funds they must prepare a budget in which they justify each sports activity separately. The budget must include special equipment, supplies, awards, pay for officials, and other items and services. Units can reduce many of their costs by being resourceful.

AWARD SYSTEM

Commanders can stimulate soldiers to participate in competitive athletics by using an award system.

Commanders can stimulate units and soldiers to participate in competitive athletics by using an award system. One type is a point-award system where teams get points based on their win/loss records and/or final league standings. This reflects the unit's standings in the overall intramural sports program. The recognition will help make units and individuals participate throughout the year. Trophies can then be given for overall performance and individual activities.

PROGRAM PLANNING

A successful program depends on sound plans and close coordination between the units involved. The intramural director should meet with subordinate commanders or a sports representative to determine what program of activities is compatible with the mission and training activities of each unit. Unless they resolve this issue, they may not get command support which, in turn, could result in forfeitures or lack of participation. The less-popular activities may not be supported because of a lack of interest.

Evaluations

Before the program is developed, leaders must study the training and availability situation at each unit level. They should include the following items in a survey to help them determine the scope of the program and to develop plans:

- *General.* Evaluate the commander's attitude, philosophy, and policy about the sports program. Understand the types of units to be served, their location, the climate, and military responsibilities.
- *Troops.* Determine the following: 1) number and types of personnel; 2) training status and general duty assignment; 3) special needs, interests, and attitudes.
- *Time available.* Coordinate the time available for the sports program with the military mission. Determine both the on-duty and off-duty time soldiers have for taking part in sports activities.
- *Equipment.* Consider the equipment that will be needed for each sport.
- *Facilities.* Determine the number, type, and location of recreational facilities both within the unit and in those controlled by units at higher levels.
- *Funds.* Determine how much each unit can spend on the intramural program.
- *Personnel.* Assess how many people are needed to run the program. The list should include a director and assistants, sports council, officials, and team captains, as well as volunteers for such tasks as setting up a playing field.
- *Coordination.* Coordinate with the units' operations sections to avoid conflict with military training schedules.
- *Activities.* The intramural director should plan a tentative program of activities based on the season, local situation, and needs and interests of the units. Both team and individual sports should be included. Some team sports are popular at all levels and need little promotional effort for success. Among these are volleyball, touch football, basketball, and softball. Some individual competitive sports have direct military value. They include boxing, wrestling, track and field, cross country, triathlon, biathlon, and swimming. While very popular, these sports are harder to organize than team sports. See Figures 9-1 and 9-2 for a list of sports activities.

SPORTS ACTIVITIES			
Team Sports			
Baseball Flag Football Softball	Basketball Water Polo Speedball	Field Hockey Pushball Tug-of-War	Football Soccer Volleyball
Field-Type-Meets			
Athletic Carnivals Physical Fitness Meet Track and Field Urban Orienteering	Cross Country Relay Carnival Water Carnival	Military Field Meets Swimming and Diving Unit Olympics	

Figure 9-1

SPORTS ACTIVITIES			
Individual Sports			
Archery Boxing Handball Marathon Track & Field Triathlon	Badminton Canoeing Judo Squash Rowing Skating	Tennis Table Tennis Horseshoes Skating Sky Diving Weightlifting	Bowling Gymnastics Modern Biathlon Mountain Climbing Skeet Shooting Swimming and Diving

Figure 9-2

ESSENTIAL ELEMENTS	
Intramural Handbook	
• Commander's foreward.	• Master calendar of activities.
• Personnel directory.	• Organization of leagues and units of competition.
• Title page.	• Command- points award system.
• Purpose.	• Facilities and their hours of operation.
• References.	• Equipment regulations.
• Objectives.	• Rules and regulations of each sport.
• Duties of the personnel.	• Reporting time for competition.
• Eligibility rules.	• Postponement of contests.
• Intramural sports council.	• Protest procedures.
• Protest and sportsmanship board.	• Awards.
• Budgets and funding.	• Records and results.
• Officials association.	• Bulletin boards and publicity.

Table 9-1

Functions

Once the evaluations have been made, the following functions should be performed:

- ***Make a handbook.*** An intramural handbook should be published at each level of command from installation to company to serve as a standing operating procedure (SOP). This handbook should include the essential elements listed in Table 9-1 above.
- ***Plan the calendar.*** Local situations and normal obstacles may conflict with the intramural program. How ever, a way can be found to provide a scheduled program for every season of the year.
- ***Choose the type of competition.*** Intramural directors should be able to choose the type of competition best suited for the sport and local circumstances. They should also know how to draw up tournaments. Unless the competition must take place in a short time, elimination tournaments should not be used. The round-robin tournament has the greatest advantage because individuals and teams are never eliminated. This type of competition is adaptable to both team and individual play. It is appropriate for small numbers of entries and league play in any sport.
- ***Make a printed schedule.*** Using scheduling forms makes this job easier. The form should include game number, time, date, court or field, and home or visiting team. Space for scores and officials is also helpful. Championship games or matches should be scheduled to take place at the best facility.

Unit Activities

The following games and activities may be included in the unit's PT program, They are large-scale activities which can combine many components of physical and motor fitness. In addition, they require quick thinking and the use of strategy. When played vigorously, they are excellent activities for adding variety to the program.

NINE-BALL SOCCER

The object of this game is for each of a team's five goalies to have one ball.

Players

There are 25 to 50 players on each team, five of whom are goalies. The other players are divided into four equal groups. The goalies play between the goal line and 5-yard line of a standard football field. The other four groups start the game between the designated 10-yard segments of the field. (See Figure 9-3.) The goalies and all other players must stay in their assigned areas throughout the game. The only exceptions are midfielder who stand between the 35- and 45-yard lines. These players may occupy both their assigned areas and the 10-yard free space at the center of the field.

The Game

The game starts with all players inside their own areas and midfielder on their own 40-yard line. The nine balls are placed as follows. Four are on each 45-yard line with at least five yards between balls. One is centered on the 50-yard line. The signal to start play is one long whistle blast. Players must pass the balls through the opposing team's defenses into the goal area using only their feet or heads. The first team whose goalies have five balls wins a point. The game then stops, and the balls are placed for the start of a new set. The first team to score five points wins.

There are no time-outs except in case of injury, which is signaled by two sharp whistle blasts. The teams change positions on the field after each set. Team members move to different zones after the set.

Rules

A ball is played along the ground or over any group or groups of players. The ball may travel any distance if it is played legally.

Goalies may use their hands in playing the ball and may give a ball to other goalies on their team. For a set to officially end, each goalie must have a ball.

If players engage in unnecessary roughness or dangerous play, the referee removes them from the game for the rest of the set and one additional set. He also removes players for the rest of the set if they step on or over a boundary or sideline or use their hands outside the goal area.

If a goalie steps on or over a boundary or sideline, the referee takes the ball being played plus another ball from the goalie's team and gives these balls to the nearest opposing player. If

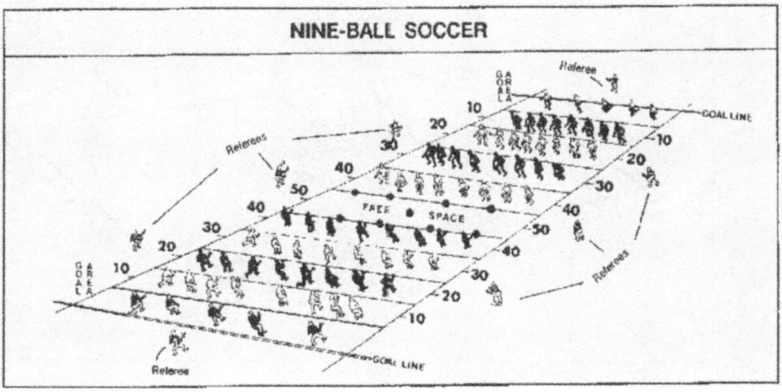

Figure 9-3

the team has no other ball in the goal area, the referee limits the penalty to the ball that is being played.

If a ball goes out of bounds, the referee retrieves it. The team that caused it to go out of bounds or over the goal line loses possession. The referee puts the ball back into play by rolling it to the nearest opposing player.

PUSHBALL

This game requires a large pushball that is five to six feet in diameter. It also requires a level playing surface that is 240 to 300 feet long and 120 to 150 feet wide. The length of the field is divided equally by a center line. Two more lines are marked 15 feet from and parallel to the end lines and extending across the entire field. (See Figure 9-4.)

Players

There are 10 to 50 soldiers on each of two teams.

The Game

The object of the game is to send the ball over the opponent's goal line by pushing, rolling, passing, carrying, or using any method other than kicking the ball.

The game begins when the ball is placed on the centerline with the opposing captains three feet away from it. The other players line up 45 feet from the ball on their half of the field. At the referee's starting whistle, the captains immediately play the ball, and their teams come to their aid.

At quarter time, the ball stays dead for two minutes where it was when the quarter ended. At halftime, the teams exchange goals, and play resumes as if the game were beginning.

A team scores a goal when it sends the ball across the opposing team's end line. A goal counts five points. The team that scores a goal may then try for an extra point. For the extra point, the ball is placed on the opposing team's 5-yard line, and the teams line

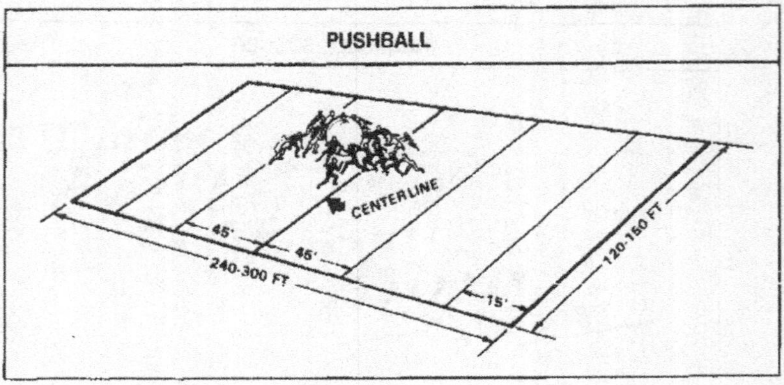

up across the field separated by the width of the ball. Only one player may place his hands on the ball. The player who just scored is directly in front of the ball. At the referee's signal, the ball is put into play for one minute. If any part of the ball is driven across the goal line in this period, the offense scores one point. The defense may not score during the extra point attempt.

The game continues until four 10-minute quarters have been played. Rest periods are allowed for two minutes between quarters and five minutes at halftime.

Rules

Players may use any means of interfering with the opponents' progress except striking and clipping. Clipping is throwing one's body across the back of an opponent's legs as he is running or standing. Force may legally be applied to all opponents whether they are playing the ball or not. A player who strikes or clips an opponent is removed from the game, and his team is penalized half the distance to its goal.

When any part of the ball goes out of bounds, it is dead. The teams line up at right angles to the sidelines. They should be six feet apart at the point where the ball went out. The referee tosses the ball between the teams.

When, for any reason, the ball is tied up in one spot for more than 10 seconds, the referee declares it dead. He returns the ball into play the same way he does after it goes out of bounds.

STRATEGY PUSHBALL

Strategy pushball is similar to pushball except that it is played on two adjacent fields, and opposing teams supply soldiers to the games on both fields. Team commanders assess the situation on the fields and distribute their soldiers accordingly. The commander decides the number of soldiers used, within limits imposed by the rules. This number may be adjusted throughout the game. Play on both fields occurs at the same time, but each game progresses independently. At the end of play, a team's points from both fields are added together to determine the overall winner.

This game requires two pushballs that are five to six feet in diameter. Pull-over vests or jerseys of two different colors are used by each team for a total of four different colors. Starters and reserves should be easily distinguishable. Starters and substitutes should wear vests of one color, while the team commander and reserves wear vests of the second color.

Players may wear any type of athletic shoes except those with metal cleats. Combat boots may be worn, but extra caution must be used to prevent injuries caused by kicking or stepping on other players. Soldiers wearing illegal equipment may not play until the problem has been corrected.

The playing area is two lined-off fields. These are 240 to 300 feet long by 120 to 150 feet wide. They are separated lengthwise by a 20-foot-wide divider strip. The length of each field is divided equally by a centerline that is parallel to the goal lines. Lines are also marked 45 feet from each side of the centerline and parallel to it. The lines extend across both fields. Dimensions may be determined locally based on available space and the number of players. The space between the fields is the team area. Each team occupies the third of the team space that immediately adjoins its initial playing field.

Time periods should be adjusted to suit weather conditions and soldiers' fitness levels.

Players

There are 25 to 40 soldiers on each team. A typical, 25-member team has the following:

- One team commander. He is responsible for overall game strategy and for determining the number and positions of players on the field.
- Sixteen starting members. Eight are on each field at all times; one is appointed field captain.
- Four reserve members. These are players the team commander designates as reinforcements.
- Three substitutes. These are replacements for starters or reserves.
- One runner. He is designated to convey messages from the team commander to field captains.

The proportion of soldiers in each category stays constant regardless of the total number on a team. Before the event, game organizers must coordinate with participating units and agree on the number on each team.

Runners serve at least one period; they may not play during that period. They are allowed on the field only during breaks in play after a dead ball or goal.

Reserves are used at any point in the game on either field and are committed as individuals or groups. They may enter or leave the playing field at any time whether the ball is in play or not. Team commanders may enter the game as reserves if they see the need for such action.

Reserves, substitutes, and starting members may be redesignated into any of the other components on a one-for-one basis only during dead balls, injury time-outs, or quarter- and half-time breaks. A reserve may become a starter by switching vests with an original starter, who then becomes a reserve.

When possible, senior NCOS and officers from higher headquarters or other units should be used as officials. Players must not question an official's

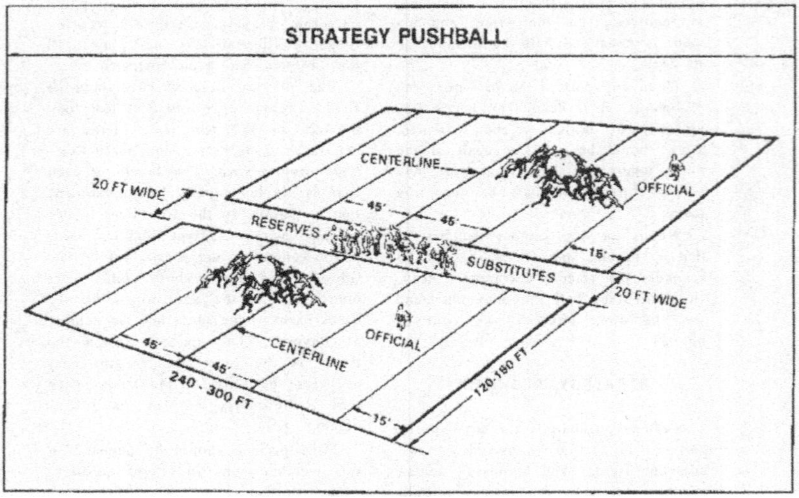

Figure 9-5

authority during play. Otherwise, the game can quickly get out of control.

Chain-of-command personnel should act as team commanders and field captains whenever possible.

The Game

The object is to propel the ball over the opponent's goal line by pushing, rolling, passing, carrying, or using any means other than kicking.

The game is officiated by two referees on each field, a chief umpire, and a scorekeeper. Referees concentrate on player actions so that they can quickly detect fouls and assess penalties. The chief umpire and scorekeeper occupy any area where they can best officiate the games. The chief umpire monitors the use of substitutes and reserves and ensures smooth progress of the games on both fields. The number of officials may be increased if teams have more than 25 players. Referees use their whistles to stop and start play except at the start and end of each quarter. The scorekeeper, who times the game with a stopwatch, starts and ends each quarter and stops play for injuries with some noisemaker other than a whistle. He may use such devices as a starter's pistol, klaxon, or air horn.

The game begins after the ball is placed on each field's center mark. Opposing field captains are three feet from the ball (six feet from the centerline). The rest of the starters are lined up 45 feet from the ball on their half of the field. (See Figure 9-5.) At the scorekeeper's signal, field captains immediately play the ball, and their teams come to their aid.

Starters may be exchanged between the fields if the minimum number of starters or substitutes per field is maintained.

Substitutes may enter the game only during breaks in play after a dead ball, goal, or time-out for injury. A substitute may not start to play until the player being replaced leaves the field.

When any part of the ball goes out of bounds, it is dead. The teams line up at right angles to the sidelines; they are 10 feet apart at the point where the ball went out of bounds. The referee places the ball between the teams at a point 15 feet inside the sideline. Play resumes when the referee blows the whistle.

When the ball gets tied up in one spot for more than 10 seconds for any reason, the referee declares it dead. He restarts play as with an out-of-bounds dead ball, except that he puts the ball on the spot where it was stopped.

Time does not stop for dead balls or goals. Play continues on one field while dead balls are restarted on the other.

At each quarter break, the ball stays on the spot where it was when the quarter ended. The next quarter, signaled by the scorekeeper, starts as it does after a ball goes out of bounds. At halftime the teams exchange goals, and play resumes as if the game were beginning.

A goal is scored when any part of the ball breaks the plane of the goal line between the sidelines. A goal counts one point. At the end of the fourth quarter, the points of each team from both fields are added together to determine the winner.

If there is a tie, a three-minute overtime is played. It is played the same as in regulation play, but only one field is used, with starting squads from both teams opposing each other. For control purposes, no more than 15 players per team are allowed on the field at once. The team with more points at the end of the overtime wins the game. If the game is still tied when time expires, the winner is the team that has gained more territory.

The game continues until four 10-minute quarters have been played. There is a 10-minute halftime between the second and third quarters. The clock stops at quarter breaks and halftime. Time-out is allowed only for serious injury. Play is then stopped on both fields.

Rules

Players may use any means of interfering with their opponents' progress, but they are penalized for striking or clipping opponents or throwing them to the ground. These penalties are enforced by the referees. Force maybe legally applied to any opponent whether or not they are playing the ball. Blocking is allowed if blockers stay on their feet and limit contact to the space between waist and shoulders. Blockers may not swing, throw, or flip their elbows or forearms. Tackling opposing soldiers who are playing the ball is allowed. The chief umpire or any referee may call infractions and impose penalties for unsportsmanlike conduct or personal fouls on either field. Penalties may also be called for infractions committed on the field or sidelines during playing time, quarter- and halftime breaks, and time-outs. Personal fouls are called for the following:

- Illegal blocking (below an opponent's waist).
- Clipping (throwing the body across the back of the opponent's legs as he is running or standing).
- Throwing an opponent to the ground (that is, lifting and dropping or slamming a player to the ground in stead of tackling cleanly).
- Spearing, tackling, or piling on an opponent who is already on the ground.
- Striking or punching with closed fist(s).
- Grasping an opponent's neck or head.
- Kicking.
- Butting heads.

Unsportsmanlike conduct is called for abusive or insulting language that the referee judges to be excessive and blatant. It is also called against a player on the sidelines who interferes with the ball or with his opponents on the field. A player who violates these rules should be removed from the game and made to run one lap around both playing fields. A penalized player leaves the team shorthanded until he completes the penalty lap and the next break in play occurs on the field from which he was removed. The penalized player or a substitute then enters the game. Referees and the chief umpire may, at their discretion, eject any player who is a chronic violator or who is judged to be dangerous to other players. Once ejected, the player must leave both the field of play and team area. Substitutes for ejected players may enter during the next break in play that follows a goal scored by either team. They enter on the field from which the players were ejected.

BROOM-BALL HOCKEY

This game is played on ice or a frozen field using hockey rules. Players wear boots with normal soles and carry broom-shaped sticks with which they hit the ball into the goals.

The object of this game is for teams to score goals through the opponent's defenses. Using only brooms, players pass the ball through the opposing team to reach its goal. The first team to score five points wins. Broom ball provides a good cardiorespiratory workout.

Players

There are 15 to 20 players on each team. One is a goalie and the others are divided into three equal groups. The goalie plays in the goal area of a standard soccer or hockey field or along the goal line if the two opposing goals are the same size. One soccer ball, or some other type of inflated ball, is used. The players need no padding.

The three groups begin the game in center field. All players must stay in their designated space throughout the game. A diagram of the field is shown at Figure 9-6.

The Game

The face-off marks the start of the game, the second half, and the restart of play after goals. Each half lasts 15 minutes. For the face-off, each player is on his own half of the field. All players, except the two centers, are outside the center circle. The referee places the ball in the center of the circle between the two centers. The signal to begin play is one long blast on the whistle. The ball must travel forward and cross the center circle before being played by another player. There are no time-outs except for injury. The time-out signal is two sharp whistle blasts.

Rules

All players, including goalies, must stay inside their legal boundaries at all times. Only goalies may use their hands to play the ball, but they must always keep control of their sticks. Other players must stay in their respective zones of play (Attack, Defense, Centerfield). The ball is played along the ground or over one or more groups of players. It may travel any distance as long as it is legally played.

The referee calls infractions and imposes penalties. Basic penalties are those called for the following:

- Unnecessary roughness or dangerous play. (The player is removed from the game; he stays in the penalty box for two minutes.)
- Ball out-of-bounds. (The team that caused it to go out loses possession, and the opposing team puts the ball back into play by hitting it to the nearest player.)
- Use of hands by a player other than a goalie. (The player must stay in the penalty box one minute.)
- Improper crossing of boundaries. (When a member of the team in possession of the ball crosses the boundary line of his zone of play, possession will be awarded to the other team.)

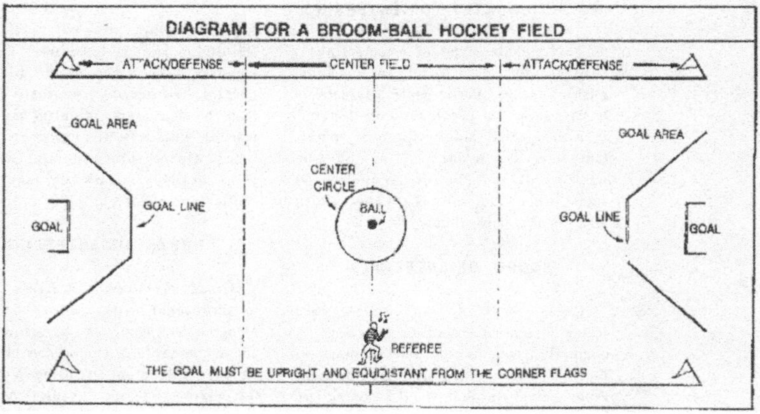

Figure 9-6

Orienteering combines- map reading, compass use, and terrain study with strategy, competition, and

Orienteering

Orienteering is a competitive form of land navigation. It combines map reading, compass use, and terrain study with strategy, competition, and exercise. This makes it an excellent activity for any training schedule.

An orienteering course is set up by placing control points or marker signs over a variety of terrain. The orienteer or navigator uses a detailed topographical map and a compass to negotiate the course. The map should be 1:25,000 scale or larger. A liquid-filled orienteering compass works best. The base of the compass is transparent plastic, and it gives accurate readings on the run. The standard military, lensatic compass will work even though it is not specifically designed for the sport.

The best terrain for an orienteering course is woodland that offers varied terrain. Several different courses can be setup in an area 2,000 to 4,000 yards square. Courses can be short and simple for training beginners or longer and more difficult to challenge the advanced competitors.

The various types of orienteering are described below.

CROSS-COUNTRY ORIENTEERING

This popular type of orienteering is used in all international and championship events. Participants navigate to a set number of check or control points in a designated order. Speed is important since the winner is the one who reaches all the control points in the right order and returns to the finish area in the least time.

SCORE ORIENTEERING

Quick thinking and strategy are major factors in score orienteering. A competitor selects the check-points to find based on point value and location. Point values throughout the course are high or low depending on how hard the markers are to reach. Whoever collects the most points within a designated time is the winner. Points are deducted for returning late to the finish area.

LINE ORIENTEERING

Line orienteering is excellent for training new orienteers. The route is premarked on the map, but checkpoints are not shown. The navagator tries to walk or run the exact map route. While negotiating the course, he looks for checkpoints or control-marker signs. The winner is determined by the time taken to run the course and the accuracy of marking the control points when they are found.

ROUTE ORIENTEERING

This variation is also excellent for beginners. The navigator follows a route that is clearly marked with signs or streamers. While negotiating the course, he records on the map the route being taken. Speed and accuracy of marking the route determine the winner.

NIGHT ORIENTEERING

Competitors in this event carry flashlights and navigate with map and compass. The night course for cross-country orienteering is usually shorter than the day course. Control points are marked with reflective material or dim lights. Open, rolling terrain, which is poor for day courses, is much more challenging at night.

URBAN ORIENTEERING

Urban orienteering is very similar to traditional types, but a compass, topographical map, and navigation skills are not needed. A course can be set up on any installation by using a map of the main post or cantonment area. Soldiers run within this area looking for coded location markers, which are numbered and marked on the map before the start. This eliminates the need for a compass. Soldiers only need a combination map-scorecard, a watch, and a pencil. (Figure 9-7 shows a sample scorecard.)

Urban orienteering adds variety and competition to a unit's PT program and is well suited for an intramural program. It also provides a good cardiovascular workout.

Participants and Rules

Urban orienteering is conducted during daylight hours to ensure safety and make the identification of checkpoint markers easy. Soldiers form two-man teams based on their APFT 2-mile-run times. Team members should have similar running ability. A handicap is given to slower teams. (See Figure 9-8.) At the assembly area, each team gets identical maps that show the

URBAN ORIENTEERING

LOCATION MARKER	POINT VALUE	LOCATION MARKER CODE	LOCATION MARKER	POINT VALUE	LOCATION MARKER CODE
1	10		26	10	
2	10		27	15	
3	15		28	5	
4	10		29	15	
5	15		30	15	
6	10		31	15	
7	25		32	25	
8	15		33	15	
9	25		34	15	
10	15		35	25	
11	15		36	15	
12	25		37	15	
13	15		38	25	
14	15		39	15	
15	25		40	25	
16	15		41	25	
17	25		42	15	
18	10		43	10	
19	10		44	15	
20	15		45	10	
21	10		46	25	
22	5		47	10	
23	15		48	15	
24	10		49	15	
25	10		50	10	

Figure 9-7

location of markers on the course. Location markers are color-coded on the map based on their point value. The markers farthest from the assembly area have the highest point values. The maps are labeled with a location number corresponding to the location marker on the course. A time limit is given, and teams finishing late are penalized. Five points are deducted for each minute a team is late. While on the course, team members must stay together and not separate to get two markers at once. A team that separates is disqualified. Any number of soldiers may participate, the limiting factors being space and the number of points on the course.

Playing the Game

Once the soldiers have been assigned a partner, the orienteering marshal briefs them on the rules and objectives of the game. He gives them their time limitations and a reminder about the overtime penalty. He also gives each team a combination map/scorecard with a two-digit number on it to identify their team. When a team reaches a location marker, it records on the scorecard the letters that correspond to its two-digit number.

Point values of each location marker are also annotated on the scorecard. When the orienteering marshal signals the start of the event, all competitors

HANDICAPS FOR URBAN ORIENTEERING

2-MILE RUN TIME	POINTS	2-MILE RUN TIME	POINTS
12:00 or faster	0	14:31-15:00	60
12:01-12:30	10	15:01-15:30	70
12:31-13:00	20	15:31-16:00	80
13:01-13:30	30	16:01-16:30	90
13:31-14:00	40	16:31-17:00	100
14:01-14:30	50	17:01+	100

Figure 9-8

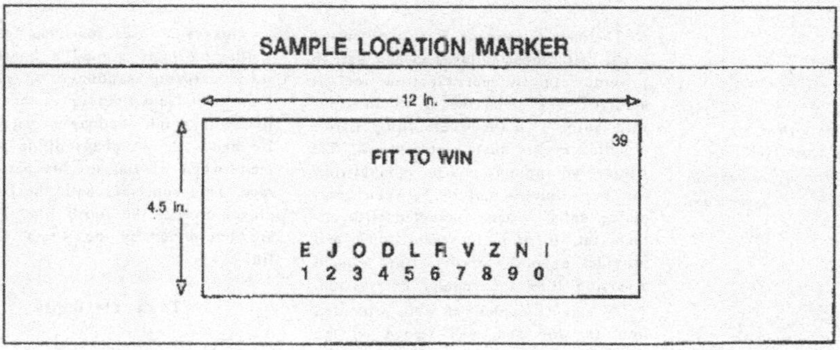

Figure 9-9

leave the assembly area at the same time. One to two hours is the optimal time for conducting the activity. A sample location marker is shown at Figure 9-9.

For this example, team number 54 found the marker. The letters corresponding to 54 are LD, so they place "LD" on line 39 of their scorecard. This line number corresponds to the location's marker number. When the location marker code is deciphered, the team moves on to the next marker of its choice. Each team goes to as many markers as possible within the allotted time. After all teams have found as many location markers as possible and have turned in their map/scorecards, the points are computed by the orienteering marshal to determine the teams' standings. He has the key to all the points and can determine each team's accuracy. Handicap points are then added. Each soldier gets points if his 2-mile-run time is slower than 12 minutes. (See Figure 9-8.) The teams' standings are displayed shortly after the activity ends.

Safety Briefing

The orienteering marshal gives a safety briefing before the event starts. He reminds soldiers to be cautious while running across streets and to emphasize that team members should always stay together.

Set Up and Materials

The course must be well thought out and set up in advance. Setting up requires some man-hours, but the course can be used many times. The major tasks are making and installing location markers and preparing map/scorecard combinations. Once the location marker numbers are marked and color coded on the maps, they are covered with combat acetate to keep them useful for a long time. Combat acetate (also called plastic sheet) can be purchased in the self-service supply center store under stock number 9330-00-618-7214.

The course organizer must decide how many location markers to make and where to put them. He should use creativity to add excitement to the course. Suggestions for locations to put point markers are as follows: at intersections, along roads in the tree line, on building corners, and along creek beds and trails. They should not be too hard to find. To help teams negotiate the course, all maps must be precisely marked to correspond with the placement of the course-location markers.

Unit Olympics incorporate athletic-events that represent all five fitness components.

Unit Olympics

The unit **Olympics** is a multifaceted event that can be tailored to any unit to provide athletic participation for all soldiers. The objective is to incorporate into a team-level competition athletic, events that represent all five fitness components. The competition can be within a unit or between competing units. When conducted with enthusiasm, it promotes team spirit and provides a good workout. It is a good diversion from the regular PT session.

A unit Olympics, if well promoted from the top and well staged by the project NCO or officer, can be a good precursor to an SDT or the EIB test.

TYPES OF EVENTS

The Olympics should include events that challenge the soldiers' muscular strength and endurance, aerobic endurance, flexibility, agility, speed, and related sports skills.

Events can be held for both individuals and teams, and they should be designed so that both male and female soldiers can take part. Each soldier should be required to do a minimum number of events. Teams should wear a distinctively marked item such as a T-shirt or arm band. This adds character to the event and sets teams apart from each other. A warm-up should precede and a cool-down should follow the events.

The following are examples of athletic events that could be included in a unit Olympics:

Push-Up Derby

This is a timed event using four-member teams. The objective is for the team to do as many correct push-ups as possible within a four-minute time limit. Only one team member does push-ups at a time. The four team members may rotate as often as desired,

Sandbag Relay

This event uses four-man teams for a running relay around a quarter-mile track carrying sandbags. One player from each team lines up at the starting line with a full sandbag *in* each hand. He hands the sandbags off to a teammate when he finishes his part of the race. This continues until the last team player crosses the finish line. Placings are determined by the teams' order of finish.

Team Flexibility

In this event, if teams are numerically equal, all members of each team should participate. If not, as many team members should participate as possible. Each team's anchor person places his foot against a wall or a curb. He stretches his other foot as far away as possible as in doing a split. The next team member puts one foot against the anchor man's extended foot and does a split-stretch. This goes on until all team members are stretched. They cover as much distance as possible keeping in contact with each other. The team that stretches farthest from the start point without a break in their chain is the winner.

Medicine-Ball Throw

This event uses four-member teams. The teams begin by throwing the ball from the same starting line. When it lands, the ball is marked for each team thrower, and the next team player throws from this spot. This is repeated until all the team's players have thrown. The team whose combined throws cover the most distance is the winner.

Job-Related Events

The organizer should use his imagination when planning activities. He may incorporate soldier skills required of an MOS. For instance, he could devise a timed land-navigation event geared toward soldiers with an MOS of 11 C. The team would carry an 81 -mm mortar (tube, tripod, and baseplate) to three different locations, each a mile apart, and set it up in a firing configuration. This type of event is excellent for fine-tuning job skills and is also physically challenging.

OPENING CEREMONY

The commander, ranking person, or ceremony host gives an inspirational speech before the opening ceremonies, welcoming competitors and wishing them good luck. The Olympics is officially opened with a torch lighting. This is followed by a short symbolic parade of all the teams. The teams are then put back into formation, and team captains lead motivating chants. The master of ceremonies (MC) announces the sequence of events and rules for each event. The games then begin.

JUDGING AND SCORING

The MC should have one assistant per team who will judge that one team during each event. Assistants give input on events that need a numerical count. The MC monitors the point accumulation of each team. Points are awarded for each event as follows:

- First = 4 points.
- Second = 3 points.
- Third = 2 points.
- Fourth = 1 point.

When two teams tie an event, the points are added together and split equally between them. After the competition ends, the totaled point scores for each team are figured. The first- through fourth-place teams are then recognized.

CHAPTER 10 .

Developing the Program

Commanders must develop prgrams that train soldiers to maximize their physical performance.

The goal of the Army's physical fitness program is to improve each soldier's physical ability so he can survive and win on the battlefield. Physical fitness includes all aspects of physical performance, not just performance on the APFT. Leaders must understand the principles of exercise, the FITT factors, and know how to apply them in order to develop a sound PT program that will improve all the fitness components. To plan PT successfully, the commander and MFT must know the training management system. (See FM 25-100.)

Commanders should not be satisfied with merely meeting the minimum requirements for physical training which is having all of their soldiers pass the APFT. They must develop programs that train soldiers to maximize their physical performance. Leaders should use incentives. More importantly, they must set the example through their own participation.

The unit PT program is the commander's program. It must reflect his goals and be based on sound, scientific principles. The wise commander also uses his PT program as a basis for building team spirit and for enhancing other training activities. Tough, realistic training is good. However, leaders must be aware of the risks involved with physical training and related activities. They should, therefore, plan wisely to minimize injuries and accidents.

Steps in Planning

STEP 1: ANALYZE THE MISSION

When planning a physical fitness program, the commander must consider the type of unit and its mission. Missions vary as do the physical requirements necessary to complete them. As stated in FM 25-100, "The wartime mission drives training." A careful analysis of the mission, coupled with the commander's intent, yields the mission-essential task list (METL) a unit must perform.

Regardless of the unit's size or mission, reasonable goals are essential. According to FM 25-100, the goals should provide a common direction for all the commander's programs and systems. An example of a goal is as follows because the exceptional physical fitness of the soldier is a critical combat-multiplier in the division, it must be our goal to ensure that our soldiers are capable of roadmarching 12 miles with a 50-pound load in less than three hours.

STEP 2: DEVELOP FITNESS OBJECTIVES

Objectives direct the unit's efforts by prescribing specific actions. The commander, as tactician, and the MFT, as physical fitness advisor, must analyze the METL and equate this to specific fitness objectives. Examples of fitness objectives are the following:

- Improve the unit's overall level of strength by ensuring that all soldiers in the unit can correctly perform at least one repetition with 50 percent of their bodyweight on the overhead press using a barbell.
- Improve the unit's average APFT score through each soldier obtaining a minimum score of 80 points on the push-up and sit-up events and 70 points on the 2-mile run.
- Decrease the number of physical training injuries by 25 percent through properly conducted training.

The commander and MFT identify and prioritize the objectives.

STEP 3: ASSESS THE UNIT

With the training objectives established, the commander and MFT are ready to find the unit's current fitness level and measure it against the desired level.

Giving a diagnostic APFT is one way to find the current level. Another way is to have the soldiers road march a certain distance within a set time while carrying a specified load. Any quantifiable, physically demanding, mission-essential task can be used as an assessment tool. Training records and reports, as well as any previous ARTEP, EDREs, and so forth, can also provide invaluable information.

STEP 4: DETERMINE TRAINING REQUIREMENTS

By possessing the unit's fitness capabilities and comparing them to the standards defined in training objectives, leaders can determine fitness training requirements. When, after extensive training, soldiers cannot reach the desired levels of fitness, training requirements may be too idealistic. Once training requirements are determined, the commander reviews higher headquarters' long- and short-range training plans to identify training events and allocations of resources which will affect near-term planning.

STEP 5: DEVELOP FITNESS TASKS

Fitness tasks provide the framework for accomplishing all training requirements. They identify what has to be done to correct all deficiencies and sustain all proficiencies. Fitness tasks establish priorities, frequencies, and the sequence for training requirements. They must be adjusted for real world constraints before they become a part of the training plan. The essential elements of fitness tasks can be cataloged into four groups:

(1) Collective tasks
(2) Individual tasks
(3) Leader tasks
(4) Resources required for training

Collective tasks. Collective tasks are the training activities performed by the unit. They are keyed to the unit's specific fitness objectives. An example would be to conduct training to develop strength and muscular endurance utilizing a sandbag circuit.

Individual tasks. Individual tasks are activities that an individual soldier must do to accomplish the collective training task. For example, to improve CR endurance the individual soldier must do ability-group running, road marching, Fartlek training, interval training, and calculate/monitor his THR when appropriate.

Leader tasks. Leader tasks are the specific tasks leaders must do in order for collective and individual training to take place. These will involve procuring resources, the setting up of training, education of individual soldiers, and the supervision of the actual training.

Resources. Identifying the necessary equipment, facilities, and training aids during the planning phase gives the trainer ample time to prepare for the training. The early identification and acquisition of resources is necessary to fully implement the training program. The bottom line is that training programs must be developed using resources which are available.

STEP 6: DEVELOP A TRAINING SCHEDULE

The fitness training schedule results from leaders' near-term planning. Leaders must emphasize the development of all the fitness components and follow the principles of exercise and the FITT factors. The training schedule shows the order, intensity, and duration of activities for PT. Figure 10-1 illustrates a typical PT session and its component parts.

There are three distinct steps in planning a unit's daily physical training activities. They are as follows:

1. Determine the minimum frequency of training. Ideally, it should in-include three cardiorespiratory and three muscular conditioning sessions each weeks. (See the FITT factors in Chapter 1.)
2. Determine the type of activity. This depends on the specific purpose of the training session. (See Figure 10-2.) For more information on this topic, see Chapters 1, 2, and 3.
3. Determine the intensity and time of the selected activity. (See the FITT factors in Chapter 1.)

Each activity period should include a warm-up, a workout that develops cardiorespiratory fitness and/or muscular endurance and strength, and a cool-down. (See Figure 10-1).

At the end of a well-planned and executed PT session, all soldiers should feel that they have been physically stressed. They should also understand the objective of the training session and how it will help them improve their fitness levels.

STEP 7: CONDUCT AND EVALUATE TRAINING

The commander and MFT now begin managing and supervising the day-to-day training. They evaluate how the training is performed by monitoring its intensity, using THR or muscle failure, along with the duration of the daily workout.

The key to evaluating training is to determine if the training being conducted will result in improvements in physical conditioning. If not, the training needs revision. Leaders should

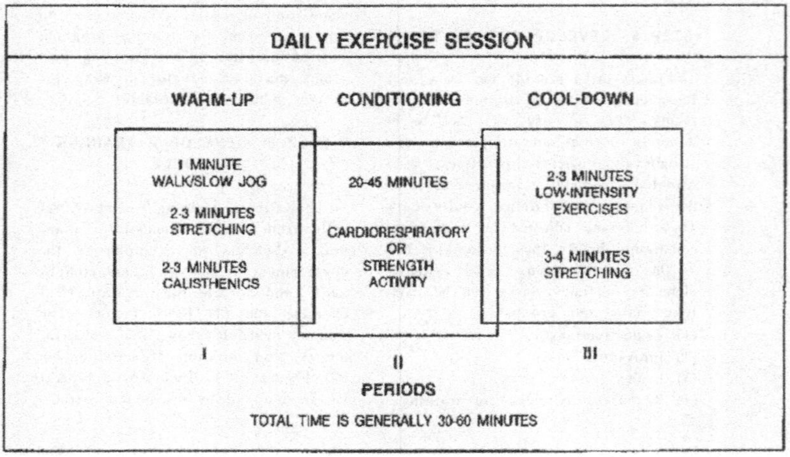

Figure 10-1

ACTIVITY SELECTION GUIDE

PURPOSE	MUSCULAR STRENGTH	MUSCULAR ENDURANCE	CARDIO-RESPIRATORY ENDURANCE	FLEXI-BILITY	BODY COMPO-SITION	SPEED/ AGILITY	COORDI-NATION	TEAM-WORK	SOLDIER SKILLS
Aerobics		X	X	X	X		X		
Bicycling		X	X		X				
Circuits	X	X	X	X	X	X	X	X	X
Competitive Activities						X	X	X	X
Calisthenics		X		X		X	X		
Cross Country Skiing	X	X	X	X	X		X		
Grass/Guerrilla Drills	X	X	X		X		X		
Obstacle Courses	X	X	X		X	X	X	X	X
Partner-Resisted Exercises	X	X					X	X	
Relays		X	X		X	X	X	X	
Rifle Drills	X	X					X		X
Road Marching	X	X	X		X				X
Running		X	X		X				
Stretching				X					
Weight Training	X	X				X	X		

Figure 10-2

not be sidetracked by PT that is all form and little substance. Such training defeats the concept of objective-based training and results in little benefit to soldiers.

Education

Teaching soldiers about physical fitness is vital. It must be an ongoing effort that uses trained experts like MFTs. Soldiers must understand why the program is organized the way it is and what the basic fitness principles are. When they know why they are training in a certain way, they are more likely to wholeheartedly take part. This makes the training more effective.

Education also helps the Army develop its total fitness concept. Total fitness should be reinforced throughout each soldier's career. Classroom instruction in subjects such as principles of exercise, diet and nutrition, tobacco cessation, and stress management should be held at regular intervals. Local "Fit to Win" coordinators (AR 600-63) can help develop classes on such subjects.

Common Errors

There are some common errors in unit programs. The most common error concerns the use of unit runs. When all soldiers must run at the same pace as with a unit run, many do not receive a training effect because they do not reach their training heart rate (THR). The least-fit soldiers of the unit may be at risk because they may be training at heart rates above their THR. Another error is exclusively using activities such as the "daily dozen." These exercises emphasize form over substance and do little to improve fitness.

Total fitness should be reinforced throughout each soldier's career by classroom instruction.

Yet another error is failing to strike a balance in a PT program between CR endurance training and muscular endurance and strength training. In addition, imbalances often stem from a lack of variety in the program which leads to boredom. The principles of exercise are described in Chapter 1, and their application is shown in the sample program below.

A Sample Program

The following sample program shows a commander's thought processes as he develops a 12-week fitness training program for his unit.

Captain Frank Jones's company has just returned from the field where it completed an ARTEP. Several injuries occurred including a broken foot, resulting from a dropped container, and three low back strains. After evaluating his unit during this ARTEP, CPT Jones concluded that its level of physical fitness was inadequate. He thought this contributed to the injuries and poor performance. The soldiers' flexibility was poor, and there was an apparent lack of prior emphasis on, and training in, good lifting techniques. This, combined with poor flexibility in the low back and hamstrings, may have contributed to the unacceptably high number of low back strains. Captain Jones decided to ask the battalion's MFT to help him develop a good unit program for the company. They went through the following steps.

7-STEP PLANNING PROCESS

ANALYZE THE MISSION

DEVELOP FITNESS OBJECTIVES

ASSESS THE UNIT

DETERMINE TRAINING REQUIREMENTS

DESIGN FITNESS TASKS

DEVELOP A TRAINING SCHEDULE

CONDUCT AND EVALUATE TRAINING

ANALYZE THE MISSION

First, they analyzed the recently completed ARTEP and reviewed the ARTEP manual to find the most physically demanding, mission-oriented tasks the unit performs. The analysis showed that, typically, the company does a tactical road march and then occupies a position. It establishes a perimeter, improves its positions, and selects and prepares alternate positions. One of the most demanding missions while in position requires soldiers to move by hand, for 15 to 30 minutes, equipment weighing up to 95 pounds. If his unit received artillery fire, it would need to be able to move to alternate positions as quickly as possible. This requires much lifting, digging, loading, unloading, and moving of heavy equipment. All of these tasks require good muscular endurance and strength and a reasonable level of cardiorespiratory endurance.

DEVELOP FITNESS OBJECTIVES

Next, CPT Jones reviewed his battalion commander's physical training guidance. It showed that the commander was aware that the unit's tasks require muscular endurance and strength and cardiorespiratory fitness. The guidance and objectives issued are as follows:

a. Units will do PT five days a week (0600-0700) when in garrison. In the field, organized PT will be at the commander's discretion.

Captain Jones determined that the major PT emphasis should be to improve muscular endurance and strength. He based this on his unit's mission, training schedule, available resources, and on his commander's guidance and objectives. With this information and the MIT's recommendations, CPT Jones developed the following fitness objectives.

- Improve the unit's overall level of muscular endurance and strength.
- Improve the unit's overall level of flexibility.
- Improve the unit's average APFT score. Each soldier will score at least 80 points on the push-up and

sit-up events and 70 points on the 2-mile run.
- Improve the unit's road marching capability so that 100 percent of the unit can complete a 12-mile road march with a 35-pound load in at least 3.5 hours.
- Decrease the number of profiles.
- Reduce tobacco use.

ASSESS THE UNIT

The next step CPT Jones accomplished was to assess his unit.

The MFT studied the results of the unit's latest APFT and came up with the following information:
- The average push-up score was 68 points.
- The average sit-up score was 72 points.
- The average number of points scored on the 2-mile run was 74.
- There were six failures, two on the 2-mile run and four on the push-up.

The MFT also recommended that the unit be assessed in the following areas: road march performance, strength, flexibility, substance abuse, and profiled soldiers.

Following the MFT's recommendations, subordinate leaders made the following assessments/determinations:
- Eighty-eight percent of the company finished the 12-mile road march with a 35-pound load in under 3 hours 30 minutes.
- A formation toe-touch test revealed that over half the company could not touch their toes while their knees were extended.
- Thirty percent of the unit uses tobacco.
- Two soldiers are in the overweight program.
- Eight percent of the unit is now on temporary profile, most from back problems.

DETERMINE TRAINING REQUIREMENTS

The next step CPT Jones accomplished was to determine the training requirements.

Training requirements are determined by analyzing the training results and the data obtained from the unit assessment. The next step is to compare this data to the standards identified in the training objectives. When performance is less than the established standard, the problem must be addressed and corrected.

Captain Jones established the following training requirements.

Units will do flexibility exercises during the warm-up and cool-down phase of every PT session. During the cool-down, emphasis on will be placed on developing flexibility in the low back, hamstrings, and hip extensor muscle groups.

Each soldier will do 8 to 12 repetitions of bent-leg, sandbag dead-lifts at least two times a week to develop strength. The section leader will supervise lifts.

Each soldier will do heavy resistance/weight training for all the muscle groups of the body two to three times a week.

Each soldier will perform timed sets of push-ups and sit-ups.

Each soldier will train at least 20 to 30 minutes at THR two to three times a week.

Road marches will be conducted at least once every other week.

Tobacco cessation classes will be established to reduce the number of tobacco users.

DESIGN FITNESS TASKS

Once all training requirements are identified, the next step is to use them to design fitness tasks which relate to the fitness objectives. In developing the fitness tasks, CPT Jones must address collective, individual, and leader tasks as well as resources required.

Fitness tasks provide the framework for accomplishing the training requirements. By accurately listing the fitness tasks that must be done and the resources required to do them, the subsequent step of developing a training schedule is greatly facilitated.

An example of designing fitness tasks is provided in Figure 10-3 by using the activities which might occur during one week of physical training.

The collective tasks for the unit are to perform the following: develop muscular endurance and strength, improve CR endurance, and improve flexibility.

The individual tasks all soldiers must perform during the week are as follows. For developing strength and muscular endurance, they must perform appropriate strength circuit exercises, PREs, sandbag circuits, to include performing bent-leg dead lifts exercises, and training for push-up/sit-up improvement. To improve cardiorespiratory endurance, they must do ability-group runs, interval training, road marching, and they must calculate their THR and monitor THR when appropriate. To improve their flexibility, they must do stretching exercises during their daily warm-up and cool-down.

The leader's tasks are to organize and supervise all strength- and muscle endurance-training sessions and CR training sessions so as to best meet all related fitness objectives. Similarly, the leader must organize and supervise all warm-up and cool-down sessions to best meet the fitness objectives for the development and maintenance of flexibility.

To provide specific examples of leaders tasks in the area of training for strength and muscle endurance, the leader will ensure the following:

- Each strength- and/or muscle endurance-training session works all the major muscle groups of the body.
- High priority is given to training those muscles and muscle groups used in mission-essential tasks.
- Areas where weaknesses exist, with respect to strength/muscle

FITNESS TASKS FOR ONE WEEK OF PHYSICAL TRAINING

COLLECTIVE	INDIVIDUAL	LEADER	RESOURCES
Improve strength and muscle endurance	Do STR CIR EX*, PRE, SNDBG CIR, SU-PU IMP	Organize & supervise STR CIR EX, PRE, & SNDBG CIR	STR RM, Gym, Sandbags, PT Field
Improve CR endurance	Do AGR, CAL/MON THR, road march, do intervals (4x440) IND AB	Organize & supervise CR workouts, CAL/MON THR, MON work/relief ratio for intervals	Track, Running Trails
Improve flexibility	Do stretching exercises	Organize & supervise activity	Gym

* A list of abbreviations appears at the end of Figure 10-4.

Figure 10-3

endurance, are targeted in all workouts.

- Problem areas related to APFT performance are addressed in appropriate workouts.
- The duration of each strength training session is 20-40 minutes.
- Soldiers train to muscle failure.
- All the principles of exercise, to include regularity, overload, recovery, progression, specificity, balance are used.

In a similar manner, the leader would ensure that the guidelines and principles outlined in this and earlier chapters are used to organize training sessions for improving CR endurance and flexibility.

The resources needed for the one-week period are as follows: a strength room, a gym, a PT field, a running track and/or running trails, and sandbags.

DEVELOP A TRAINING SCHEDULE

The next step was to develop a fitness training schedule (shown at Figure 10-4). It lists the daily activities and their intensity and duration.

12-WEEK TRAINING PLAN

JULY

MONDAY	TUESDAY	WEDNESDAY	THURSDAY	FRIDAY
START ASSESSMENT* ACT: AGR** INT: 70% HRR*** DUR: 20 MIN	FINISH ASSESSMENT ACT: PLT 1 & 2 STR CIR; PLT 3 & 4 SNDBG CIR/PU-SU IMP INT: MF/MF DUR: 30/4 MIN	ACT: AGR/LINE SOCCER INT: 70% HRR/NA DUR: 20/30 MIN	ACT: PRE/PU-SU IMP INT: MF DUR: 35/4 MIN	ACT: ROAD MARCH, 5 MLE W 35 LBS IN 90 MIN
ACT: PLT 1 & 2 WT STR CIR; PLT 3 & 4 SNDBG CIR/PU-SU IMP INT: MF DUR : 30-35/4 MIN	ACT: AGR/ PAR COURSE INT: 70% HRR DUR: 20/15-20 MIN	FLIP - FLOP MONDAY's WORKOUT	ACT: AGR/GDR INT: 70% HRR DUR: 20/15-20 MIN	ACT: PRE/PU-SU IMP INT: MF DUR: 35/4 MIN
ACT: FIXED CIR I INT: 70% HRR DUR: 30-40 MIN	ACT: PRE/PU-SU IMP INT: MF DUR: 40/5 MIN	ACT: AGR/GDR INT: 70% HRR DUR: 22/20 MIN	ACT: SNDBG CIR/PU-SU IMP INT: MF DUR: 35-40/5 MIN	ACT: ROAD MARCH, 5 MLE W 35 LBS IN 90 MIN
ACT: PLT 1 & 2 STR CIR; PLT 3 & 4 SNDBG CIR/PU-SU IMP INT: MF DUR: 30-40/5 MIN	ACT: AGR INT: 70% HRR DUR: 25 MIN	FLIP - FLOP MONDAY'S WORKOUT	ACT: AGR INT: 70% HRR DUR: 25 MIN	ACT: OBS CRS/ PRE/PU-SU IMP INT: MF DUR: 25/20/5 MIN

* Initially, assessments must be made of each soldier's level of physical fitness. Particularly important is assessing a soldier's strength and muscular endurance by determining his 8-12 RM for each resistance exercise he will be doing. As mentioned in the Phases of Conditioning section in Chapter 3, this will take two weeks and should be planned for accordingly. The other components of fitness should also be addressed as the need arises.

** A list of abbreviations and acronyms appears at the end of this training plan.

*** Those soldiers with a fairly good level of CR fitness (that is, the average soldier) should exercise at about 70 percent HRR. Those with very high levels of CR fitness may benefit most from training at around 80 to 85 percent HRR during a CR training workout.

Figure 10-4

12-WEEK TRAINING PLAN

AUGUST				
MONDAY	**TUESDAY**	**WEDNESDAY**	**THURSDAY**	**FRIDAY**
ACT: AGR INT: 70% HRR DUR: 27 MIN	ACT: PLT 1 & 2 SNDBG CIR; PLT 3 & 4 PRE/PU-SU IMP INT: MF DUR: 40/6	ACT: INTERVALS INT: 8 x 440 IND AB DUR: 45 MIN	FLIP - FLOP TUESDAY'S WORKOUT	ACT: ROAD MARCH, 7.5 MLE W 35 LBS IN 2.5 HOURS
ACT: PLT 1 & 2 SNDBG CIR; PLT 3 & 4 STR CIR/PU SU IMP INT: MF DUR: 40/6 MIN	ACT: LAST MAN-UP RUN IN AG/ PAR CRS INT: 70-80% HRR*/70%HRR DUR: 30/20 MIN	FLIP - FLOP MONDAY'S WORKOUT	ACT: AGR/FIT-NESS RELAYS INT: 70% HRR/NA DUR: 30/20 MIN	ACT: PLT 1 & 2 SNDBG CIR; PLT 3 & 4 PRE/ PU-SU IMP INT: MF DUR: 40/6 MIN
IN FIELD: PLAN FOR	IN FIELD PLAN FOR	IN FIELD: PLAN FOR	IN FIELD: PLAN FOR	IN FIELD: PLAN FOR
ACT: LAST MAN-UP RUN IN AG INT: 70-90% HRR* DUR: 32 MIN	ACT: PRE/PU-SU IMP INT: MF DUR: 40/7 MIN	ACT: FARTLEK IN AG INT: 60-90% HRR* DUR: 32 MIN	ACT: PRE/PU-SU IMP INT: MF DUR: 40/8 MIN	ACT: ROAD MARCH, 10 MLE W 35 LBS IN 3.5 HOURS
ACT: PRE/PU-SU IMP INT: MF DUR: 35/10 MIN	ACT: INTERVALS INT: 8 x 440 IND AB DUR: 45 MIN	ACT: PU-SU, PULL-UP IMP INT: MF DUR: 45 MIN	ACT: FARTLEK IN AG INT: 60-90% HRR* DUR: 35 MIN	ACT: PU-SU, PULL-UP IMP INT: MF DUR: 45 MIN LC: APFT FOR GRADERS

* During the Last-Man-Up and Fartlek running, the heart rate will vary depending on whether it is taken during the slower or the faster portion of the run. The smaller and larger numbers provided for percent HRR should set the lower and upper limits, respectively, for a soldier's heart rate during this type of training. During interval running, the soldier should concern himself with running at the appropriate pace; he should not monitor THR during interval work.

Figure 10-4 (continued)

12-WEEK TRAINING PLAN

SEPTEMBER

MONDAY	TUESDAY	WEDNESDAY	THURSDAY	FRIDAY
ACT: DEVEL-OPMENTAL STRETCHING INT: SLIGHT TENSION, NOT PAIN DUR: 20-30 MIN	ACT: APFT DUR: NA	ACT: UNIT OLYMPICS, PART I DUR: NA	ACT: UNIT OLYMPICS, PART II DUR: NA	ACT: APFT & OLYMPIC AWARDS CEREMONY/ UNIT RUN INT: NA/CD DUR: NA/30-40 MIN
ACT: PLT 1 & 2 STR CIR; PLT 3, & 4 PRE INT: MF DUR: 40 MIN	ACT: PLT 1 & 2 AGR; PLT 3 & 4 EX TO MUSIC INT: 70% HRR DUR: 35/45 MIN	FLIP - FLOP MONDAY'S WORKOUT	ACT: ROAD MARCH, 10 MLE W 35 LBS IN 3 HOURS	ACT: PLT 1 & 2 SNDBG CIR; PLT 3 & 4 PRE INT: MF DUR: 40 MIN
ACT: AGR INT: 70% HRR DUR: 35 MIN	ACT: PRE/PU-SU IMP INT: MF DUR: 40/8 MIN	ACT. FIXED CIRCUIT/RELAYS INT: 70% HRR/NA DUR: 20/20 MIN	ACT: UPPER BODY PRE/PU-SU IMP INT: MF DUR: 30/8 MIN	ACT: ROAD MARCH, 12 MLE W 35 LBS IN UNDER 4 HOURS
ACT: PLT 1 & 2 LOG DRILLS; PLT 3 & 4 SNDBG CIR/PU-SU IMP INT: MF DUR: 30/8 MIN	ACT: PLT 1 & 2 FIXED CIR; PLT 3 & 4 AQUATICS INT: 70% HRR/NA DUR: 30 MIN	FLIP - FLOP MONDAY'S WORKOUT	FLIP - FLOP TUESDAY'S WORKOUT	ACT: PRE/PU-SU IMP INT: MF DUR: 35/8 MIN

NOTES

1. Push-ups and sit-ups are done as part of each strength workout. In the above sessions, they have been placed near the end of the workout. However, they can occasionally be done before the strength workout for variety. An example of a beginning PU-SU improvement workout lasting about three minutes follows:

   a. Perform one timed set of push-ups for 50 seconds. Follow this immediately with one 50-second, timed set of sit-ups. For all timed sets, each soldier must perform as many repetitions of the exercise as he can during the alloted time period.

   b. Perform a second set of push-ups for 40 seconds. Follow this immediately with a timed set of sit-ups of equal duration.

   As the soldier adapts to this, the difficulty of the session can be increased by adding more timed sets and/or by decreasing the rest interval between like or unlike sets of exercises. For example, the rest period between timed sets of push-ups and sit-ups can be decreased. Also, all of the timed sets for push-ups may be done back-to-back (as can the sit-ups), the rest interval between these timed sets of push-ups can be progressively reduced to make the workout more demanding. Many more options exist for increasing the difficulty of, and adding variety to, these sessions.

2. Activities are planned for the FTX; duration is determined on site.
3. The unit's olympic events include the following:
   a. Ammo-box shuttle-run (fastest time by section).
   b. Biceps, barbell curl (most reps with 60 lbs., total by section).
   c. Leg press (most weight lifted by section).
   d. Standing-toe touch (most soldiers touching toes by section/must hold five seconds).
   e. Highest APFT score by section.

Figure 10-4 (continued)

ABBREVIATIONS AND ACRONYMS

ACT	activity
AG	ability groups
AGR	ability group run
ANA ACT	anaerobic activity
AOTR	assessment of training requirements
CAL/MON	calculate/monitor
CD	commander's decision
CFA	competitive fitness activities
CIR	circuit
CR	cardiorespiratory training
DL	dead-lift (bent-leg)
EX	exercise
GDR	grass drills
GUD	guerilla drills
HRR	heart rate reserve
IMP	improvement
IND AB	at the individual's ability
INT	intensity
LBS	pounds
LC	leader's class
MIN	minute(s)
MF	muscle failure (due to fatigue)
MLE	mile(s)
MS/E	muscle strength/endurance
NA	not applicable
OBS CRS	obstacle course
PLT	platoon
PRE	partner-resisted exercise(s)
PT FLD	physical training field
PU-SU IMP	push-up, sit-up improvement
R	run
SNDBG	sandbag
STR CIR EX	strength circuit exercise
STR RM	strength room
THR	training heart rate
TNG	training
W	with
WT STR CIR	strength circuit with weights
% HRR	percent of heart rate reserve
2-MR	2-mile run

Figure 10-4 (continued)

CONDUCT AND EVALUATE TRAINING

Conducting and evaluating training is the final phase of the training process. This phase includes the evaluation of performance, assessment of capabilities, and feedback portions of the training management cycle. These portions of the cycle must be simultaneous and continuous. To be effective, the evaluation process must address why weaknesses exist, and it must identify corrective actions to be taken. Evaluations should address the following:

- Assessment of proficiency in mission-essential tasks.
- Status of training goals and objectives.
- Status of training in critical individual and collective tasks.
- Shortfalls in training.
- Recommendations for next training cycle (key in on correcting weaknesses).
- Results of educational programs.

Using the Principles of Exercise

As CPT Jones developed his program, he made sure he used the seven principles of exercise. He justified his program as follows:

- Balance. This program is balanced because all the fitness components are addressed. The emphasis is on building muscular endurance and strength in the skeletal muscular system because of the many lifting tasks the unit must do. The program also trains cardiorespiratory endurance and flexibility, and warm-up and cool-down periods are included in every workout.
- Specificity. The unit's fitness goals are met. The sand-bag lifting and weight training programs help develop muscular endurance and strength. The movements should, when possible, stress muscle groups used in their job-related lifting tasks. Developmental stretching should help reduce work-related back injuries. The different types of training in running will help ensure that soldiers reach a satisfactory level of CR fitness and help each soldier score at least 70 points on the APFT's 2-mile run. Soldiers do push-ups and sit-ups at least two or three times a week to improve the unit's performance in these events. The competitive fitness activities will help foster teamwork and cohesion, both of which are essential to each section's functions.
- Overload. Soldiers reach overload in the weight circuit by doing each exercise with an 8- to 12-RM lift for a set time and/or until they reach temporary muscle failure. For the cardiorespiratory workout, THR is calculated initially using 70 percent of the HRR. They do push-ups and sit-ups in multiple, timed sets with short recovery periods to ensure that muscle failure is reached. They also do PREs to muscle failure.
- Progression. To help soldiers reach adequate overload as they improve, the program is made gradually more difficult. Soldiers progress in their CR workout by increasing the time they spend at THR up to 30 to 45 minutes per session and by maintaining THR. They progress on the weight training circuit individually. When a soldier can do an exercise for a set time without reaching muscle failure, the weight is increased so that the soldier reaches muscle failure between the 8th and 12th repetition again. Progression in push-ups and sit-ups involves slowly increasing the duration of the work intervals.
- Variety. There are many different activities for variety. For strength and muscular endurance training the soldiers use weight circuits, sandbag circuits, and PREs. Ability group runs, intervals, Par courses,

Fartlek running, and guerrilla drills are all used for CR training. Varied stretching techniques, including static, partner-assisted, and contract-relax, are used for developmental stretching.

- Regularity. Each component of fitness is worked regularly. Soldiers will spend at least two to three days a week working each of the major fitness components. They will also do push-ups and sit-ups regularly to help reach their peak performance on the APFT.
- Recovery. The muscular and cardiorespiratory systems are stressed in alternate workouts. This allows one system to recover on the day the other is working hard.

Conclusion

CPT Jones's step-by-step process of developing a sound PT program for his unit is an example of what each commander should do in developing his own unit program.

Good physical training takes no more time to plan and execute than does poor training. When commanders use a systematic approach to develop training, the planning process bears sound results and the training will succeed.

CHAPTER II.

Physical Training During Initial Entry Training

Soldiers report to initial entry training (IET) ranging widely in their levels of physical fitness. Because of this, there are special considerations when designing a physical training program for IET soldiers. Physical training involves safely training and challenging all soldiers while improving their fitness level to meet required standards. The regulations which govern the conduct of physical training in IET and explain the graduation requirements are TRADOC Reg. 350-6 and AR 350-15.

The mission of physical training in IET is twofold: to safely train soldiers to meet the graduation requirements of each course and to prepare soldiers to meet the physical demands of their future assignments.

Program Development

All physical training programs in IET must do the following: 1) progressively condition and toughen soldiers for military duties; 2) develop soldiers' self-confidence, discipline, and team spirit; 3) develop healthy life-styles through education; and, 4) improve physical fitness to the highest levels possible in all five components of physical fitness (cardiorespiratory endurance, muscular strength, muscular endurance, flexibility, and body composition).

Because each IET school is somewhat different, commanders must examine the graduation requirements for the course and establish appropriate fitness objectives. They can then design a program that attains these objectives. The seven principles of exercise outlined in Chapter I are universal, and they apply to all PT programs including those in IET. Commanders of initial entry training should look beyond the graduation requirements of their own training course to ensure that their soldiers are prepared for the physical challenges of their future assignments. This means developing safe training programs which will produce the maximum physical improvement possible.

MFTs are skilled at assessing soldiers* capabilities. They use the five components of physical fitness in designing programs to reach the training objectives established by the commander. They also know how to conduct exercise programs that are effective and safe. MFTs are not, however, trained to diagnose or treat injuries.

The commander's latitude in program development varies with the length and type of the IET course. For example, commanders of basic combat training (BCT) may do a standard PT program at one installation, while AIT commanders may design their own programs. Regardless of the type of course, all leaders must strive to train their soldiers to attain the highest level of physical fitness possible. This means using the established principles of exercise to develop a safe physical training program.

Safety Considerations

Overuse injuries are common in IET. However, they can be avoided by carefully following the exercise principles of "recovery" and "progression."

Research suggests that soldiers are more prone to injuries of the lower extremities after the third week of IET. High-impact activities, such as road marching and running on hard surfaces, should be carefully monitored during at this time. During this period, fixed circuits and other activities that develop CR fitness are good, low-impact alternatives.

Properly fitted, high-quality running shoes are important, especially when PT sessions require running on hard surfaces. Court shoes, like basketball or tennis shoes, are not designed to absorb the repetitive shock of running. Activities such as running obstacle courses and road marching require combat boots to protect and support the feet and ankles. Naturally, common sense dictates a reasonable break-in period for new combat boots, especially before long marches.

Examples of recommended PT sessions and low-risk exercises are in Chapter 7. Specific health and safety considerations are in TRADOC Reg. 350-6, paragraph 4-2.

Road Marching

One road march should be conducted weekly with the difficulty of the marches progressing gradually throughout IET.

In the first two weeks of IET, soldiers can be expected to road march up to 5 kilometers with light loads. Loads should be restricted to the standard LCE, kevlar helmet, and weapon. Bones, ligaments, and tendons respond slowly to training and may be injured if the load and/or duration are increased too quickly.

After the initial adaptations in the early weeks of IET, soldiers can be expected to carry progressively heavier loads including a rucksack. By he start of the fourth week, they should be accustomed to marching in boots, and their feet should be less prone to blistering. By the sixth week, the load may be increased to 40 pounds including personal clothing and equipment. At no time during IET or one-station unit training (OSUT) should loads exceed 40 pounds.

A sample regimen for road marches during IET is at Figure 11-1.

SAMPLE ROAD MARCH PROGRAM

WEEK	DISTANCE	EQUIPMENT	REMARKS
1	5 km	LCE, kevlar helmet, weapon	
2	5 km	SAME AS WEEK 1	
3	7 km	LCE, kevlar helmet, weapon, and 10-pound rucksack	With all equipment, the total load is 30 pounds.
4	7 km	SAME AS WEEK 3	SAME TOTAL LOAD AS WEEK 3
5	7 km	SAME AS WEEK 3	SAME TOTAL LOAD AS WEEK 3
6	10 km	LCE, hevlar helmet, weapon, and 20-pound rucksack	With all equipment, the total load is 40 pounds.
7	10 km	SAME AS WEEK 6	SAME TOTAL LOAD AS WEEK 6
8	10 km	SAME AS WEEK 6	SAME TOTAL LOAD AS WEEK 6

Note: The total load carried (to include the LCE, kevlar helmet, weapon, and ruck load) should not exceed that shown in the remarks column. If the road marches are to or from non-tactical training, they need not be tactical road marches.

Figure 11-1

CHAPTER 12

Environmental Considerations

In today's Army, soldiers may deploy anywhere in the world. They may go into the tropical heat of Central America, the deserts of the Middle East, the frozen tundra of Alaska, or the rolling hills of Western Europe. Each environment presents unique problems concerning soldiers' physical performance. Furthermore, physical exertion in extreme environments can be life-threatening. While recognizing such problems is important, preventing them is even more important. This requires an understanding of the environmental factors which affect physical performance and how the body responds to those factors.

Temperature Regulation

The body constantly produces heat, especially during exercise. To maintain a constant normal temperature, it must pass this heat on to the environment. Life-threatening circumstances can develop if the body becomes too hot or too cold. Body temperature must be maintained within fairly narrow limits, usually between 74 and 110 degrees Fahrenheit. However, hypothermia and heat injuries can occur within much narrower limits. Therefore, extreme temperatures can have a devastating effect on the body's ability to control its temperature.

Overheating is a serious threat to health and physical performance. During exercise, the body can produce heat at a rate 10 to 20 times greater than during rest. To survive, it must get rid of the excess heat.

The four ways in which the body can gain or lose heat are the following:

- Conduction-the transfre of heat from a warm object to a cool one that is touching it. (Warming boots by putting them on is an example.)
- Convection-the transfer of heat by circulation or movement of air. (Using a fan on a hot day is an example.)
- Radiation-the transfer of heat by electromagnetic waves. (Sitting under a heat lamp is an example.)
- Evaporation- the transfer of heat by changing a liquid into a gas. (Evaporating sweat cooling the skin is an example.)

Heat moves from warm to cool areas. During exercise, when the body is extremely warm, heat can be lost by a combination of the four methods. Sweating, however, is the body's most important means for heat loss, especially during exercise. Any condition that slows or blocks the transfer of heat from the body by evaporation causes heat storage which results in an increase in body temperature.

The degree to which evaporative cooling occurs is also directly related to the air's relative humidity (a measure of the amount of water vapor in the air). When the relative humidity is 100 percent, the air is completely saturated at its temperature. No more water can evaporate into the surrounding air. As a result, sweat does not evaporate, no cooling effect takes place, and the body temperature increases. This causes even more sweating. During exercise in the heat, sweat rates of up to two quarts per hour are not uncommon.

Adapting to differing environmental conditions is called acclimatization.

If the lost fluids are not replaced, dehydration can occur. This condition, in turn, can result in severe heat injuries.

Thus, in hot, humid conditions when a soldier's sweat cannot evaporate, there is no cooling effect through the process of evaporation. High relative humidities combined with high temperatures can cause serious problems. Weather of this type occurs in the tropics and equatorial regions such as Central America and southern Asia. These are places where soldiers have been or could be deployed.

Heat Injuries and Symptoms

The following are common types of heat injuries and their symptoms.

- Heat cramps-muscles cramps of the abdomen, legs, or arms.
- Heat exhaustion-headache, excessive sweating, dizziness, nausea, clammy skin.
- Heat stroke-hot, dry skin, cessation of sweating, rapid pulse, mental confusion, unconsciousness.

To prevent heat injuries while exercising, trainers must adjust the intensity to fit the temperature and humidity. They must ensure that soldiers drink enough water before and during the exercise session. Body weight is a good gauge of hydration. If rapid weight loss occurs, dehydration should be suspected. Plain water is the best replacement fluid to use. Highly concentrated liquids such as soft drinks and those with a high sugar content may hurt the soldier's performance because they slow the absorption of water from the stomach.

To prevent heat injuries, the following hydration guidelines should be used:

- Type of drink: cool water (45 to 55 degrees F).
- Before the activity: drink 13 to 20 ounces at least 30 minutes before.
- During the activity: drink 3 to 6 ounces at 15 to 30 minute intervals.
- After the activity: drink to satisfy thirst, then drink a little more.

Acclimatization to Hot, Humid Environments

Adapting to differing environmental conditions is called acclimatization. Soldiers who are newly introduced to a hot, humid climate and are moderately active in it can acclimatize in 8 to 14 days. Soldiers who are sedentary take much longer. Until they are acclimatized, soldiers are much more likely to develop heat injuries.

A soldier's ability to perform effectively in hot, humid conditions depends on both his acclimatization and level of fitness. The degree of heat stress directly depends on the relative workload. When two soldiers do the same task, the heat stress is less for the soldier who is in better physical condition, and his performance is likely to be better. Therefore, it is important to maintain high levels of fitness.

Increased temperatures and humidity cause increased heart rates. Consequently, it takes much less effort to elevate the heart rate into the training zone, but the training effect is the same. These facts underscore the need to use combat-development running and to monitor heart rates when running, especially in hot, humid conditions.

Some important changes occur as a result of acclimatization to a hot climate. The following physical adaptations help the body cope with a hot environment

- Sweating occurs at a lower body temperature.
- Sweat production is increased.
- Blood volume is increased.
- Heart rate is less at any given work rate.

Exercising in Cold Environments

Contrary to popular belief, there are few real dangers in exercising at temperatures well below freezing. Since the body produces large amounts of heat during exercise, it has little trouble maintaining a normal temperature. There is no danger of freezing the lungs. However, without proper precautions, hypothermia, frostbite, and dehydration can occur.

HYPOTHERMIA

If the body's core temperature drops below normal, its ability to regulate its temperature can become impaired or lost. This condition is called hypothermia. It develops because the body cannot produce heat as fast as it is losing it. This can lead to death. The chance of a soldier becoming hypothermic is a major threat any time he is exposed to the cold.

Hypothermia develops when the body cannot produce heat as fast as it is losing it.

Some symptoms of hypothermia are shivering, loss of judgment, slurred speech, drowsiness, and muscle weakness.

During exercise in the cold, people usually produce enough heat to maintain normal body temperature. As they get fatigued, however, they slow down and their bodies produce less heat. Also, people often overdress for exercise in the cold. This makes the body sweat. The sweat dampens the clothing next to the skin making it a good conductor of heat. The combination of decreased heat production and increased heat loss can cause a rapid onset of hypothermia.

Some guidelines for dressing for cold weather exercise are shown in Figure 12-1.

GUIDELINES FOR DRESSING FOR EXERCISE IN THE COLD

Clothing for cold weather should protect, insulate, and ventilate.

- Protect by covering as large an area of the body as possible.
- Insulation will occur by trapping air which has been warmed by the body and holding it near the skin.
- Ventilate by allowing a two-way exchange of air through the various layers of clothing.

Clothing should leave your body slightly cool rather than hot.

Clothing should also be loose enough to allow movement.

Clothing soaked with perspiration should be removed if reasonably possible.

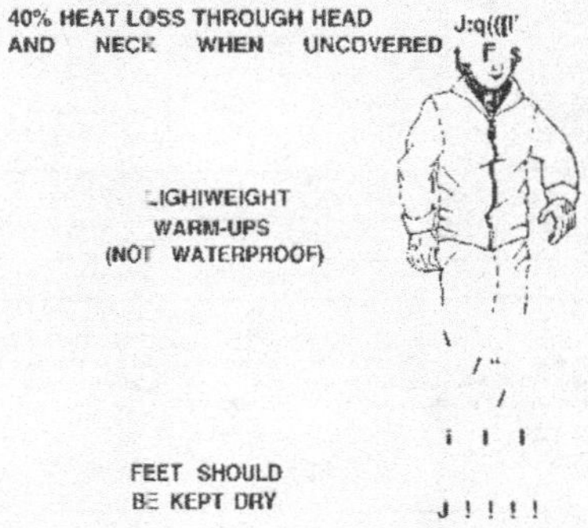

Figure 12-1

FROSTBITE

Frostbite is the freezing of body tissue. It commonly occurs in body parts located away from the core and exposed to the cold such as the nose, ears, feet, hands, and skin. Severe cases of frostbite may require amputation.

Factors which lead to frostbite are cold temperatures combined with windy conditions. The wind has a great cooling effect because it causes rapid convective heat transfer from the body. For a given temperature, the higher the wind speed, the greater the cooling effect. Figure 12-2 shows how the wind can affect cooling by providing information on windchill factors,

A person's movement through the air creates an effect similar to that caused by wind. Riding a bicycle at 15 mph is the same as standing in a 15-mph wind. If, in addition, there is a 5-mph headwind, the overall effect is equivalent to a 20-mph wind. Therefore, an exercising soldier must be very cautious to avoid getting frostbite. Covering exposed parts of the body will substantially reduce the risks.

DEHYDRATION

Dehydration can result from losing body fluids faster than they are replaced. Cold environments are often dry, and water may be limited. As a result, soldiers may in time become dehydrated. While operating in extremely cold climates, trainers should check the body weights of the soldiers regularly and encourage them to drink liquids whenever possible.

WINDCHILL FACTOR

Cooling Power of Wind on Exposed Flesh Expressed as an Equivalent Temperature (under calm conditions)

Estimated wind speed (in mph)	Actual Thermometer Reading (°F)											
	50	40	30	20	10	0	−10	−20	−30	−40	−50	−60
	EQUIVALENT CHILL TEMPERATURE (°F)											
Calm	50	40	30	20	10	0	−10	−20	−30	−40	−50	−60
5	48	37	27	16	6	−5	−15	−26	−36	−47	−57	−68
10	40	28	16	4	−9	−24	−33	46	58	70	83	95
15	36	22	9	5	−18	−32	−45	−58	−72	−85	−99	−112
20	32	18	4	−10	−25	−39	−53	−67	−82	−96	−110	−124
25	30	16	0	−15	−29	−44	−59	−74	−88	−104	−118	−133
30	28	13	−2	−18	33	−48	−63	79	94	109	125	140
35	27	11	−4	−21	35	51	−67	−82	98	113	129	−145
40	26	10	6	−21	−37	−53	−69	−85	−100	−116	−132	−148
(wind speed greater than 40mph have little additional effect)	LITTLE DANGER In 0.5 hr with dry skin Maximum danger of false sense of security				INCREASING DANGER Danger from freezing of exposed flesh within one minute			GREAT DANGER Flesh may freeze within 30 seconds				
	Trenchfoot and immersion foot may occur at any point on this chart.											

INSTRUCTIONS

MEASURE local temperature and wind speed if possible. If not, ESTIMATE. Enter table at closest 5° F interval along the top and with appropriate wind speed along left side. Intersection gives approximate equivalent chill temperature. That is, the temperature that would cause the same rate of cooling under calm conditions. Note that regardless of cooling rate, you do not cool below the actual air temperature unless wet.

Figure 12-2

Acclimatization to High Altitudes

Elevations below 5,000 feet have little noticeable effect on healthy people. However, at higher elevations the atmospheric pressure is reduced, and the body tissues get less oxygen. This means that soldiers cannot work or exercise as well at high altitudes. The limiting effects of high elevation are often most pronounced in older soldiers and persons with low levels of fitness.

Due to acclimatization, the longer a soldier remains at high altitude, the better his performance becomes. Generally, however, he will not perform as well as at sea level and should not be expected to. For normal activities, the time required to acclimatize depends largely on the altitude. In order to insure that soldiers who are newly assigned to altitudes above 5,000 feet are not at a disadvantage, it is recommended that 30 days of acclimatization, including regular physical activity, be permitted before they are administered a record APFT.

Before acclimatization is complete, people at high altitudes may suffer acute mountain sickness. This includes such symptoms as headache, rapid pulse, nausea, loss of appetite, and an inability to sleep. The primary treatment is further acclimatization or returning to a lower altitude.

Once soldiers are acclimatized to altitudes above 5,000 feet, deacclimatization will occur if they spend 14 or more days at lower altitudes. For this reason, soldiers should be permitted twice the length of their absence, not to exceed 30 days, to reacclimatize before being required to take a record APFT. A period of 30 days is adequate for any given reacclimatization.

Air Pollution and Exercise

Pollutants are substances in the environment which lower the environment's quality. Originally, air pollutants were thought to be only byproducts of the industrial revolution. However, many pollutants are produced naturally. For example, volcanoes emit sulfur oxides and ash, and lightning produces ozone.

There are two classifications of air pollutants - primary and secondary. Primary pollutants are produced directly by industrial sources. These include carbon monoxide (CO), sulfur oxides (SO), hydrocarbons, and particulate (ash). Secondary pollutants are created by the primary pollutant's interaction with the environment. Examples of these include ozone **(03)**, aldehydes, and sulfates. Smog is a combination of primary and secondary pollutants.

Some pollutants have negative effects on the body. For example, carbon monoxide binds to hemoglobin in the red blood cells and reduces the amount of oxygen carried in the blood. Ozone and the oxides irritate the air passageways in the lungs, while other pollutants irritate the eyes.

When exercisers in high-pollution areas breathe through the mouth, the nasal mucosa's ability to remove impurities is bypassed, and many pollutants can be inhaled. This irritates the respiratory tract and makes the person less able to perform aerobically.

Pollutants can irritate the respiratory tract and make the person less able to perform aerobically.

The following are some ways to deal with air pollution while exercising:

- Avoid exposure to pollutants before and during exercise, if possible.
- In areas of high ozone concentration, train early in the day and after dark.
- Avoid exercising near heavily traveled streets and highways during rush hours.
- Consult your supporting preventive-medicine activity for advice in identifying or defining training restrictions during periods of heavy air pollution.

CHAPTER **13**

Injuries

Most injuries can be prevented by designing a well-balanced PT program.

Injuries are not an uncommon occurrence during intense physical training. It is, nonetheless, a primary responsibility of all leaders to minimize the risk of injury to soldiers. Safety is always a major concern.

Most injuries can be prevented by designing a well-balanced PT program that does not overstress any body parts, allows enough time for recovery, and includes a warm-up and cool-down. Using strengthening exercises and soft, i,,,i surfaces for stretching and run- .:., .]., helps prevent injuries. If, however, injuries do occur, they should be recognized and properly treated in a timely fashion. If a soldier suspects that he is injured, he should stop what he is doing, report the injury, and seek medical help.

Many common injuries are caused by overuse, that is, soldiers often exercise too much and too often and with too rapid an increase in the workload. Most overuse injuries can be treated with rest, ice, compression, and elevation (RICE). Following any required first aid, health-care personnel should evaluate the injured soldier.

Typical Injuries Associated with Physical Training

Common injuries associated with exercise are the following:

t Abrasion (strawberry) - the rubbing off of skin by friction.
- Dislocation - "the displacement of one or more bones of a joint from their natural positions.
- Hot spot - a hot or irritated feeling of the skin which occurs just before a blister forms. These can be prevented by using petroleum jelly over friction-prone areas.
- Blister - a raised spot on the skin filled with liquid. These can generally be avoided by applying lubricants such as petroleum jelly to areas of friction, keeping footwear (socks, shoes, boots) in good repair, and wearing the proper size of boot or shoe.
- Shinsplints - a painful injury to the soft tissues and bone in the shin area. These are generally caused by wearing shoes with inflexible soles or inadequate shock absorption, running on the toes or on hard surfaces, and/or having calf muscles with a limited range of motion.
- Sprain - a stretching or tearing of the ligament(s) at a joint.
- Muscle spasm (muscle cramp) - a sudden, involuntary contraction of one or more muscles.
- Contusion - a bruise with bleeding into the muscle tissue.
- Strain - a stretching or tearing of the muscles.
- Bursitis - an inflammation of the bursa (a sack-like structure where tendons pass over bones). This occurs at a joint and produces pain when the joint is moved or touched. Sometimes swelling occurs.
- Tendinitis - an inflammation of a tendon that produces pain when the attached muscle contracts. Swelling may not occur.
- Stress fractures of the feet.
- Tibial stress fractures - overuse injuries which seem like shinsplints except that the pain is in a specific area.
- Knee injuries - caused by running on uneven surfaces or with worn out shoes, overuse, and improper body alignment. Soldiers who have problems with their knees can benefit from doing leg exercises that strengthen the front (quadriceps) and rear (hamstrings) thigh muscles.
- Low back problems - caused by poor running, sitting, or lifting techniques, and by failing to stretch the back and hip-flexor muscles and to strengthen the abdominal muscles.

The most common running injuries occur in the feet, ankles, knees, and legs. Although they are hard to eliminate, much can be done to keep them to a minimum. Preventive measures include proper warm-up and cool-down along with stretching exercises. Failure to allow recovery between hard bouts of running can lead to overtraining and can also be a major cause of injuries. A well-conditioned soldier can run five to six times a week. However, to do this safely, he should do two things: gradually build up to running that frequently and vary the intensity of the running sessions to allow recovery between them.

Many running injuries can be prevented by wearing proper footwear. Soldiers should train in running shoes. These are available in a wide range of prices and styles. They should fit properly and have flexible, multi-layered soles with good arch and heel support. Shoes made with leather and nylon uppers are usually the most comfortable. See Appendix E for more information on running shoes.

Many running injuries can be prevented by wearing proper

Since injuries can also be caused by running on hard surfaces, soldiers should, if possible, avoid running on concrete. Soft, even surfaces are best for injury prevention. Whenever possible, soldiers should run on grass paths, dirt paths, or park trails. However, with adequate footwear and recovery periods, running on roads and other hard surfaces should pose no problem.

Common running injuries include the following:

- Black toenails.
- Ingrown toenails.
- Stress fractures of the feet.
- Ankle sprains and fractures.
- Achilles tendinitis (caused by improper stretching and shoes that do not fit.
- Upper leg and groin injuries (which can usually be prevented by using good technique in stretching and doing strengthening exercises).

Tibial stress fractures, knee injuries, low back problems, shinsplints, and blisters, which were mentioned earlier, are also injuries which commonly occur in runners.

Other Factors

Proper clothing can also help prevent injuries. Clothes used for physical activity should be comfortable and fit loosely. A T-shirt or sleeveless undershirt and gym shorts are best in warm weather. In cold weather, clothing may be layered according to personal preference. For example, soldiers can wear a BDU, sweat suit, jogging suit, or even Army-issued long underwear. In very cold weather, soldiers may need gloves or mittens and ear-protecting caps. Rubberized or plastic suits should never be worn during exercise. They cause excessive sweating which can lead to dehydration and a dangerous increase in body temperature.

Army Regulation 385-55 (paragraph B- 12, C) prohibits the use of headphones or earphones while walking, jogging, skating, or bicycling on the roads and streets of military installations. However, they may be worn on tracks and running trails.

Road safety equipment is required on administative-type walks, marches, or runs which cross highways, roads, or tank trails or which are conducted on traffic ways. If there is reduced visibility, control personnel must use added caution to ensure the safety of their soldiers.

CHAPTER 14

Army Physical Fitness Test

All soldiers in the Active Army, Army National Guard, and Army Reserve must take the Army Physical Fitness Test (APFT) regardless of their age. The APFT is a three-event physical performance test used to assess muscular endurance and cardiorespiratory (CR) fitness. It is a simple way to measure a soldier's ability to effectively move his body by using his major muscle groups and CR system. Performance on the APFT is strongly linked to the soldier's fitness level and his ability to do fitness-related tasks. An APFT with alternate test events is given to soldiers with permanent profiles and with temporary profiles greater than three months' duration.

The APFT is a three-event physical performance test used to assess muscular endurance and cardiorespiratory (CR) fitness.

While the APFT testing is an important tool in determining the physical readiness of individual soldiers and units, it should not be the sole basis for the unit's physical fitness training. Commanders at every level must ensure that fitness training is designed to develop physical abilities in a balanced way, not just to help soldiers do well on the APFT.

Commanders should use their unit's APFT results to evaluate its physical fitness level. APFT results may indicate a need to modify the fitness programs to attain higher fitness levels. However, mission-essential tasks, not the APFT, should drive physical training.

Performance on the APFT is strongly linked to the soldier's fitness level and his ability to do fitness-related tasks.

Additional physical performance tests and standards which serve as prerequisites for Airborne/Ranger/Special Forces/SCUBA qualification are provided in DA Pam 351-4.

Methods of Evaluation

Commanders are responsible for ensuring that their soldiers are physically fit (AR 350- 15). There are several ways they can assess fitness including the following

- Testing. This is an efficient way to evaluate both the individual's and the unit's physical performance levels.
- Inspection. This evaluates training procedures and indicates the sound ness of the unit's physical fitness program.
- Observation. This is an ongoing way to review training but is not as reliable as testing as an indicator of the unit's level of fitness.
- Medical examination. This detects individual disabilities, health-re-related problems, and physical problems.

Over-Forty Cardiovascular Screening Program

The Army's over-40 cardiovascular screening program (CVSP) does the following:

- Identifies soldiers with a risk of coronary heart disease.
- Provides guidelines for safe, regular CR exercise.
- Gives advice and help in controlling heart-disease risk factors.
- Uses treadmill testing only for high-risk soldiers who need it.

All soldiers, both active and reserve component, must take the APFT for record regardless of age unless prohibited by a medical profile. For soldiers who reached age 40 on or after 1 January 1989, there is no requirement for clearance in the cardiovascular screening program before taking a record APFT. Soldiers who reached age 40 before 1 January 1989 must be cleared through the cardiovascular screening program before taking a record APFT. Prior to their CVSP evaluation, however, they may still take part in physical training to include diagnostic APFTs unless profiled or contraindications to exercise exist. All soldiers must undergo periodic physical examinations in accordance with AR 40-50! and NGR 40-501. These include screening for cardiovascular risk factors.

Overview

As stated, APFT events assess muscular endurance and CR fitness. The lowest passing APFT standards reflect the minimum acceptable fitness level for all soldiers, regardless of MOS or component. When applied to a com - mand, APFT results show a unit's overall level of physical fitness. However, they are not all-inclusive, overall measures of physical-combat readiness. To assess this, other physical capabilities must be measured. The APFT does, however, give a commander a sound measurement of the general fitness level of his unit.

Service schools, agencies, and units may set performance goals which are above the minimum APFT standards in accordance with their missions (AR 350- 15). Individual soldiers are also encouraged to set for themselves a series of successively higher APFT performance goals. They should always strive to improve themselves physically and never be content with meeting minimum standards. Competition on the APFT among soldiers or units can also be used to motivate them to improve their fitness levels.

Testing is not a substitute for a regular, balanced exercise program. Diagnostic testing is important in monitoring training progress but, when done too often, may decrease motivation and waste training time.

The test period is defined as *the* period of time which elapses from starting to finishing the three events. It must not take more than two hours. Soldiers must do all three events in the same test period.

Test Administration

The APFT must be administered properly and to standard in order to accurately evaluate a soldier's physical fitness and to be fair to all soldiers. (Test results are used for personnel actions.)

Individual soldiers are not authorized to administer the APFT to themselves for the purpose of satisfying a unit's diagnostic or record APFT requirement.

REQUIRED EQUIPMENT

The OIC or NCOIC at the test site must have a copy of FM 21-20 on hand. The supervisor of each event must have the event instructions and standards. Scorers should have a clipboard and an ink pen to record the results on the soldiers' scorecards.

Two stopwatches are needed. They must be able to measure time in both minutes and seconds.

Runners must wear numbers or some other form of identification for the 2-mile run. The numbers may be stenciled or pinned onto pullover vests or sleeveless, mesh pullovers or attached to the runners themselves.

Soldiers should wear clothing that is appropriate for PT such as shorts, T-shirts, socks, and running shoes (not tennis shoes). They should not wear basketball shoes or other types of court shoes. BDUs may be worn but may be a hindrance on some events.

Anything that gives a soldier an unfair advantage is not permitted during the APFT. Wearing devices such as weight belts or elastic bandages may or may not provide an advantage. However, for standardization, such additional equipment is not authorized unless prescribed by medical personnel. The only exception is gloves. They may be worn in cold weather when approved by the local commander.

Each soldier needs a DA Form 705, Army Physical Fitness Test Scorecard. The soldier fills in his name, social security number, grade, age, and sex.

(See Figure 14-1.) The unit will complete the height and weight data.

Scorers record the raw score for each event and initial the results. If a soldier fails an event or finds it difficult to perform, the scorer should write down the reasons and other pertinent information in the comment block. After the entire APFT has been completed, the event scorer will convert raw scores to point scores using the scoring standards on the back of the scorecards. (See Figure 14-1.)

See page 14-8.1 for instructions on completing DA Form 705.

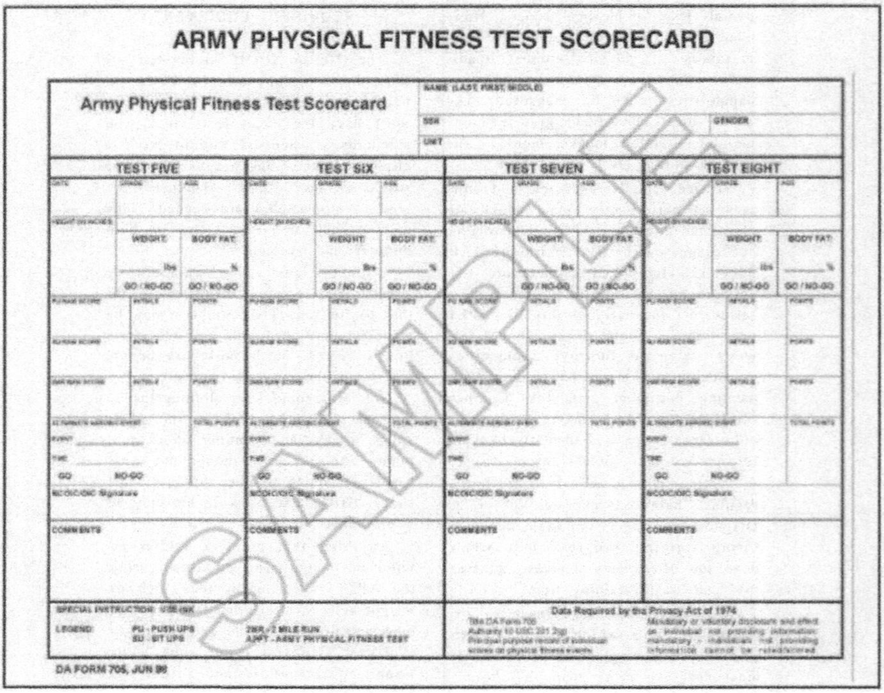
ARMY PHYSICAL FITNESS TEST SCORECARD

Army Physical Fitness Test Scorecard

NAME (LAST, FIRST, MIDDLE)

SSN | GENDER

UNIT

TEST FIVE	TEST SIX	TEST SEVEN	TEST EIGHT
WEIGHT ___ lbs / BODY FAT ___ %	WEIGHT ___ lbs / BODY FAT ___ %	WEIGHT ___ lbs / BODY FAT ___ %	WEIGHT ___ lbs / BODY FAT ___ %
GO / NO-GO	GO / NO-GO	GO / NO-GO	GO / NO-GO
GO NO-GO	GO NO-GO	GO NO-GO	GO NO-GO
NCOIC/OIC Signature	NCOIC/OIC Signature	NCOIC/OIC Signature	NCOIC/OIC Signature
COMMENTS	COMMENTS	COMMENTS	COMMENTS

SPECIAL INSTRUCTION: USE INK

LEGEND: PU - PUSH UPS, SU - SIT UPS, 2MR - 2 MILE RUN, APFT - ARMY PHYSICAL FITNESS TEST

Data Required by the Privacy Act of 1974

DA FORM 705, JUN 98

'Figure 14-1

PUSH-UP STANDARDS

Age Group	17-21		22-26		27-31		32-36		37-41		Age Group	42-46		47-51		52-56		57-61		62+		Age Group
Repetitions	M	F	M	F	M	F	M	F	M	F	Repetitions	M	F	M	F	M	F	M	F	M	F	Repetitions
77					100						77											77
76					99						76											76
75			100		98		100				75											75
74			99		97		99				74											74
73			98		96		98		100		73											73
72			97		95		97		99		72											72
71	100		95		94		96		98		71											71
70	99		94		93		95		97		70											70
69	97		93		92		94		96		69											69
68	96		92		91		93		95		68											68
67	94		91		89		92		94		67											67
66	93		90		88		91		93		66	100										66
65	92		89		87		90		92		65	99										65
64	90		87		86		89		91		64	98										64
63	89		86		85		88		90		63	97										63
62	88		85		84		87		89		62	95										62
61	86		84		83		86		88		61	94										61
60	85		83		82		85		87		60	93										60
59	83		82		81		84		86		59	92		100								59
58	82		81		80		83		85		58	91		99								58
57	81		79		79		82		84		57	90		98								57
56	79		78		78		81		83		56	89		97		100						56
55	78		77		77		79		82		55	88		96		99						55
54	77		76		76		78		81		54	87		94		98						54
53	75		75		75		77		79		53	86		93		97		100				53
52	74		74		74		76		78		52	84		92		96		99				52
51	72		73		73		75		77		51	83		91		94		98				51
50	71		71		72	100	74		76		50	82		89		93		97		100		50
49	70		70		71	99	73		75		49	81		88		92		95		99		49
48	68		69		69	98	72		74		48	80		87		91		94		98		48
47	67		68		68	96	71		73		47	79		86		90		93		96		47
46	66		67	100	67	95	70		72		46	78		85		89		92		95		46
45	64		66	99	66	94	69	100	71		45	77		84		88		91		94		45
44	63		65	97	65	93	68	98	70		44	76		82		87		90		93		44
43	61		63	96	64	92	67	97	69		43	74		81		86		89		92		43
42	60	100	62	94	63	90	66	96	68		42	73		80		85		87		91		42
41	59	99	61	93	62	89	65	95	67		41	72		79		83		86		89		41
40	57	97	60	92	61	88	64	93	66	100	40	71		78		82		85		88		40
39	56	95	59	90	60	87	63	92	65	99	39	70		76		81		84		87		39
38	54	93	58	89	59	85	62	91	64	97	38	69		75		80		83		86		38
37	53	91	57	88	58	84	61	89	63	96	37	68	100	74		79		82		85		37
36	52	90	55	86	57	83	60	88	62	94	36	67	99	73		78		81		84		36
35	50	88	54	85	56	82	59	87	61	93	35	66	97	72		77		79		82		35
34	49	86	53	83	55	81	58	85	60	91	34	64	96	71	100	76		78		81		34
33	48	84	52	82	54	79	57	84	59	90	33	63	94	69	99	74		77		80		33
32	46	83	51	81	53	78	56	83	58	88	32	62	92	68	97	73		76		79		32
31	45	81	50	79	52	77	55	81	57	87	31	61	90	67	95	72	100	75		78		31
30	43	79	49	78	50	76	54	80	56	85	30	60	89	66	93	71	98	74		76		30
29	42	77	47	77	49	75	53	79	55	84	29	59	87	65	92	70	96	73		75		29
28	41	76	46	75	48	73	52	77	54	82	28	58	86	64	90	69	95	71	100	74		28
27	39	74	45	74	47	72	51	76	53	81	27	57	84	62	88	68	93	70	98	73		27
26	38	72	44	72	46	71	50	75	52	79	26	56	82	61	87	67	91	69	96	72		26
25	37	70	43	71	45	70	49	73	51	78	25	54	81	60	85	66	89	68	94	71	100	25
24	35	69	42	70	44	68	48	72	50	76	24	53	79	59	83	64	87	67	92	69	98	24
23	34	67	41	68	43	67	47	71	49	75	23	52	78	58	82	63	85	66	90	68	95	23
22	32	65	39	67	42	66	46	69	48	73	22	51	76	56	80	62	84	65	88	67	93	22
21	31	63	38	66	41	65	45	68	47	72	21	50	74	55	78	61	82	63	86	66	91	21
20	30	62	37	64	40	64	44	67	46	70	20	49	73	54	77	60	80	62	84	65	89	20
19	28	60	36	63	39	62	43	65	45	69	19	48	71	53	75	59	78	61	82	64	87	19
18	27	58	35	61	38	61	42	64	44	67	18	47	70	52	73	58	76	60	80	62	84	18
17	26	57	34	60	37	60	41	63	43	66	17	46	68	51	72	57	75	59	78	61	82	17
16	24	55	33	59	36	59	39	61	42	64	16	44	66	49	70	56	73	58	76	60	80	16
15	23	53	31	57	35	58	38	60	41	63	15	43	65	48	68	54	71	57	74	59	78	15
14	21	51	30	56	34	56	37	59	39	61	14	42	63	47	67	53	69	55	72	58	76	14
13	20	50	29	54	33	55	36	58	38	60	13	41	62	46	65	52	67	54	70	56	73	13
12	19	48	28	52	32	54	35	56	37	59	12	40	60	45	63	51	65	53	68	55	71	12
11	17	46	27	50	31	52	34	54	36	57	11	39	58	44	62	50	64	52	66	54	69	11
10	16	44	26	49	29	50	33	52	35	56	10	38	57	42	60	49	62	51	64	53	67	10
9	14	43	25	48	28	49	32	50	34	54	9	37	55	41	58	48	60	50	62	52	64	9
8	13	41	23	47	27	49	31	49	33	53	8	36	54	40	57	47	58	49	60	51	62	8
7	12	39	22	46	26	48	30	48	32	51	7	34	52	39	55	46	56	47	58	49	60	7
6	10	37	21	45	25	47	29	48	31	50	6	33	50	38	53	44	55	46	56	48	58	6
5	9	36	20	43	24	46	28	47	30	48	5	32	49	36	52	43	53	45	54	47	56	5
4	8	34	19	42	23	44	27	45	29	47												
3	6	32	18	41	22	43	26	44	28	45												
2	5	30	17	39	21	42	25	43	27	44												
1	3	29	15	38	20	41	24	41	26	42												
Repetitions	M	F	M	F	M	F	M	F	M	F	Repetitions	M	F	M	F	M	F	M	F	M	F	Repetitions
Age Group	17-21		22-26		27-31		32-36		37-41		Age Group	42-46		47-51		52-56		57-61		62+		Age Group

Scoring standards are used to convert raw scores to point scores after test events are completed. Male point scores are indicated by the M at the top and bottom of the shaded column. Female point scores are indicated by the F at the top and bottom of the unshaded column. To convert raw scores to point scores, find the number of repetitions performed in the left-hand column. Next, move right along that row and locate the intersection of the soldier's appropriate age column. Record that number in the Push-Up points block on the front of the scorecard.

'Figure 14-1 (<ontinued)

SIT-UP STANDARDS												
AGE GROUP	17-21	22-26	27-31	32-36	37-41	AGE GROUP	42-46	47-51	52-56	57-61	62+	AGE GROUP
Repetitions	M/F	M/F	M/F	M/F	M/F	Repetitions	M/F	M/F	M/F	M/F	M/F	Repetitions
82			100			82						82
81			99			81						81
80		100	98			80						80
79		99	97			79						79
78	100	97	96			78						78
77	98	96	95			77						77
76	97	95	94	100	100	76						76
75	95	93	92	99	99	75						75
74	94	92	91	98	98	74						74
73	92	91	90	96	97	73						73
72	90	89	89	95	96	72	100					72
71	89	88	88	94	95	71	99					71
70	87	87	87	93	94	70	98					70
69	86	85	86	92	93	69	97					69
68	84	84	85	91	92	68	96					68
67	82	83	84	89	91	67	95					67
66	81	81	83	88	89	66	94	100	100			66
65	79	80	82	87	88	65	93	99	99			65
64	78	79	81	86	87	64	92	98	98	100		64
63	76	77	79	85	86	63	91	97	97	99	100	63
62	74	76	78	84	85	62	90	96	96	98	99	62
61	73	75	77	82	84	61	89	94	95	97	98	61
60	71	73	76	81	83	60	88	93	94	96	97	60
59	70	72	75	80	82	59	87	92	93	95	96	59
58	68	71	74	79	81	58	86	91	92	94	95	58
57	66	69	73	78	80	57	85	90	91	92	94	57
56	65	68	72	76	79	56	84	89	89	91	92	56
55	63	67	71	75	78	55	83	88	88	90	91	55
54	62	65	70	74	77	54	82	87	87	89	90	54
53	60	64	69	73	76	53	81	86	86	88	89	53
52	58	63	68	72	75	52	80	84	85	87	88	52
51	57	61	66	71	74	51	79	83	84	86	87	51
50	55	60	65	69	73	50	78	82	83	85	86	50
49	54	59	64	68	72	49	77	81	82	84	85	49
48	52	57	63	67	71	48	76	80	81	83	84	48
47	50	56	62	66	69	47	75	79	80	82	83	47
46	49	55	61	65	68	46	74	78	79	81	82	46
45	47	53	60	64	67	45	73	77	78	79	81	45
44	46	52	59	62	66	44	72	76	77	78	79	44
43	44	51	58	61	65	43	71	74	76	77	78	43
42	42	49	57	60	64	42	70	73	75	76	77	42
41	41	48	56	59	63	41	69	72	74	75	76	41
40	39	47	55	58	62	40	68	71	73	74	75	40
39	38	45	54	56	61	39	67	70	72	73	74	39
38	36	44	52	55	60	38	66	69	71	72	73	38
37	34	43	51	54	59	37	65	68	69	71	72	37
36	33	41	50	53	58	36	64	67	68	70	71	36
35	31	40	49	52	57	35	63	66	67	69	70	35
34	30	39	48	50	56	34	62	64	66	68	69	34
33	28	37	47	49	55	33	61	63	65	66	68	33
32	26	36	46	48	54	32	60	62	64	65	66	32
31	25	35	45	47	53	31	59	61	63	64	65	31
30	23	33	44	46	52	30	58	60	62	63	64	30
29	22	32	43	45	50	29	57	59	61	62	63	29
28	20	31	42	44	49	28	56	58	60	61	62	28
27	18	29	41	42	48	27	55	57	59	60	61	27
26	17	28	39	41	47	26	54	56	58	59	60	26
25	15	27	38	40	46	25	53	54	57	58	59	25
24	14	25	37	39	45	24	52	53	56	57	58	24
23	12	24	36	38	44	23	51	52	55	56	57	23
22	10	23	35	36	43	22	50	51	54	55	56	22
21	9	21	34	35	42	21	49	50	53	54	55	21
Repetitions	M/F	M/F	M/F	M/F	M/F	Repetitions	M/F	M/F	M/F	M/F	M/F	Repetitions
AGE GROUP	17-21	22-26	27-31	32-36	37-41	AGE GROUP	42-46	47-51	52-56	57-61	62+	AGE GROUP

Scoring standards are used to convert raw scores to point scores after test events are completed. To convert raw scores to point scores, find the number of repetitions performed in the left-hand column. Next, move right along that row and locate the intersection of the soldier's appropriate age column. Record that number in the Sit-Up points block on the front of the scorecard.

•Figure 14-1 (continued)

2-MILE RUN STANDARDS

AGE GROUP	17-21		22-26		27-31		32-36		37-41		AGE GROUP	42-46		47-51		52-56		57-61		62+		AGE GROUP
Time	M	F	M	F	M	F	M	F	M	F	Time	M	F	M	F	M	F	M	F	M	F	Time
12:54											12:54											12:54
13:00	100		100								13:00											13:00
13:06	99		99								13:06											13:06
13:12	97		98								13:12											13:12
13:18	96		97		100		100				13:18											13:18
13:24	94		96		99		99				13:24											13:24
13:30	93		94		98		98				13:30											13:30
13:36	92		93		97		97		100		13:36											13:36
13:42	90		92		96		96		99		13:42											13:42
13:48	89		91		95		95		98		13:48											13:48
13:54	88		90		94		95		97		13:54											13:54
14:00	86		89		92		94		97		14:00											14:00
14:06	85		88		91		93		96		14:06	100										14:06
14:12	83		87		90		92		95		14:12	99										14:12
14:18	82		86		89		91		94		14:18	98										14:18
14:24	81		84		88		90		93		14:24	97		100								14:24
14:30	79		83		87		89		92		14:30	97		99								14:30
14:36	78		82		86		88		91		14:36	96		98								14:36
14:42	77		81		85		87		91		14:42	95		98		100						14:42
14:48	75		80		84		86		90		14:48	94		97		99						14:48
14:54	74		79		83		85		89		14:54	93		96		98						14:54
15:00	72		78		82		85		88		15:00	92		95		98						15:00
15:06	71		77		81		84		87		15:06	91		95		97						15:06
15:12	70		76		79		83		86		15:12	90		94		96						15:12
15:18	68		74		78		82		86		15:18	90		93		95		100				15:18
15:24	67		73		77		81		85		15:24	89		92		95		99				15:24
15:30	66		72		76		80		84		15:30	88		91		94		98				15:30
15:36	64	100	71	100	75		79		83		15:36	87		91		93		97				15:36
15:42	63	99	70	99	74		78		82		15:42	86		90		92		97		100		15:42
15:48	61	98	69	98	73	100	77		81		15:48	85		89		91		96		99		15:48
15:54	60	98	68	97	72	99	76	100	80		15:54	84		88		91		95		98		15:54
16:00	59	96	67	96	71	98	75	99	80		16:00	83		87		90		94		97		16:00
16:06	57	94	66	95	70	97	75	99	79		16:06	83		87		89		93		96		16:06
16:12	56	93	64	94	69	97	74	98	78		16:12	82		86		88		92		95		16:12
16:18	54	92	63	93	68	96	73	97	77		16:18	81		85		87		91		94		16:18
16:24	53	90	62	92	66	95	72	97	76		16:24	80		84		87		90		93		16:24
16:30	52	89	61	91	65	94	71	96	75		16:30	79		84		86		90		93		16:30
16:36	50	88	60	90	64	93	70	95	74		16:36	78		83		85		89		92		16:36
16:42	49	87	59	89	63	92	69	94	74		16:42	77		82		84		88		91		16:42
16:48	48	85	58	88	62	91	68	94	73		16:48	77		81		84		87		90		16:48
16:54	46	84	57	87	61	91	67	93	72		16:54	76		80		83		86		89		16:54
17:00	45	83	56	86	60	90	66	92	71	100	17:00	75		80		82		85		88		17:00
17:06	43	82	54	85	59	89	65	92	70	99	17:06	74		79		81		84		87		17:06
17:12	42	81	53	84	58	88	65	91	69	99	17:12	73		78		80		83		86		17:12
17:18	41	79	52	83	57	87	64	90	68	98	17:18	72		77		80		83		85		17:18
17:24	39	78	51	82	56	86	63	90	67	97	17:24	71	100	76		79		82		84		17:24
17:30	38	77	50	81	55	86	62	89	67	96	17:30	70	99	76		78		81		83		17:30
17:36	37	76	49	80	54	85	61	88	66	96	17:36	70	98	75	100	77		80		82		17:36
17:42	35	75	48	79	53	84	60	88	65	95	17:42	69	98	74	99	76		79		81		17:42
17:48	34	73	47	78	51	83	59	87	64	94	17:48	68	97	73	99	76		78		80		17:48
17:54	32	72	46	77	50	82	58	86	63	94	17:54	67	97	73	98	75		77		80		17:54
18:00	31	71	44	76	49	81	57	85	63	93	18:00	67	96	72	97	74		77		79		18:00
18:06	30	70	43	75	48	80	56	85	62	92	18:06	66	96	71	97	73		76		78		18:06
18:12	28	68	42	74	47	80	55	84	61	92	18:12	64	95	70	96	73		75		77		18:12
18:18	27	67	41	73	46	79	55	83	60	91	18:18	63	94	69	96	72		74		76		18:18
18:24	26	66	40	72	45	78	54	83	59	90	18:24	63	94	69	95	71		73		75		18:24
18:30	24	65	39	71	44	77	53	82	58	89	18:30	62	93	68	94	70		72		74		18:30
18:36	23	64	38	70	43	76	52	81	57	89	18:36	61	92	67	94	69		71		73		18:36
18:42	21	62	37	69	42	76	51	81	57	88	18:42	60	92	66	93	69		70		72		18:42
18:48	20	61	36	68	41	74	50	80	56	87	18:48	59	91	65	92	68		70		71		18:48
18:54	19	60	34	67	39	74	49	79	55	87	18:54	58	90	65	92	67		69		70		18:54
19:00	17	59	33	66	38	73	48	79	54	86	19:00	57	90	64	91	66	100	68		69		19:00
19:06	16	58	32	65	37	72	47	78	53	85	19:06	57	89	63	91	65	99	67		68		19:06
19:12	14	56	31	64	36	71	46	77	52	85	19:12	56	89	62	90	65	99	66		67		19:12
19:18	13	55	30	63	35	70	45	77	51	84	19:18	55	88	62	89	64	98	65		67		19:18
19:24	12	54	29	62	34	69	45	76	51	83	19:24	54	87	61	89	63	97	64		66		19:24
19:30	10	53	28	61	33	69	44	75	50	82	19:30	53	87	60	88	62	96	63		65		19:30
19:36	9	52	27	60	32	68	43	74	49	82	19:36	52	86	59	87	62	96	63		64		19:36
19:42	8	50	26	59	31	67	42	74	48	81	19:42	51	85	58	87	61	95	62	100	63		19:42
19:48	6	49	24	58	30	66	41	73	47	80	19:48	50	85	58	86	60	94	61	99	62		19:48
19:54	5	48	23	57	29	65	40	72	46	80	19:54	50	84	57	86	59	93	60	98	61		19:54
20:00	3	47	22	56	28	64	39	72	45	79	20:00	49	83	56	85	58	93	59	98	60	100	20:00
20:06	[illegible]	45	21	55	[illegible]	63	38	71	[illegible]	78	20:06	48	[illegible]	55	84	58	[illegible]	58	97	59	[illegible]	20:06

'Figure 14-1 (continued)

Time	M	F	M	F	M	F	M	F	M	F	Time	M	F	M	F	M	F	M	F	M	F	Time
20:06	2	45	21	55	26	63	38	71	45	78	20:06	48	83	55	84	59	92	58	97	59	99	20:06
20:12	1	44	20	54	25	63	37	70	44	78	20:12	47	82	55	84	57	91	57	96	58	98	20:12
20:18	0	43	19	53	24	62	36	70	43	77	20:18	46	82	54	83	56	90	57	95	57	98	20:18
20:24		42	18	52	23	61	35	69	42	76	20:24	45	81	53	82	55	90	56	95	56	97	20:24
20:30		41	17	51	22	60	35	68	41	75	20:30	44	80	52	82	55	89	55	94	55	96	20:30
20:36		39	16	50	21	59	34	68	40	75	20:36	43	80	51	81	54	88	54	93	54	95	20:36
20:42		38	14	49	20	58	33	67	40	74	20:42	43	79	51	81	53	87	53	92	53	94	20:42
20:48		37	13	48	19	57	32	66	39	73	20:48	42	78	50	80	52	87	52	91	53	94	20:48
20:54		36	12	47	18	57	31	66	38	73	20:54	41	78	49	79	51	86	51	91	52	93	20:54
21:00		35	11	46	17	56	30	65	37	72	21:00	40	77	48	79	51	85	50	90	51	92	21:00
21:06		33	10	45	16	55	29	64	36	71	21:06	39	77	47	78	50	84	50	89	50	91	21:06
21:12		32	9	44	15	54	28	63	35	71	21:12	38	76	47	77	49	84	49	88	49	90	21:12
21:18		31	8	43	14	53	27	63	34	70	21:18	37	75	46	77	48	83	48	87	48	90	21:18
21:24		30	7	42	12	52	26	62	34	69	21:24	37	75	45	76	47	82	47	87	47	89	21:24
21:30		28	6	41	11	51	25	61	33	68	21:30	36	74	44	76	47	81	46	86	46	88	21:30
21:36		27	4	40	10	51	25	61	32	68	21:36	35	73	44	75	46	81	45	85	45	87	21:36
21:42		26	3	39	9	50	24	60	31	67	21:42	34	73	43	74	45	80	44	84	44	86	21:42
21:48		25	2	38	8	49	23	59	30	66	21:48	33	72	42	74	44	79	43	84	43	86	21:48
21:54		24	1	37	7	48	22	59	29	66	21:54	32	71	41	73	44	79	43	83	42	85	21:54
22:00		22	0	36	6	47	21	58	28	65	22:00	31	71	40	72	45	78	42	82	41	84	22:00
22:06		21		35	5	46	20	57	28	64	22:06	30	70	40	72	42	77	41	81	40	83	22:06
22:12		20		34	4	46	19	57	27	64	22:12	30	70	39	71	41	76	40	80	40	82	22:12
22:18		19		33	3	45	18	56	26	63	22:18	29	69	38	71	40	76	39	80	39	82	22:18
22:24		18		32	2	44	17	55	25	62	22:24	28	68	37	70	40	75	38	79	38	81	22:24
22:30		16		31	1	43	16	54	24	61	22:30	27	68	36	69	39	74	37	78	37	80	22:30
22:36		15		30	0	42	15	54	23	61	22:36	26	67	35	69	38	73	37	77	36	79	22:36
22:42		14		29		41	15	53	23	60	22:42	25	66	35	68	37	73	36	76	35	78	22:42
22:48		13		28		40	14	52	22	59	22:48	24	65	34	67	36	72	35	76	34	78	22:48
22:54		12		27		40	13	52	21	59	22:54	23	65	33	67	36	71	34	75	33	77	22:54
23:00		10		26		39	12	51	20	58	23:00	23	64	33	66	35	70	33	74	32	76	23:00
23:06		9		25		38	11	50	19	57	23:06	22	64	32	66	34	70	32	73	31	75	23:06
23:12		8		24		37	10	49	18	56	23:12	21	63	31	65	33	69	31	73	30	74	23:12
23:18		7		23		36	9	49	17	56	23:18	20	63	30	64	33	68	30	72	29	74	23:18
23:24		6		22		35	8	48	17	55	23:24	19	62	29	64	32	67	30	71	28	73	23:24
23:30		4		21		34	7	48	16	54	23:30	18	61	29	63	31	67	29	70	27	72	23:30
23:36		3		20		34	6	47	15	54	23:36	17	61	28	62	30	66	28	69	27	71	23:36
23:42		2		19		33	5	46	14	53	23:42	17	60	27	62	29	65	27	69	26	70	23:42
23:48		1		18		32	5	46	13	52	23:48	16	59	26	61	29	64	26	68	25	70	23:48
23:54		0		17		31	4	45	12	[illegible]	23:54	15	59	25	61	28	64	25	67	24	69	23:54
24:00				16		30	3	44	11	51	24:00	14	58	25	60	27	63	24	66	23	68	24:00
24:06				15		29	2	43	11	50	24:06	13	57	24	59	26	62	23	65	22	67	24:06
24:12				14		29	1	43	10	49	24:12	12	57	23	58	25	61	23	65	21	66	24:12
24:18				13		28	0	42	9	49	24:18	11	56	22	58	25	61	22	64	20	65	24:18
24:24				12		27		41	8	48	24:24	10	56	22	57	24	60	21	63	19	65	24:24
24:30				11		26		41	7	47	24:30	10	55	21	57	23	59	20	62	18	64	24:30
24:36				10		25		40	6	47	24:36	9	54	20	56	22	59	19	62	17	63	24:36
24:42				9		24		39	6	46	24:42	8	54	19	56	22	58	18	61	16	62	24:42
24:48				8		23		39	5	45	24:48	7	53	18	55	21	57	17	60	15	62	24:48
24:54				7		23		38	4	45	24:54	6	52	18	54	20	56	17	59	14	61	24:54
25:00				6		22		37	3	44	25:00	5	52	17	54	19	56	16	58	13	60	25:00
25:06				5		21		37	2	43	25:06	4	51	16	53	19	55	15	58	13	59	25:06
25:12				4		20		36	1	42	25:12	3	50	15	52	18	54	14	57	12	58	25:12
25:18				3		19		35	0	42	25:18	3	50	15	52	17	53	13	56	11	58	25:18
25:24				2		18		34		41	25:24	2	49	14	51	16	53	12	55	10	57	25:24
25:30				1		17		34		40	25:30	1	49	13	51	15	52	11	55	9	56	25:30
25:36				0		17		33		40	25:36	0	48	12	50	15	51	10	54	8	55	25:36
25:42						16		32		39	25:42		47	11	49	14	50	10	53	7	54	25:42
25:48						15		32		38	25:48		47	11	49	13	50	9	52	6	54	25:48
25:54						14		31		38	25:54		46	10	48	12	49	8	51	5	53	25:54
26:00						13		30		37	26:00		45	9	47	11	48	7	51	4	52	26:00
26:06						12		30		36	26:06		45	8	47	11	47	6	50	3	51	26:06
26:12						11		29		35	26:12		44	7	46	10	47	5	49	2	50	26:12
26:18						11		29		35	26:18		43	7	46	9	46	4	48	1	50	26:18
26:24						10		28		34	26:24		43	6	45	8	45	3	47	0	49	26:24
26:30						9		27		33	26:30		42	5	44	7	44	3	47	0	48	26:30
Time	M	F	M	F	M	F	M	F	M	F	Time	M	F	M	F	M	F	M	F	M	F	Time
Age Group	17-21		22-26		27-31		32-36		37-41		Age Group	42-46		47-51		52-56		57-61		62+		Age Group

Scoring standards are used to convert raw scores to point scores after test events are completed. Male point scores are indicated by the M at the top and bottom of the shaded column. Female point scores are indicated by the F at the top and bottom of the unshaded column. To convert raw scores to point scores, find the number of repetitions performed in the left-hand column. Next, move right along that row and locate the intersection of the soldier's appropriate age column. **In all cases, when a time falls between two point values, the lower point value is used.** Record that number in the 2MR points block on the front of the scorecard.

'Figure 14-1 (continued)

SUPERVISION

The APFT must be properly supervised to ensure that its objectives are met. Proper supervision ensures uniformity in the following:

- Scoring the test.
- Training of supervisors and scorers.
- Preparing the test and controlling performance factors.

The goal of the APFT is to get an accurate evaluation of the soldiers' fitness levels. Preparations for administering an accurate APFT include the following:

- Selecting and training supervisors and scorers.
- Briefing and orienting administrators and participants.
- Securing a location for the events.

Commanders must strictly control those factors which influence test performance. They must ensure that events, scoring, clothing, and equipment are uniform. Commanders should plan testing which permits each soldier to perform to his maximal level. They should also ensure the following:

- Soldiers are not tested when fatigued or ill.
- Soldiers do not have tiring duties just before taking the APFT.
- Weather and environmental conditions do not inhibit performance.
- Safety is the first consideration.

Duties of Test Personnel

Testers must be totally familiar with the instructions for each event and trained to administer the tests. Correctly supervising testees and laying out the test area are essential duties. The group administering the test must include the following:

- OIC or NCOIC.
- Event supervisor, scorers, and a demonstrator for each event.
- Support personnel (safety, control, and medical as appropriate). There should be no less than one scorer for each 15 soldiers tested. Twelve to 15 scorers are required when a company-sized unit is tested.

OIC OR NCOIC

The OIC or NCOIC does the following:

- Administers the APFT.
- Procures all necessary equipment and supplies.
- Arranges and lays out the test area.
- Trains the event supervisors, scorers, and demonstrators. (Training video tape No. 21-191 should be used for training those who administer the APFT.)
- Ensures the test is properly administered and the events are explained, demonstrated, and scored according to the test standards in this chapter.
- Reports the results after the test.

EVENT SUPERVISORS

Event supervisors do the following:

- Administer the test events.
- Ensure that necessary equipment is on hand.
- Read the test instructions, and have the events demonstrated.
- Supervise the scoring of events, and ensure that they are done correctly.
- Rule on questions and scoring discrepancies for their event.

SCORERS

Scorers do the following:

- Supervise the performance of testees.
- Enforce the test standards in this chapter.
- Count the number of correctly performed repetitions aloud.
- Record the correct, raw score on each soldier's scorecard, and initial the scorecard block.
- Perform other duties assigned by the OIC or NCOIC.

Scorers must be thoroughly trained to maintain uniform scoring standards. They do not participate in the test.

The goal of the APFT is to get an accurate evaluation of the soldier's fitness levels.

Instructions for Completing DA Form 705, Army Physical Fitness Scorecard, June 1998.

NAME Print soldier's last name, first name and middle initial in NAME block.

SSN Print soldier's social security number in SSN block.

GENDER Print **M** for male or **F** for female in GENDER block.

UNIT Print soldier's unit designation in UNIT block.

DATE Print date the APFT is administered in DATE block.

GRADE Print soldier's grade in GRADE block.

AGE Print soldier's age on the date the APFT is administered in AGE block.

HEIGHT Print soldier's height in HEIGHT block. Height will be rounded to the nearest inch. If the height fraction is less than 1/2 inch, round down to the nearest whole number in inches. If the height fraction is greater than 1/2 inch, round up to the next highest whole number in inches.

WEIGHT Print soldier's weight in WEIGHT block. Weight will be recorded to the nearest pound. If the weight fraction is less than 1/2 pound, round down to the nearest pound. If the weight fraction is 1/2 pound or greater, round up to the nearest pound. Circle **GO** if soldier meets screening table weight IAW AR 600-9. Circle **NO-GO** if soldier exceeds screening table weight IAW AR 600-9.

BODY FAT If soldier exceeds screening table weight, print the soldier's body fat in the BODY FAT block. Percent body fat is recorded from DA Form 5500-R, Body Fat Content Worksheet, Dec 85, for male soldiers and DA Form 5501-R, Body Fat Content Worksheet, Dec 85, for female soldiers. Circle **GO** if soldier meets percent body fat for their age and gender IAW AR 600-9. Circle **NO-GO** if soldier exceeds percent body fat for their age and gender IAW AR 600-9. If soldier does not exceed screening table weight or does not appear to have excessive body fat IAW AR 600-9, print N/A (not applicable) in the BODY FAT block.

PU RAW SCORE The event scorer records the number of correctly performed repetitions of the push-up in the PU RAW SCORE block and prints his or her initials in the INITIALS block.

SU RAW SCORE The event scorer records the number of correctly performed repetitions of the sit-up in the SU RAW SCORE block and prints his or her initials in the INITIALS block.

2MR RAW SCORE The event scorer records the two-mile run time in the 2MR RAW SCORE block. The time is recorded in minutes and seconds. The event scorer then determines the point value for the two-mile run using the scoring standards on the reverse side of the scorecard. The point value is recorded in the 2MR POINTS block and the event scorer prints his or her initials in the INITIALS block. In all cases when a point value falls between two point values, the lower point value is used and recorded. The two-mile run event scorer also determines the point value for push-ups and sit-ups using the scoring standards on the reverse side of the scorecard. The point values are recorded in the appropriate push-up and sit-up POINTS block and the event scorer prints his or her initials in the INITIALS block. The two-mile run event scorer totals the points from the three events and records the total APFT score in the TOTAL POINTS block.

ALTERNATE AEROBIC EVENT The event scorer prints the alternate aerobic event administered (800-yard swim, 6.2-mile-stationary bicycle ergometer, 6.2-mile-bicycle test or 2.5-mile walk) in the ALTERNATE AEROBIC EVENT block. The time the soldier completes the alternate aerobic event is recorded in minutes and seconds in the ALTERNATE AEROBIC EVENT block. The standards for the alternate aerobic event tests are listed in FM 21-20, Chapter 14, Figure 14-9. Scoring for all alternate aerobic events is on a **GO** or **NO-GO** basis. No point values are awarded. Circle **GO** if the soldier completes the alternate aerobic event within the required time or less. Circle **NO-GO** if the soldier fails to complete the alternate aerobic event within the required time. The alternate aerobic event scorer also determines the point value for push-ups and or sit-ups using the scoring standards on the reverse side of the scorecard. The point values are recorded in the appropriate push-up and or sitip POINTS block and the event scorer prints his or her initials in the 2MR INITIALS block. The alternate aerobic event scorer totals the points from the push-up and or sit-up events and records the total APFT score in the TOTAL POINTS block.

NCOIC/OIC Signature The NCOIC/OIC checks all test scores for accuracy and signs their name in the NCOIC/OIC Signature block.

COMMENTS The event supervisor, event scorer, NCOIC, or OIC may record comments appropriate to the APFT in the COMMENTS block. Appropriate comments may include: weather conditions, injury during APFT and or appeals.

(THIS PAGE INTENTIONALLY LEFT BLANK.)

SUPPORT PERSONNEL

Safety and control people should be at the test site, depending on local policy and conditions. Medical personnel may also be there. However, they do not have to be on site to have the APFT conducted. At a minimum, the OIC or NCOIC should have a plan, known to all test personnel, for getting medical help if needed.

Test Site

The test site should be fairly flat and free of debris. It should have the following:

- An area for stretching and warming up.
- A soft, flat, dry area for performing push-ups and sit-ups.
- A flat, 2-mile running course with a solid surface and no more than a three-percent grade. (Commanders must use good judgement; no one is expected to survey terrain.)
- No significant hazards, (for example, traffic, slippery road surfaces, heavy pollution).

When necessary or expedient, a quarter-mile running track can "be used. It can be marked with a series of stakes along the inside edge. When the track is laid out, a horizontal midline 279 feet, 9 3/4 inches long must be marked in the center of a clear area. A 120-foot circle is marked at both ends of this line. The track is formed when the outermost points of the two circles are connected with tangent lines. (See Figure 14-2.)

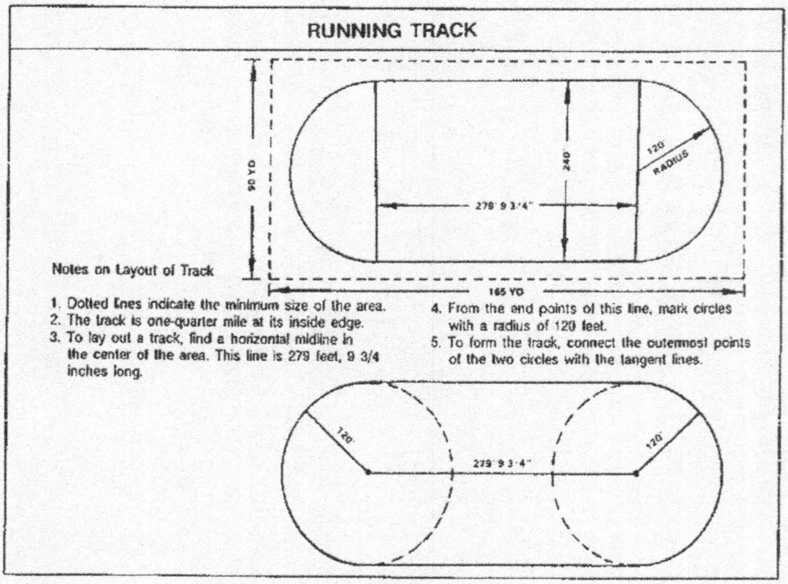

Figure 14-2

A 400-meter track may be used in place of the standard quarter-mile (440-yard) track for the 2-mile run. However, one lap run on a 400-meter track is 92 inches shorter than one lap on a 440-yard track. Eight laps on a 400-meter track is 736 inches shorter than eight laps (2 miles) on a 440-yard track. Therefore, soldiers who run the 2-mile event on a 400-meter track must run eight laps plus an additional 61 feet, 4 inches.

Test Procedures

On test day, soldiers are assembled in a common area and briefed by the test OIC or NCOIC about the purpose and organization of the test. The OIC or NCOIC then explains the scorecard, scoring standards, and sequence of events.

The instructions printed here in large type must be read to the soldiers: "YOU ARE ABOUT TO TAKE THE ARMY PHYSICAL FITNESS TEST, A TEST THAT WILL MEASURE YOUR MUSCULAR ENDURANCE AND CARDIORESPIRATORY FITNESS. THE RESULTS OF THIS TEST WILL GIVE YOU AND YOUR COMMANDERS AN INDICATION OF YOUR STATE OF FITNESS AND WILL ACT AS A GUIDE IN DETERMINING YOUR PHYSICAL TRAINING NEEDS. LISTEN CLOSELY TO THE TEST INSTRUCTIONS, AND DO THE BEST YOU CAN ON EACH OF THE EVENTS."

If scorecards have not already been issued, they are handed out at this time. The OIC or NCOIC then says the following: "IN THE APPROPRIATE SPACES, PRINT IN INK THE PERSONAL INFORMATION REQUIRED ON THE SCORECARD." (If scorecards have been issued to the soldiers and filled out before they arrive at the test site, this remark is omitted.)

The OIC or NCOIC pauses briefly to give the soldiers time to check the information. He then says the following: "YOU ARE TO CARRY THIS CARD WITH YOU TO EACH EVENT. BEFORE YOU BEGIN, HAND THE CARD TO THE SCORER. AFTER YOU COMPLETE THE EVENT, THE SCORER WILL RECORD YOUR RAW SCORE, INITIAL THE CARD, AND RETURN IT TO YOU." (At this point, the scoring tables are explained so everyone understands how raw scores are converted to point scores.) Next, the OIC or NCOIC says the following "EACH OF YOU WILL BE ASSIGNED TO A GROUP. STAY WITH YOUR TEST GROUP FOR THE ENTIRE TEST. WHAT ARE YOUR QUESTIONS ABOUT THE TEST AT THIS POINT?"

Groups are organized as required and given final instructions including what to do after the final event. The test is then given.

RETAKING OF EVENTS

Soldiers who start an event incorrectly must be stopped by the scorer before they complete 10 repetitions and told what their errors are. They are then sent to the end of the line to await their turn to retake the event.

A soldier who has problems such as muscle cramps while performing an event may rest if he does not assume an illegal position in the process. If he continues, he receives credit for all correctly done repetitions within the two-minute period. If he does not continue, he gets credit for the number of correct repetitions he has performed up to that time. If he has not done 10 correct repetitions, he is sent to the end of the line to retake that event. He may not retake the event if he has exceeded 10 repetitions. Soldiers who are unable to perform 10 correct repetitions because of low fitness levels may not retake an event.

TEST FAILURES

Soldiers who stop to rest in an authorized rest position continue to receive credit for correct repetitions performed after their rest. Soldiers who rest in an unauthorized rest position will have their performance in that event immediately terminated.

The records of soldiers who fail a record APFT for the first time and those who fail to take the APFT within the required period (AR 350-15, paragraph 11) must be flagged I AW AR 600-8-2 (Reference B).

RETESTING

Soldiers who fail any or all of the events must retake the entire APFT. In case of test failure, commanders may allow soldiers to retake the test as soon as the soldiers and commanders feel they are ready. Soldiers without a medical profile will be retested not-later-than three months following the initial APFT failure in accordance with AR 350-15, paragraph 11.

Test Sequence

The test sequence is the push-up, sit-up, and 2-mile run (or alternate, aerobic event). The order of events cannot be changed. There are no exceptions to this sequence.

Soldiers should be allowed no less than 10 minutes, but ideally no more than 20 minutes, to recover between each event. The OIC or NCOIC determines the time to be allotted between events, as it will depend on the total number of soldiers who are participating in the APFT. If many soldiers are to be tested, staggered starting times should be planned to allow the proper intervals between events. Under no circumstances is the APFT valid if a soldier cannot begin and end all three events in two hours or less.

The following paragraphs describe the equipment, facilities, personnel, instructions, administration, timing techniques, and scorers' duties for the pushup, sit-up, and 2-mile-run events.

PUSH-UPS

Push-ups measure the endurance of the chest, shoulder, and triceps muscles. (See Figure 14-3.)

Equipment

One stopwatch is needed along with one clipboard and pen for each scorer. The event supervisor must have the following the instructions in this chapter on how to conduct the event and one copy of the push-up scoring standards (DA Form 705).

Facilities

There must be at least one test station for every 15 soldiers to be tested. Each station is 6 feet wide and 15 feet deep.

Personnel

One event supervisor must beat the test site and one scorer at each station. The event supervisor may not be the event scorer.

Instructions

The event supervisor must read the following: "THE PUSH-UP EVENT MEASURES THE ENDURANCE OF THE CHEST, SHOULDER, AND TRICEPS MUSCLES. ON THE COMMAND 'GET SET,' ASSUME THE FRONT-LEANING REST POSITION BY PLACING YOUR HANDS WHERE THEY ARE COMFORTABLE FOR YOU. YOUR FEET MAY BE TOGETHER OR UP TO 12 INCHES APART. WHEN VIEWED FROM THE SIDE, YOUR BODY SHOULD FORM A GENERALLY STRAIGHT LINE FROM YOUR SHOULDERS TO YOUR ANKLES.

PUSH-UPS

Figure 14-3

ON THE COMMAND 'GO,' BEGIN THE PUSH-UP BY BENDING YOUR ELBOWS AND LOWERING YOUR ENTIRE BODY AS A SINGLE UNIT UNTIL YOUR UPPER ARMS ARE AT LEAST PARALLEL TO THE GROUND. THEN, RETURN TO THE STARTING POSITION BY RAISING YOUR ENTIRE BODY UNTIL YOUR ARMS ARE FULLY EXTENDED. YOUR BODY MUST REMAIN RIGID IN A GENERALLY STRAIGHT LINE AND MOVE AS A UNIT WHILE PERFORMING EACH REPETITION. AT THE END OF EACH REPETITION, THE SCORER WILL STATE THE NUMBER OF REPETITIONS YOU HAVE COMPLETED CORRECTLY. IF YOU FAIL TO KEEP YOUR BODY GENERALLY STRAIGHT, TO LOWER YOUR WHOLE BODY UNTIL YOUR UPPER ARMS ARE AT LEAST PARALLEL TO THE GROUND, OR TO EXTEND YOUR ARMS COMPLETELY, THAT REPETITION WILL NOT COUNT, AND THE SCORER WILL REPEAT THE NUMBER OF THE LAST CORRECTLY PERFORMED REPETITION. IF YOU FAIL TO PERFORM THE FIRST TEN PUSH-UPS CORRECTLY, THE SCORER WILL TELL YOU TO GO TO YOUR KNEES AND WILL EXPLAIN TO YOU WHAT YOUR MISTAKES ARE. YOU WILL THEN BE SENT TO THE END OF THE LINE TO BE RETESTED. AFTER THE FIRST 10 PUSH-UPS HAVE BEEN PERFORMED AND COUNTED, HOWEVER, NO RESTARTS ARE ALLOWED. THE TEST WILL CONTINUE, AND ANY INCORRECTLY PERFORMED PUSH-UPS WILL NOT BE COUNTED. AN ALTERED, FRONT-LEANING REST POSITION IS THE ONLY AUTHORIZED REST POSITION. THAT IS, YOU MAY

SAG IN THE MIDDLE OR FLEX YOUR BACK. WHEN FLEXING YOUR BACK, YOU MAY BEND YOUR KNEES, BUT NOT TO SUCH AN EXTENT THAT YOU ARE SUPPORTING MOST OF YOUR BODY WEIGHT WITH YOUR LEGS. IF THIS OCCURS, YOUR PERFORMANCE WILL BE TERMINATED. YOU MUST RETURN TO, AND PAUSE IN, THE CORRECT STARTING POSITION BEFORE CONTINUING. IF YOU REST ON THE GROUND OR RAISE EITHER HAND OR FOOT FROM THE GROUND, YOUR PERFORMANCE WILL BE TERMINATED. YOU MAY REPOSITION YOUR HANDS AND/OR FEET DURING THE EVENT AS LONG AS THEY REMAIN IN CONTACT WITH THE GROUND AT ALL TIMES. CORRECT PERFORMANCE IS IMPORTANT. YOU WILL HAVE TWO MINUTES IN WHICH TO DO AS MANY PUSH-UPS AS YOU CAN. WATCH THIS DEMONSTRATION." (The exercise is then demonstrated. See Figure 14-4 for a list of points that need to be made during the demonstration.) "WHAT AREYOUR QUESTIONS?"

Administration

After reading the instructions, the supervisor answers questions. Then he moves the groups to their testing stations. The event supervisor cannot be ready to begin. Successive groups do the event until all soldiers have completed it.

Timing Techniques

The event supervisor is the timer. He calls out the time remaining every 30 seconds and every second for the last 10 seconds of the two minutes. He ends the event after two minutes by the command "Halt!"

Scorers' Duties

Scorers must allow for differences in the body shape and structure of each soldier. The scorer uses each soldier's starting position as a guide throughout the event to evaluate each repetition. The scorer should talk to the soldier before the event begins and have him do a few repetitions as a warm-up and reference to ensure he is doing the exercise correctly.

ADDITIONAL POINTS TO DEMONSTRATE FOR THE PUSH-UP EVENT

The following points must be clarified during the demonstration:

The soldier's chest may touch the ground (mat or floor) during the push-up as long as the contact does not provide him an advantage. He cannot use the ground to bounce off of or momentarily rest on. However, penalizing a soldier for touching the ground with the chest is unfair. Some soldiers have a large chest or abdomen or are otherwise developed in a way which makes touching the ground unavoidable when they are in the correct down position. Do not count those repetitions in which bouncing off the ground has given the soldier an unfair advantage- Do not count those repetitions in which the long bone of the upper arm does not reach a position parallel to the ground,

Soldiers may reposition their hands during the push-up event as long as the hands remain in contact with the at all times. The hands can be repositioned either forward, inward, outward, or backward. If a soldier repositions his hands too far backward, the legal front-leaning rest position may be violated.

If a mat is used, the entire body must be on the mat.

- In the rest position, a soldier may sag in the middle or flex his back in the altered front-leaning rest position; however, he may not readjust his hands backward and/or bend his knees to such a point that when he bends at the waist and/or knees, he supports most of his body weight with his legs. If this occurs, the soldier's performance in the event will be terminated.
- The feet may not be braced during the push-up event. Test administrators must ensure that a non-slip surface is available.
- Soldiers may do the push-up event on their fists. This may be necessary due.to a prior injury. There is no unfair advantage to be gained by doing so.
- Soldiers may not cross their feet while doing the push-up event. This ensures as much standardization as possible and avoids violation of the proper front-leaning rest position, which is the only authorized starting position for this event.
- Soldiers may not take any part of the APFT in bare feet.
- Soldiers should not wear glasses while performing the push-up event.

Figure 14-4

The scorer may either sit or kneel about three feet from the testee's shoulder at a 45-degree angle in front of it. The scorer's head should be about even with the testee's shoulder when the latter is in the front-leaning rest position. Each scorer determines for himself if he will sit or kneel when scoring. He may not lie down or stand while scoring. He counts out loud the number of correct repetitions completed and repeats the number of the last correct push-up if an incorrect one is done. Scorers tell the testees what they do wrong as it occurs during the event. A critique of the performance is done following the test.

When the soldier completes the event, the scorer records the number of correctly performed repetitions, initials the scorecard, and returns it to the soldier.

SIT-UPS

This event measures the endurance of the abdominal and hip-flexor muscles. (See Figure 14-5.)

Equipment

One stopwatch is needed along with one clipboard and pen for each scorer. The event supervisor must have the following: the instructions in this chapter on how to conduct the event and one copy of the sit-up scoring standards (DA Form 705).

Facilities

Each station is 6 feet wide and 15 feet deep. Ensure that no more than 15 soldiers are tested at a station.

Personnel

One event supervisor must be at the test site and one scorer at each station. The event supervisor may not be the event scorer.

Instructions

The event supervisor must read the following: "THE SIT-UP EVENT MEASURES THE ENDURANCE OF THE ABDOMINAL AND HIP-FLEXOR MUSCLES. ON THE

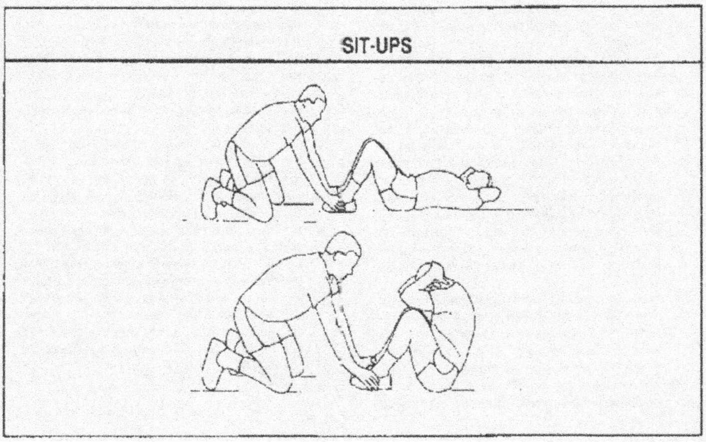

Figure 14-5

COMMAND "GET SET", ASSUME THE STARTING POSITION BY LYING ON YOUR BACK WITH YOUR KNEES BENT AT A 90-DEGREE ANGLE. YOUR FEET MAY BE TOGETHER OR UP TO 12 INCHES APART. ANOTHER PERSON WILL HOLD YOUR ANKLES WITH THE HANDS ONLY. NO OTHER METHOD OF BRACING OR HOLDING THE FEET IS AUTHORIZED. THE HEEL IS THE ONLY PART OF YOUR FOOT THAT MUST STAY IN CONTACT WITH THE GROUND. YOUR FINGERS MUST BE INTERLOCKED BEHIND YOUR HEAD AND THE BACKS OF YOUR HANDS MUST TOUCH THE GROUND. YOUR ARMS AND ELBOWS NEED NOT TOUCH THE GROUND. ON THE COMMAND "GO", BEGIN RAISING YOUR UPPER BODY FORWARD TO, OR BEYOND, THE VERTICAL POSITION. THE VERTICAL POSITION MEANS THAT THE BASE OF YOUR NECK IS ABOVE THE BASE OF YOUR SPINE. AFTER YOU HAVE REACHED OR SURPASSED THE VERTICAL POSITION, LOWER YOUR BODY UNTIL THE BOTTOM OF YOUR SHOULDER BLADES TOUCH THE GROUND. YOUR HEAD, HANDS, ARMS, OR ELBOWS DO NOT HAVE TO TOUCH THE GROUND. AT THE END OF EACH REPETITION, THE SCORER WILL STATE THE NUMBER OF SIT-UPS YOU HAVE CORRECTLY COMPLETED. A REPETITION WILL NOT COUNT IF YOU FAIL TO REACH THE VERTICAL POSITION, FAIL TO KEEP YOUR FINGERS INTERLOCKED BEHIND YOUR HEAD, ARCH OR BOW YOUR BACK AND RAISE YOUR BUTTOCKS OFF THE

ADDITIONAL POINTS TO DEMONSTRATE FOR THE SIT-UP EVENT

The following points must be clarified during the demonstration:

- To minimize stress to the neck, it is recommended that the soldier keep his chin curled downward and touching the top of his chest throughout the performance of the sit-up event.
- From the starting (down) position, or during any phase of the sit-up, the soldier may not use his hands or arms to pull himself up to push off the ground (floor or mat) in order to help himself attain the up position. Any of these procedures can give the violator an unfair advantage. They also violate the intent of the event. The sit-up event will be terminated immediately for those soldiers who, by pushing or pulling, use their arms to assist themselves in attaining the up position.
- If a mat is used, the entire body, including the feet and head, must be on the mat at the start.
- From the starting (down) position, or during any phase of the sit-up, the soldier may not swing his hands or arms in order to help himself attain the up position. If this occurs, that repetition does not count.
- The soldier may wiggle to attain the up position. This gives him no advantage.
- While in the up position, the soldier may not help himself stay in that position by using the elbows or any part of the arms to lock on to or brace against the legs. The elbows can go either inside or outside the knees. However, to push or pull them into the sides or tops of the knees to get extra leverage and rest gives an unfair advantage to that soldier. Therefore, soldiers who use this technique will be warned once for the first violation and immediately terminated if the violation continues or recurs.
- During the performance of the sit-up event, the fingers must be interlocked and behind the head. As long as any of the fingers are overlapping to any degree, the fingers are considered to be interlocked.
- If either foot breaks contact with the ground during a repetition, that repetition will not count. Both heels must stay in contact with the ground (floor or mat) during the performance of the event. The scorer should ensure that the holder has the soldier's feet properly secured. The scorer tells the soldier if his heel(s) is raised from the ground and that the repetition will not count.

Figure 14-6

GROUND TO RAISE YOUR UPPER BODY, OR LET YOUR KNEES EXCEED A 90-DEGREE ANGLE. IF A REPETITION DOES NOT COUNT, THE SCORER WILL REPEAT THE NUMBER OF YOUR LAST CORRECTLY PERFORMED SIT-UP. THE UP POSITION IS THE ONLY AUTHORIZED REST POSITION. IF YOU STOP AND REST IN THE DOWN (STARTING) POSITION, THE EVENT WILL BE TERMINATED. AS LONG AS YOU MAKE A CONTINUOUS PHYSICAL EFFORT TO SIT UP, THE EVENT WILL NOT BE TERMINATED. YOU MAY NOT USE YOUR HANDS OR ANY OTHER MEANS TO PULL OR PUSH YOURSELF UP TO THE UP (RESTING) POSITION OR TO HOLD YOURSELF IN THE REST POSITION. IF YOU DO SO, YOUR PERFORMANCE IN THE EVENT WILL BE TERMINATED. CORRECT PERFORMANCE IS IMPORTANT. YOU WILL HAVE TWO MINUTES TO PERFORM AS MANY SIT-UPS AS YOU CAN. WATCH THIS DEMONSTRATION." (The exercise is then demonstrated. See Figure 14-6 for a list of points that need to be made during the demonstration.) "WHAT ARE YOUR QUESTIONS?"

Administration

After reading the instructions, the supervisor answers questions. He then moves the groups to their testing stations. The event supervisor cannot be a scorer. At this point, the testing is ready to begin. Successive groups do the event until all soldiers have completed it.

Timing Techniques

The event supervisor is the timer. He calls out the time remaining every 30 seconds and every second for the last 10 seconds of the two minutes. He ends the event after two minutes by the command "Halt!"

Scorers' Duties

The scorer may either kneel or sit about three feet from the testee's hip. The scorer's head should be about even with the testee's shoulder when the latter is in the vertical (up) position. Each scorer decides for himself whether to sit or kneel down when scoring. He may not lie down or stand while scoring. The scorer counts aloud the number of correctly performed sit-ups and repeats the number of the last correctly performed repetition if an incorrect one is done. Scorers tell the testees what they are doing wrong as it occurs during the event. A critique of his performance is given to each soldier after the event. When the soldier completes the event, the scorer records the number of correctly performed sit-ups, initials the scorecard, and returns it to the soldier.

When checking for correct body position, the scorer must be sure that at a 90-degree angle is formed at each knee by the soldier's upper and lower leg. The angle to be measured is not the one formed by the lower leg and the ground. If, while performing the sit-up event, this angle becomes greater than 90 degrees, the scorer should instruct the testee and holder to reposition the legs to the proper angle and obtain compliance before allowing the testee's performance to continue. The loss of the proper angle does not terminate the testee's performance in the event. When the soldier comes to the vertical position, the scorer must be sure that the base of the soldier's neck is above or past the base of the spine. A soldier who simply touches his knees with his elbows may not come to a completely vertical position. The scorer must ensure that the holder uses only his hands to brace the exerciser's feet.

TWO-MILE RUN

This event tests cardiorespiratory (aerobic) endurance and the endurance of the leg muscles. (See Figure 14-7.)

Equipment

Two stopwatches for the event supervisor, one clipboard and pen for each scorer, copies of the event's instructions and standards, and numbers for the testees are needed.

Facilities

There must be a level area with no more than a three-degree slope on which a measured course has been marked. An oval-shaped track of known length may be used. If a road course is used, the start and finish and one-mile (half way) point must be clearly marked.

Personnel

One event supervisor and at least one scorer for every 15 runners are required.

Instructions

The event supervisor must read the following: "THE TWO-MILE RUN IS USED TO ASSESS YOUR AEROBIC FITNESS AND YOUR LEG MUSCLES' ENDURANCE. You MUST COMPLETE THE RUN WITHOUT ANY PHYSICAL HELP. AT THE START, ALL SOLDIERS WILL LINE UP BEHIND THE STARTING LINE. ON THE COMMAND 'GO,' THE CLOCK WILL START. YOU WILL BEGIN RUNNING AT YOUR OWN PACE. TO RUN THE REQUIRED TWO MILES, YOU MUST COMPLETE (describe the number of laps, start and finish points, and course layout). YOU ARE BEING TESTED ON YOUR ABILITY TO COMPLETE THE 2-MILE COURSE IN THE SHORTEST TIME POSSIBLE. ALTHOUGH WALKING IS AUTHORIZED, IT IS STRONGLY DISCOURAGED. IF YOU ARE PHYSICALLY HELPED IN ANY WAY (FOR EXAMPLE, PULLED, PUSHED, PICKED UP, AND/OR CARRIED) OR LEAVE THE DESIGNATED RUNNING COURSE FOR ANY

Figure 14-7

REASON, YOU WILL BE DISQUALIFIED. (IT IS LEGAL TO PACE A SOLDIER DURING THE 2-MILE RUN. AS LONG AS THERE IS NO PHYSICAL CONTACT WITH THE PACED SOLDIER AND IT DOES NOT PHYSICALLY HINDER OTHER SOLDIERS TAKING THE TEST, THE PRACTICE OF RUNNING AHEAD OF, ALONG SIDE OF, OR BEHIND THE TESTED SOLDIER, WHILE SERVING AS A PACER, IS PERMITTED. CHEERING OR CALLING OUT THE ELAPSED TIME IS ALSO PERMITTED.) THE NUMBER ON YOUR CHEST IS FOR IDENTIFICATION. YOU MUST MAKE SURE IT IS VISIBLE AT ALL TIMES. TURN IN YOUR NUMBER WHEN YOU FINISH THE RUN. THEN, GO TO THE AREA DESIGNATED FOR THE COOL-DOWN AND STRETCH. DO NOT STAY NEAR THE SCORERS OR THE FINISH LINE AS THIS MAY INTERFERE WITH THE TESTING. WHAT ARE YOUR QUESTIONS ON THIS EVENT?"

Administration

After reading the instructions, the supervisor answers questions. He then organizes the soldiers into groups of no more than 10. The scorer for each group assigns a number to each soldier in the group. At the same time, the scorer collects the scorecards and records each soldier's number.

Timing Techniques

The event supervisor is the timer. He uses the commands "Get set" and "Go." Two stopwatches are used in case one fails. As the soldiers near the finish line, the event supervisor calls off the time in minutes and seconds (for example, "Fifteen-thirty, fifteen-thirty-one, fifteen-thirty -two," and so on).

Scorers' Duties

The scorers observe those runners in their groups, monitor their laps (if appropriate), and record their times as they cross the finish line. (It is often helpful to record the soldiers' numbers and times on a separate sheet of paper or card. This simplifies the recording of finish times when large groups of soldiers are simultaneously tested.) After all runners have completed the run, the scorers determine the point value for each soldier's run time, record the point values on the scorecards, and enter their initials in the scorers' blocks. In all cases, when a time falls between two point values, the lower point value is used and recorded. For example, if a female soldier, age 17 to 21, runs the two miles in 15 minutes and 19 seconds, the score awarded is 95 points.

At this time, the scorers for the 2-mile run also convert the raw scores for the push-up and sit-up events by using the scoring standards on the back side of the scorecard. They enter those point values on the scorecards and determine the total APFT score for each soldier before giving the scorecards to the test's OIC or NCOIC. After the test scores have been checked, the test's OIC or NCOIC signs all scorecards and returns them to the unit's commander or designated representative.

Test Results

The soldier's fitness performance for each APFT event is determined by converting the raw score for each event to a point score.

Properly interpreted, performance on the APFT shows the following:

- Each soldier's level of physical fitness.
- The entire unit's level of physical fitness.
- Deficiencies in physical fitness.
- Soldiers who need special attention.

(Leaders must develop special programs to improve the performance of soldiers who are below the required standards.)

Commanders should not try to determine the individual's or the unit's strengths and weaknesses in fitness by using only the total scores. A detailed study of the results on each event is more important. For a proper analysis of the unit's performance, event scores should be used. They are corrected for age and sex. Therefore, a female's 80-point push-up score should be considered the same as a male's 80-point push-up score. Using the total point value or raw scores may distort the interpretation.

Scores Above Maximum

Even though some soldiers exceed the maximum score on one or more APFT events, the official, maximum score on the APFT must remain at 300 (100 points per event). Some commanders, however, want to know unofficial point scores to reward soldiers for their extra effort.

Only those soldiers who score 100 points in all three events are eligible to determine their score on an extended scale. To fairly determine the points earned, extra points are awarded at the same rate as points obtained for scores at or below the 100 point level. Each push-up and sit-up beyond the maximum is worth one point as is every six-second decrease in the run time. Take, for example, the following case shown in Figure 14-8. A male soldier performs above the maximum in the 17-21 age group by doing 87 push-ups and 98 sit-ups and by running the two miles in 11 minutes and 12 seconds. His score would be calculated as follows:

CALCULATION FOR AGE 17-21 MALE

PUSH-UPS	
Actual	87
Maximum	82
Additional points	+5
Points (official)	100
Points (unofficial)	105
SIT-UPS	
Actual	98
Maximum	92
Additional points	+6
Points (official)	100
Points (unofficial)	106
2-MILE RUN	
Actual	11:12
Maximum	11:54
	:42
ADDITIONAL POINTS (42 sec/6 = 7)	+7
Points (official)	100
Points (unofficial)	107

Thus, the unofficial total for this soldier in the three events is determined in the following way:

PUSH-UPS	105
SIT-UPS	106
2-MILE RUN	+ 107
UNOFFICIAL TOTAL	318

Figure 14-8

The calculations on the previous page, give the soldier a total score of 318 points. This method lets the commander easily determine the scores for performances that are above the maximum. He may recognize soldiers for their outstanding fitness achievements, not only on the APFT but also for other, unofficial fitness challenges. Using this method ensures that each soldier has an equal chance to be recognized for any of the tested fitness components. Commanders may also establish their own incentive programs and set their own unit's standards (AR 350-15).

Temporary Profiles

A soldier with a temporary profile must take the regular three-event APFT after the profile has expired. (Soldiers with temporary profiles of more than three months may take an alternate test as determined by the commander with input from health-care personnel.) Once the profile is lifted, the soldier must be given twice the time of the profile (but not more than 90 days) to train for the APFT. For example, if the profile period was 7 days, the soldier has 14 days to train for the APFT after the profile period ends. If a normally scheduled APFT occurs during the profile period, the soldier should be given a mandatory make-up date.

Permanent Profiles

A permanently profiled soldier is given a physical training program by the profiling officer using the positive profile form DA 3349 (see Appendix B). The profiling officer gives the unit's commander a list of physical activities that are suitable for the profiled soldier. He also indicates the events and/or alternate aerobic event that the soldier will do on the APFT. This recommendation, made after consultation with the profiled soldier, should address the soldier's abilities and preference and the equipment available. (See DA Form 3349, Physical Profile, referenced in AR 40-501.)

The profiled soldier must perform all the regular APFT events his medical profile permits. Each soldier must earn at least 60 points on the regular events to receive a "go." He must also complete the alternate event in a time equal to or less than the one listed for his age group. For example, a soldier whose profile forbids only running will do the push-up and sit-up events and an alternate aerobic event. He must get at least a minimum passing score on each event to earn a "go" for the test. A soldier whose profile prevents two or more APFT events must complete the 2-mile run or an alternate aerobic event to earn a "go" on the test. Soldiers who cannot do any of the aerobic events due to a profile cannot be tested. Such information will be recorded in their official military record.

The standards for alternate events are listed in Figure 14-9. Scoring for all alternate events is on a go/no go basis. Soldiers who do push-up and sit-up events but who take an alternate aerobic event are not awarded promotion points for APFT performance.

Alternate Events

Alternate APFT events assess the aerobic fitness and muscular endurance of soldiers with permanent medical profiles or long-term (greater than three months) temporary profiles who cannot take the regular, three-event APFT.

The alternate aerobic APFT events are the following:

- 800-yard-swim test.
- 6.2-mile-stationary- bicycle ergometer test with a resistance setting of 2 kiloponds (2 kilograms) or 20 newtons.
- 6.2-mile-bicycle test on a conventional bicycle using one speed.
- 2.5-mile-walk test.

ALTERNATE TEST STANDARDS BY EVENT, SEX, AND AGE

EVENT	SEX	AGE 17-21	22-26	27-31	32-36	37-41	42-46	47-51	52-56	57-61	62+
800-YARD SWIM	Men	20:00	20:30	21:00	21:30	22:00	22:30	23:00	24:00	24:30	25:00
	Women	21:00	21:30	22:00	22:30	23:00	23:30	24:00	25:00	25:30	26:00
6.2-MILE BIKE (Stationary and track)	Men	24:00	24:30	25:00	25:30	26:00	27:00	28:00	30:00	31:00	32:00
	Women	25:00	25:30	26:00	26:30	27:00	28:00	30:00	32:00	33:00	34:00
2.5-MILE WALK	Men	34:00	34:30	35:00	35:30	36:00	36:30	37:00	37:30	38:00	38:30
	Women	37:00	37:30	38:00	38:30	39:00	39:30	40:00	40:30	41:00	41:30

•Figure 14-9

800-YARD-SWIM TEST

This event is used to assess cardio-respiratory (aerobic) fitness. (See Figure 14-10.)

Equipment

Two stopwatches, one clipboard and pen for each scorer, one copy each of the test instructions and standards, and appropriate safety equipment are needed.

Facilities

A swimming pool at least 25 yards long and 3 feet deep, or an approved facility, is needed.

Personnel

One event supervisor and at least one scorer for every soldier to be tested are required. Appropriate safety, control, and medical personnel must also be present.

Instructions

The event supervisor must read the following statement: "THE 800-YARD SWIM IS USED TO ASSESS YOUR LEVEL OF AEROBIC FITNESS. YOU WILL BEGIN IN THE WATER; NO DIVING IS ALLOWED. AT THE START, YOUR BODY MUST BE IN CONTACT WITH THE WALL OF THE POOL. ON THE COMMAND 'GO,' THE CLOCK WILL START. YOU SHOULD THEN BEGIN SWIMMING AT YOUR OWN PACE, USING ANY STROKE OR COMBINATION OF STROKES YOU WISH. YOU MUST SWIM (tell the number) LAPS TO COMPLETE THIS DISTANCE. YOU MUST TOUCH THE WALL OF THE POOL AT EACH END OF THE POOL AS YOU TURN. ANY TYPE OF TURN IS AUTHORIZED. YOU WILL BE SCORED ON YOUR ABILITY TO COMPLETE THE SWIM IN A TIME EQUAL TO, OR LESS THAN, THAT LISTED FOR YOUR AGE AND SEX. WALKING ON THE BOTTOM TO RECUPERATE IS AUTHORIZED. SWIMMING GOGGLES ARE PERMITTED, BUT NO OTHER EQUIPMENT IS AUTHORIZED. WHAT ARE YOUR QUESTIONS ABOUT THIS EVENT?"

Figure 14-10

Administration

After reading the instructions, the event supervisor answers only related questions. He assigns one soldier to each lane and tells the soldiers to enter the water. He gives them a short warm-up period to acclimate to the water temperature and loosen up. Above all, the event supervisor must be alert to the safety of the testees throughout the test.

Timing Techniques

The event supervisor is the timer. He uses the commands "Get set" and "Go." Two stopwatches are used in case one fails. As the soldiers near the finish, the event supervisor begins calling off the elapsed time in minutes and seconds (for example, "Nineteen-eleven, nineteen-twelve, nineteen-thirteen," and so on). The time is recorded when each soldier touches the end of the pool on the final lap or crosses a line set as the 800-yard mark.

Scorers' Duties

Scorers must observe the swimmers assigned to the. They must be sure that each swimmer touches the bulkhead at every turn. The scorers record each soldier's time in the 2-mile-run block on the scorecard and use the comment block to identify the time as an 800-yard-swim time. If the pool length is measured in meters, the scorers convert the exact distance to yards. To convert meters to yards, multiply the number of meters by 39.37 and divide the product by 36; that is, (meters x 39.37)/36 - yards. For example, 400 meters equals 437.4 yards; that is, (400 x 39.37)/36 = 437.4 yards.

6.2-MILE STATIONARY-BICYCLE ERGOMETER TEST

This event is used to assess the soldier's cardiorespiratory and leg-muscle endurance. (See Figure 14-11.)

Equipment

Two stopwatches, one clipboard and pen for each scorer, a copy of the test instructions and standards, and one stationary bicycle ergometer are needed. The ergometers should measure resistance in kiloponds or newtons. The bicycle should be one that can be used for training and testing. Its seat and

6.2 MILE STATIONARY-BICYCLE ERGOMETER TEST

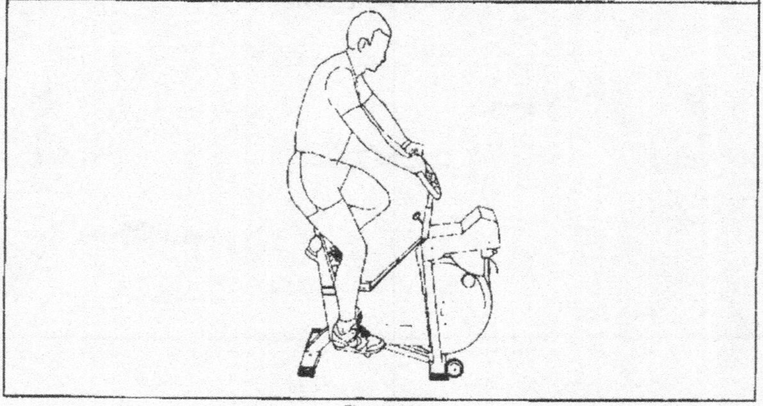

Figure 14-11

handlebars must be adjustable to let the soldier fully extend his legs when pedaling. It should have an adjustable tension setting and an odometer. The resistance is usually set by a tension strap on a weighted pendulum connected to the flywheel. See Appendix D for guidance on using various types of stationary bikes.

Facilities

The test site can be any place where there is an approved bicycle ergometer. This could be the post's fitness facility or the hospital's therapy clinic. Each test station must be two yards wide and four yards deep.

Personnel

One event supervisor and at least one scorer for every three soldiers to be tested are required. Appropriate safety, control, and medical personnel should also be present.

Instructions

The event supervisor must read the following: "THE 6.2-MILE STATIONARY-BICYCLE ERGOMETER EVENT TESTS YOUR CARDIORESPIRATORY FITNESS AND LEG MUSCLE ENDURANCE. THE ERGOMETER'S RESISTANCE MUST BE SET AT TWO KILOPOUNDS (20 NEWTONS). ON THE COMMAND 'GO,' THE CLOCK WILL START, AND YOU WILL BEGIN PEDALING AT YOUR OWN PACE WHILE MAINTAINING THE RESISTANCE INDICATOR AT TWO POUNDS. YOU WILL BE SCORED ON YOUR ABILITY TO COMPLETE 6.2 MILES (10 KILOMETERS), AS SHOWN ON THE ODOMETER, IN A TIME EQUAL TO OR LESS THAN THAT LISTED FOR YOUR AGE AND SEX. WHAT ARE YOUR QUESTIONS ABOUT THIS EVENT?"

Administration

After reading the instructions, the event supervisor answers any related questions. Each soldier is given a short warm-up period and allowed to adjust the seat and handlebar height.

Timing Techniques

The event supervisor is the timer. He uses the commands "Get set" and "Go." Two stopwatches are used in case one fails. As the soldiers pedal the last two-tenths of the test distance, the event supervisor should start calling off the time in minutes and seconds (for example, "Twenty-thirty-one, twenty -thirty -two, twenty-thirty-three," and so on). He calls the time remaining every 30 seconds for the last two minutes of the allowable time and every second during the last ten seconds.

Scorers' Duties

Scorers must ensure that the bicycle ergometer is functioning properly. They must then make sure that the bicycle ergometers' tension settings have been calibrated and are accurate and that the resistance of the ergometers has been set at two kiloponds (20 newtons). The scorers must observe the soldiers throughout the event. From time to time the scorer may need to make small adjustments to the resistance control to ensure that a continuous resistance of exactly 2 kiloponds (20 newtons) is maintained throughout the test. At the end of the test, they record each soldier's time on the scorecard in the 2-mile-run block, initial the appropriate block, and note in the comment block that the time is for a 6.2-mile stationary-bicycle ergometer test.

6.2-MILE BICYCLE TEST

This event is used to assess the soldier's cardiorespiratory and leg-muscle endurance.

Equipment

Two stopwatches, one clipboard and pen for each scorer, a copy of the test instructions and standards, and numbers are needed. Although one-speed bicycles are preferred for this event, multispeed bicycles may be used. If a multispeed bicycle is used, measures must be taken to ensure that only one gear is used throughout the test. (This can usually be done by taping the gear shifters at the setting preferred by the testee.)

Facilities

A relatively flat course with a uniform surface and no obstacles must be used. It must also be clearly marked. Soldiers should not be tested on a quarter-mile track, and they should never be out of the scorers' sight. The course should be completely free of runners and walkers.

Personnel

One event supervisor and at least one scorer for every 10 soldiers are required. Safety, control, and medical personnel should also be present as appropriate.

Instructions

The event supervisor must read the following: "THE 6.2-MILE BICYCLE TEST IS USED TO ASSESS YOUR CARDIORESPIRATORY FITNESS AND LEG MUSCLES' ENDURANCE. YOU MUST COMPLETE THE 6.2

MILES WITHOUT ANY PHYSICAL HELP FROM OTHERS. YOU MUST KEEP YOUR BICYCLE IN ONE GEAR OF YOUR CHOOSING FOR THE ENTIRE TEST. CHANGING GEARS IS NOT PERMITTED AND WILL RESULT IN DISQUALIFICATION. TO BEGIN, YOU WILL LINE UP BEHIND THE STARTING LINE. ON THE COMMAND 'GO,' THE CLOCK WILL START, AND YOU WILL BEGIN PEDALING AT YOUR OWN PACE. TO COMPLETE THE REQUIRED DISTANCE OF 6.2 MILES, YOU MUST COMPLETE (describe the number of laps, start and finish points, and course layout). YOU WILL BE SCORED ON YOUR ABILITY TO COMPLETE THE DISTANCE-OF 6.2 MILES (10 KILOMETERS) IN A TIME EQUAL TO OR LESS THAN THAT LISTED FOR YOUR AGE AND SEX. IF YOU LEAVE THE DESIGNATED COURSE FOR ANY REASON, YOU WILL BE DISQUALIFIED. WHAT ARE YOUR QUESTIONS ABOUT THIS EVENT?"

Administration

After reading the instructions, the event supervisor answers any related questions. He then organizes the soldiers into groups of no more than ten and assigns each group to a scorer. Scorers assign numbers to the soldiers in their groups and record each soldier's number on the appropriate scorecard.

Timing Techniques

The event supervisor is the timer. He uses the commands "Get set" and "Go." Two stopwatches are used in case one fails. As soldiers near the end of the 6.2-mile ride, the event supervisor starts calling off the time in

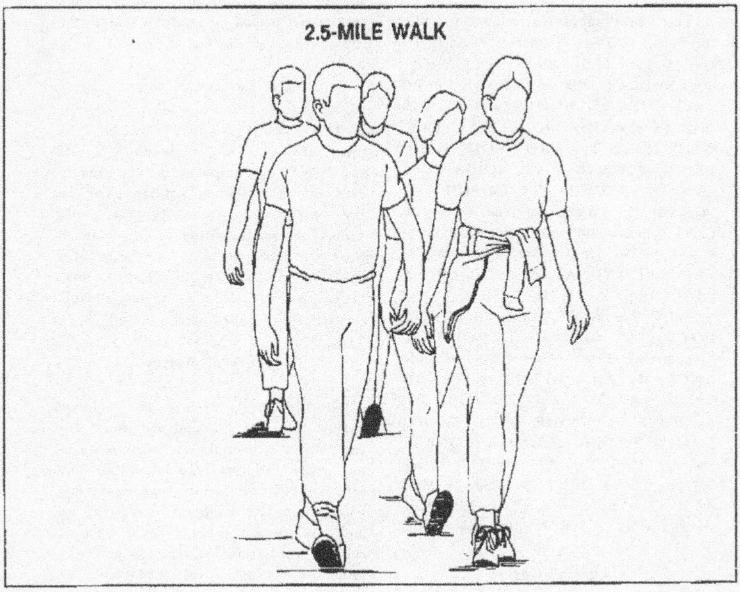

minutes and seconds (for example, "Thirty-twenty-one, thirty- twenty-two, thirty -twenty-three," and so on).

Scorers' Duties

When the event is over, scorers record each soldier's time in the 2-mile-run block. They initial the appropriate block and note in the comment block that the time is for a 6.2-mile-bicycle test and whether or not the testee met the required standards for his age and sex.

2.5-MILE WALK

This event serves to assess cardiorespiratory and leg-muscle endur-

Equipment

Two stopwatches, one clipboard and pen for each scorer, numbers, and copies of the test instructions and standards are needed.

Facilities

This event uses the same course as the 2-mile run.

Personnel

One event supervisor and at least one scorer for every three soldiers to be tested are required. Appropriate safety, control, and medical personnel should be present.

Instructions

The event supervisor must read the following: "THE 2.5-MILE WALK IS USED TO ASSESS YOUR CARDIORESPIRATORY FITNESS AND LEG-MUSCLE ENDURANCE. ON THE COMMAND 'GO,' THE CLOCK WILL START, AND YOU WILL BEGIN WALKING AT YOUR OWN PACE. YOU MUST COMPLETE (describe the number of laps, start and finish points, and course layout). ONE FOOT MUST BE IN CONTACT WITH THE GROUND AT ALL TIMES. IF YOU BREAK INTO A RUNNING STRIDE AT ANY TIME OR HAVE BOTH FEET OFF THE GROUND AT THE SAME TIME, YOUR PERFORMANCE IN THE EVENT WILL BE TERMINATED. YOU WILL BE SCORED ON YOUR ABILITY TO COMPLETE THE 2.5-MILE COURSE IN A TIME EQUAL TO OR LESS THAN THAT LISTED FOR YOUR AGE AND SEX. WHAT ARE YOUR QUESTIONS ABOUT THIS EVENT?"

Administration

After reading the instructions, the event supervisor answers any related questions. He then divides the soldiers into groups of no more than three and assigns each group to a scorer. Each soldier is issued a number which the scorer records on the scorecard.

Timing Techniques

The event supervisor is the timer. He uses the commands "Get set" and "Go." Two stopwatches are used in case one fails. As the soldiers near the end of the 2.5-mile walk, the event supervisor starts calling off the elapsed time in minutes and seconds (for example, "Thirty-three-twenty-two, thirty -three -twenty -three, thirty-three-twenty -four," and so on).

Scorers' Duties

Scorers must observe the soldiers during the entire event and must ensure that the soldiers maintain a walking stride. Soldiers who break into any type of running stride will be terminated from the event and given a "no go." When the event is over, scorers record the time in the 2-mile-run block on the scorecard, initial the appropriate block, and note in the comment block that the time is for a 2.5-mile walk and whether or not the testee received a "go" or "no go."

APPENDIX A

PHYSIOLOGICAL DIFFERENCES BETWEEN THE SEXES

Soldiers vary in their physical makeup. Each body reacts differently to varying degrees of physical stress, and no two bodies react exactly the same way to the same physical stress. For everyone to get the maximum benefit from training, leaders must be aware of these differences and plan the training to provide maximum benefit for everyone. They must also be aware of the physiological differences between men and women. While leaders must require equal efforts of men and women during the training period, they must also realize that women have physiological limitations which generally preclude equal performance. The following paragraphs describe the most important physical and physiological differences between men and women.

SIZE

The average 18- year-old man is 70.2 inches tall and weighs 144.8 pounds, whereas the average woman of the same age is 64.4 inches tall and weighs 126.6 pounds. This difference in size affects the absolute amount of physical work that can be performed by men and women.

MUSCLES

Men have 50 percent greater total muscle mass, based on weight, than do women. A woman who is the same size as her male counterpart is generally only 80 percent as strong. Therefore, men usually have an advantage in strength, speed, and power over women.

FAT

Women carry about 10 percentage points more body fat than do men of the same age. Men accumulate fat primarily in the back, chest, and abdomen; women gain fat in the buttocks, arms, and thighs. Also, because the center of gravity is lower in women than in men, women must overcome more resistance in activities that require movement of the lower body.

BONES

Women have less bone mass than men, but their pelvic structure is wider. This difference gives men an advantage in running efficiency.

HEART SIZE AND RATE

The average woman's heart is 25 percent smaller than the average man's. Thus, the man's heart can pump more blood with each beat. The larger heart size contributes to the slower resting heart rate (five to eight beats a minute slower) in males. This lower rate is evident both at rest and at any given level of submaximal exercise. Thus, for any given work rate, the faster heart rate means that most women will become fatigued sooner than men.

FLEXIBILITY

Women generally are more flexible than men.

LUNGS

The lung capacity of men is 25 to 30 percent greater than that of women. This gives men still another advantage in the processing of oxygen and in doing aerobic work such as running.

RESPONSE TO HEAT

A woman's response to heat stress differs somewhat from a man's. Women sweat less, lose less heat through evaporation, and reach higher body temperatures before sweating starts. Nevertheless, women can adapt to heat stress as well as men. Regardless of gender, soldiers with a higher level of physical fitness generally better tolerate, and adapt more readily to, heat stress than do less fit soldiers.

OTHER FACTORS

Knowing the physiological differences between men and women is just the first step in planning physical training for a unit. Leaders need to understand other factors too.

Women can exercise during menstruation; it is, in fact, encouraged. However, any unusual discomfort, cramps, or pains while menstruating should be medically evaluated.

Pregnant soldiers cannot be required to exercise without a doctor's approval. Generally, pregnant women may exercise until they are close to childbirth if they follow their doctors' instructions. The Army agrees with the position of the American College of obstetricians and Gynecologists regarding exercise and pregnancy. This guidance is available from medical authorities and the U.S. Army Physical Fitness School (USAPFS). The safety and health of the mother and fetus are primary concerns when dealing with exercise programs.

Vigorous activity does not harm women's reproductive organs or cause menstrual problems. Also, physical fitness training need not damage the breasts. Properly fitted and adjusted bras, however, should be worn to avoid potential injury to unsupported breast tissue that may result from prolonged jarring during exercise.

Although female soldiers must sometimes be treated differently from males, women can reach high levels of physical performance. Leaders must use common sense to help both male and female soldiers achieve acceptable levels of fitness. For example, ability-group running alleviates gender-based differences between men and women. Unit runs, however, do not.

APPENDIX B

POSITIVE PROFILE FORM

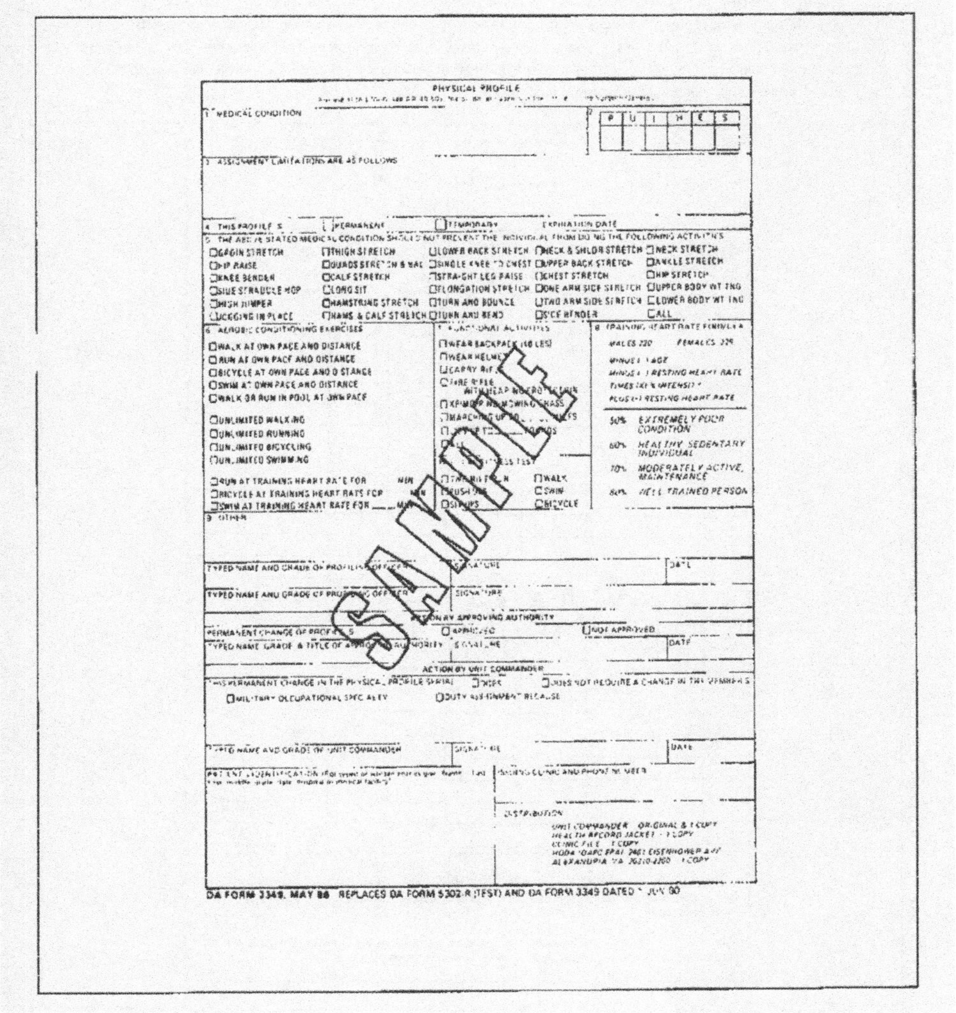

PHYSICAL PROFILE

1. MEDICAL CONDITION

2.

P	U	L	H	E	S

3. ASSIGNMENT LIMITATIONS ARE AS FOLLOWS

4. THIS PROFILE IS ☐ PERMANENT ☐ TEMPORARY EXPIRATION DATE

5. THE ABOVE STATED MEDICAL CONDITION SHOULD NOT PREVENT THE INDIVIDUAL FROM DOING THE FOLLOWING ACTIVITIES

☐ GROIN STRETCH	☐ THIGH STRETCH	☐ LOWER BACK STRETCH	☐ NECK & SHLDR STRETCH	☐ NECK STRETCH
☐ HIP RAISE	☐ QUADS STRETCH & VAL	☐ SINGLE KNEE TO CHEST	☐ UPPER BACK STRETCH	☐ ANKLE STRETCH
☐ KNEE BENDER	☐ CALF STRETCH	☐ STRAIGHT LEG RAISE	☐ CHEST STRETCH	☐ HIP STRETCH
☐ SIDE STRADDLE HOP	☐ LONG SIT	☐ ELONGATION STRETCH	☐ ONE ARM SIDE STRETCH	☐ UPPER BODY WT TNG
☐ HIGH JUMPER	☐ HAMSTRING STRETCH	☐ TURN AND BOUNCE	☐ TWO ARM SIDE STRETCH	☐ LOWER BODY WT TNG
☐ JOGGING IN PLACE	☐ HAMS & CALF STRETCH	☐ TURN AND BEND	☐ SIDE BENDER	☐ ALL

6. AEROBIC CONDITIONING EXERCISES

☐ WALK AT OWN PACE AND DISTANCE
☐ RUN AT OWN PACE AND DISTANCE
☐ BICYCLE AT OWN PACE AND DISTANCE
☐ SWIM AT OWN PACE AND DISTANCE
☐ WALK OR RUN IN POOL AT OWN PACE

☐ UNLIMITED WALKING
☐ UNLIMITED RUNNING
☐ UNLIMITED BICYCLING
☐ UNLIMITED SWIMMING

☐ RUN AT TRAINING HEART RATE FOR ___ MIN
☐ BICYCLE AT TRAINING HEART RATE FOR ___ MIN
☐ SWIM AT TRAINING HEART RATE FOR ___ MIN

7. FUNCTIONAL ACTIVITIES

☐ WEAR BACKPACK (40 LBS)
☐ WEAR HELMET
☐ CARRY RIFLE
☐ FIRE RIFLE
WITH HEAD PROTECTION
☐ MARCHING UP TO ___ MILES
☐ LIFT UP TO ___ POUNDS
☐ ALL

FITNESS TEST

☐ ☐ WALK
☐ PUSH-UPS ☐ SWIM
☐ SIT-UPS ☐ BICYCLE

8. TRAINING HEART RATE FORMULA

MALES 220 FEMALES 226
MINUS (-) AGE
MINUS (-) RESTING HEART RATE
TIMES (X) % INTENSITY
PLUS (+) RESTING HEART RATE

50% EXTREMELY POOR CONDITION
60% HEALTHY SEDENTARY INDIVIDUAL
70% MODERATELY ACTIVE, MAINTENANCE
80% WELL TRAINED PERSON

9. OTHER

TYPED NAME AND GRADE OF PROFILING OFFICER	SIGNATURE	DATE
TYPED NAME AND GRADE OF PROFILING OFFICER	SIGNATURE	

ACTION BY APPROVING AUTHORITY

PERMANENT CHANGE OF PROFILE IS ☐ APPROVED ☐ NOT APPROVED

TYPED NAME, GRADE & TITLE OF APPROVING AUTHORITY	SIGNATURE	DATE

ACTION BY UNIT COMMANDER

THIS PERMANENT CHANGE IN THE PHYSICAL PROFILE SERIAL ☐ DOES ☐ DOES NOT REQUIRE A CHANGE IN THE MEMBER'S
☐ MILITARY OCCUPATIONAL SPEC ALTY ☐ DUTY ASSIGNMENT RECAUSE

TYPED NAME AND GRADE OF UNIT COMMANDER	SIGNATURE	DATE

PATIENT IDENTIFICATION

ISSUING CLINIC AND PHONE NUMBER

DISTRIBUTION

UNIT COMMANDER - ORIGINAL & 1 COPY
HEALTH RECORD JACKET - 1 COPY
CLINIC FILE - 1 COPY
HQDA (DAPC-EPAI) 2461 EISENHOWER AVE
ALEXANDRIA, VA 22310-4200 - 1 COPY

DA FORM 3349, MAY 86 REPLACES DA FORM 5302-R (TEST) AND DA FORM 3349 DATED 1 JUN 80

Figure B-1

PHYSICAL FITNESS LOG

Soldiers can use a physical fitness log to record their fitness goals. The log will serve as a diary of how well they achieve them. Fitness goals are determined before the training begins. The results should closely parallel or exceed the unit's goals. While this is not a requirement, the log may also be used by commanders and supervisors as a record of physical fitness training. Figure C-l shows an example of a physical fitness log that could be reproduced locally.

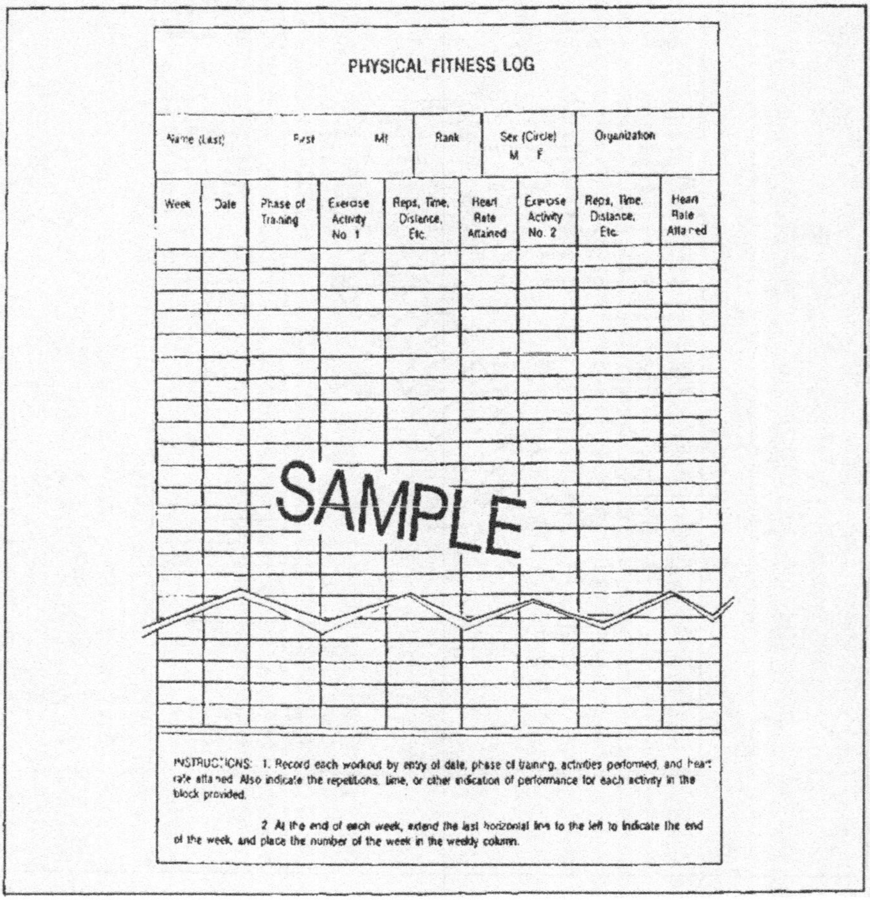
PHYSICAL FITNESS LOG

Name (Last)	First	MI	Rank	Sex (Circle) M F	Organization

Week	Date	Phase of Training	Exercise Activity No. 1	Reps, Time, Distance, Etc.	Heart Rate Attained	Exercise Activity No. 2	Reps, Time, Distance, Etc.	Heart Rate Attained

SAMPLE

INSTRUCTIONS: 1. Record each workout by entry of date, phase of training, activities performed, and heart rate attained. Also indicate the repetitions, time, or other indication of performance for each activity in the block provided.

2. At the end of each week, extend the last horizontal line to the left to indicate the end of the week, and place the number of the week in the weekly column.

Figure C-1

APPENDIX D

STATIONARY BICYCLE TEST

Only stationary bicycles which can be calibrated and which have mechanically adjustable resistances may be used to test profiled soldiers on the 6.2-mile (I O-kilometer), alternate APFT event. Therefore, the event supervisor or scorer must be sure that the stationary bicycle can be accurately adjusted to ensure that the soldier pedals against the correct resistance (force) of 2 kiloponds or 20 newtons. If the stationary bicycle cannot be properly calibrated and adjusted, the soldier may end up pedalling against a resistance which is too great or not great enough. In either case, the test would not provide an accurate indication of the soldier's level of cardiorespiratory fitness.

The best type of stationary bicycle for testing has the following features:

- Calibration adjustment.
- Adjustable resistance displayed in kiloponds or newtons.
- Odometer which accurately measures the distance traveled in either miles or tenths of miles or in kilometers and tenths of kilometers.

Examples of stationary bicycles which meet the above criteria are the mechanically braked Bodyguard 990 and Monark 868. Such bicycles can be used to accurately measure a person's rate of work or the total amount of work. They are often called bicycle ergometers.

If the stationary bicycle has an odometer, the soldier must pedal 6.2 miles (10.0 kilometers or 10,000 meters) against a resistance set at 2 kiloponds or 20 newtons. The test is completed when the soldier pedals 6.2 miles (10.0 kilometers). He receives a "Go" if he is below or at the time allotted for his particular age group and gender. Care should be taken to ensure that, when using a stationary bicycle which measures distance in kilometers, the test is ended at 10 kilometers, not 6.2 kilometers.

There are many electrically operated, stationary bicycles (EOSBS) on the market and in gymnasiums on Army installations. Most of them are designed for physical fitness training. Only a limited number of EOSB models are designed to accurately assess a person's energy expenditure during exercise. Such EOSBS are relatively expensive and are generally found in medical and scientific laboratories. Very few, if any, are found in gymnasiums on Army installations.

Because most of the more common training EOSBS were not designed to accurately assess energy expenditure, they should not be used for the alternate, cardiorespiratory APFT event.

For the sake of accuracy and ease of administration, soldiers designated to be tested on either of the two bicycle protocols should be tested using a moving bicycle IAW the guidelines provided elsewhere in this field manuel. If the mechanical y- braked Bodyguard 990 or Monark 868 is used, however, the tester must ensure that the equipment has been properly calibrated prior to each test.

TABLE D-1								
MALES								
AGE (YEARS)	17-21	22-26	27-31	32-36	37-41	42-46	47-51	52+
TIME ALLOTTED (MINUTES)	24.0	24.5	25.0	25.5	26.0	27.0	28.0	30.0
CALORIES/MIN.	9.8	9.7	9.5	9.3	9.2	8.9	8.6	8.2
CALORIES/HR.	590	580	570	560	550	535	520	490
N.-METERS/SEC. or WATTS	139	136	133	131	128	124	119	111
TOTAL CALORIES EXPENDED	236	237	238	239	240	241	242	245

TABLE D-2								
FEMALES								
AGE YEARS	17-21	22-26	27-31	32-36	37-41	42-46	47-51	52+
TIME ALLOTTED (MINUTES)	25.0	25.5	26.0	26.5	27.0	28.0	30.0	32.0
CALORIES/MIN.	9.5	9.3	9.2	9.0	8.9	8.6	8.2	7.8
CALORIES/HR.	570	560	550	545	535	520	490	465
N.-METERS/SEC. or WATTS	133	131	128	126	124	120	110	104
TOTAL CALORIES	237.5	238	239	240	240.5	242	245	248

APPENDIX E

SELECTING THE RIGHT RUNNING SHOE

Choosing a running shoe that is suitable for your particular type of foot can help you avoid some common running- related injuries. It can also make running more enjoyable and let you get more mileage out of your shoes.

Shoe manufacturers are aware that, anatomically, feet usually fall into one of three categories. Some people have "floppy" feet that are very "loose- jointed." Because feet like this are too mobile, they "give" when they hit the ground. These people need shoes that are built to control the foot's motion. At the other extreme are people with "rigid" feet. These feet are very tight-jointed and do not yield enough upon impact. To help avoid impact-related injuries, these people need shoes that cushion the impact of running. Finally, the third type, or normal foot, falls somewhere between mobile and rigid. This type of foot can use any running shoe that is stable and properly cushioned. Use the chart at Figure E-1 to help you determine what kind of foot you have. Then, read the information on special features you should look for in a shoe.

When shopping for running shoes, keep the following in mind:

- Expect to spend between $30 and $100 for a pair of good shoes.
- Discuss your foot type, foot problems, and shoe needs with a knowledgeable salesperson.
- Check the PX for available brands and their prices before shopping at other stores.
- Buy a training shoe, not a racing shoe.
- When trying on shoes, wear socks that are as similar as possible to those in which you will run. Also, be sure to try on both shoes.
- Look at more than one model of shoe.
- Choose a pair of shoes that fit both feet well while you are standing.
- Ask if you can try running in the shoes on a non-carpeted surface. This gives you a feel for the shoes.
- Carefully inspect the shoes for defects that might have been missed by quality control. Do the following:
  - -Place the shoes on a flat surface and check the heel from behind to see that the heel cup is perpendicular to the sole of the shoe.
  - -Feel the seams inside the shoe to determine if they are smooth, even, and well-stitched.
  - -Check for loose threads or extra glue spots; they are usually signs of poor construction.

The shoes' ability to protect you from injury decreases as the mileage on them increases. Record the number of miles you run with them on a regular basis, and replace the shoes when they have accumulated 500 to 700 miles even if they show little wear.

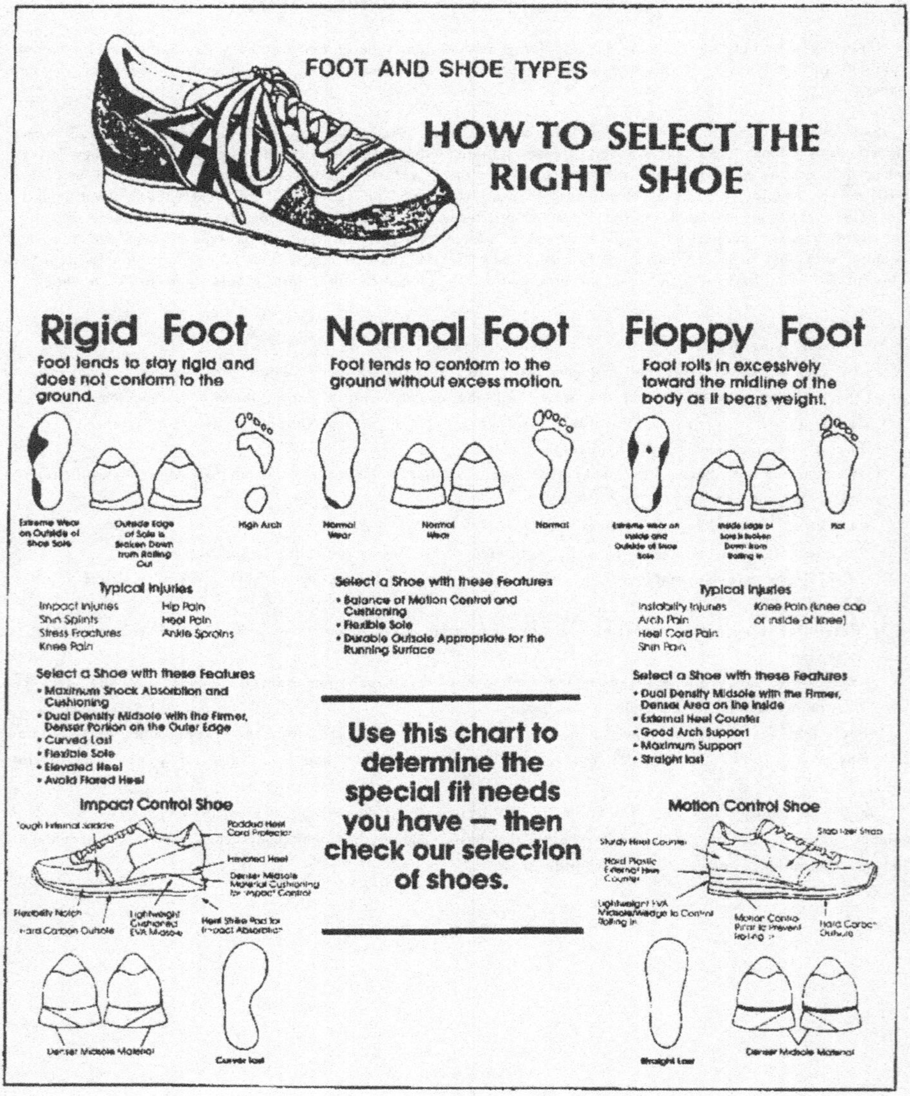
FOOT AND SHOE TYPES
HOW TO SELECT THE RIGHT SHOE
Rigid Foot
Foot tends to stay rigid and does not conform to the ground.
High Arch
Typical Injuries
Impact Injuries
Shin Splints
Stress Fractures
Knee Pain
Hip Pain
Heel Pain
Ankle Sprains
Select a Shoe with these Features
• Maximum Shock Absorbtion and Cushioning
• Dual Density Midsole with the Firmer, Denser Portion on the Outer Edge
• Curved Last
• Flexible Sole
• Elevated Heel
• Avoid Flared Heel
Impact Control Shoe
Denser Midsole Material
Curved last
Normal Foot
Foot tends to conform to the ground without excess motion.
Normal Wear
Normal Wear
Normal
Select a Shoe with these Features
• Balance of Motion Control and Cushioning
• Flexible Sole
• Durable Outsole Appropriate for the Running Surface
Use this chart to determine the special fit needs you have — then check our selection of shoes.
Floppy Foot
Foot rolls in excessively toward the midline of the body as it bears weight.
Flat
Typical Injuries
Instability Injuries
Arch Pain
Heel Cord Pain
Shin Pain
Knee Pain (knee cap or inside of knee)
Select a Shoe with these Features
• Dual Density Midsole with the Firmer, Denser Area on the Inside
• External Heel Counter
• Good Arch Support
• Maximum Support
• Straight last
Motion Control Shoe
Straight Last
Denser Midsole Material

Figure E-1

APPENDIX F

CALCULATION OF V0,max

This appendix gives a step-by-step example of how a soldier can calculate VOjmax using his all-out, 2-mile-run time. This lets interested soldiers compare their fitness levels with others such as athletes whose VO,max values are published in magazines or journals.

The two equations below convert the 2-mile-run times of males and females to maximum oxygen uptake values. The VC^max values obtained are shown as the maximum amount of oxygen in milliliters used per kilogram of the person's body weight in one minute during maximum aerobic exercise. VOgmax values are generally expressed more succinctly as ml CVkg x min.

For males, the following equation is used to calculate V0,max:

V0,max = 99.7 - [3.35 x (2-mile-run time in decimal form)].

For females, the following equation is used:

V0,max • 72.9 - [1.77 x (2-mile-run time in decimal form)].

The example below shows how to use the equation for males. The data is for a 21-year-old male whose all-out, 2-mile-run time is 12 minutes and 36 seconds.

STEP 1. Express the 2-mile-run time as a decimal, and insert it into the equation. When 12 minutes 36 seconds is written as a decimal, it becomes 12.60 minutes. (To determine what fraction of a minute 36 seconds is, divide 36 seconds by the number of seconds in one minute, that is, 60 seconds. Thus, 36/60 = 0.60. This fraction is added to the minute value to give 12.60 as the run time expressed as a decimal.) After putting the decimal form into the equation, the equation should resemble the one below.

VOj,max = 99.7 -[3.35 x (12.60)]

STEP 2. Multiply the decimal form of the 2-mile-run time by 3.35. In this case, we get [3.35 x (12.60)], which equals 42.21. At this point, the equation should resemble the one below.

V0,max = 99.7 -[42.21].

STEP 3. Subtract the product obtained in Step 2 from 99.7. For our example, we make the following subtraction: 99.7-[42.21], This gives a value of 57.49. Thus, the equation should look like the one below.

VCumax = 57.49

This calculation reveals that a male whose all-out, 2-mile-run time is 12 minutes 36 seconds will have a V0,max of approximately 57.49 ml Q2/kg x min.

To determine how this value or others translates into fitness ratings, refer to Table F-1. It presents information for finding one's level of CR fitness based on V0,max. By matching a soldier's value for maximum oxygen uptake with those in the table corresponding to his age group and sex, one gets an adjectival rating (fair, good, superior, etc.).

$\dot{V}O_2max$ AND CR FITNESS CLASSIFICATIONS

CATEGORY	SEX	AGE 20-29	30-39	40-49	50-59	60+
SUPERIOR	MALE	54.0+*	52.5+	50.4+	47.1+	45.2+
	FEMALE	46.8+	43.9+	41.0+	36.8+	37.5+
EXCELLENT	MALE	48.2-51.4	46.8-50.4	44.1-48.2	41.0-45.3	38.1-42.5
	FEMALE	41.0-44.2	38.5-41.0	36.3-39.5	32.1-35.2	31.2-35.2
GOOD	MALE	44.2-47.0	42.4-45.3	39.9-43.9	36.7-39.5	33.6-36.7
	FEMALE	36.7-39.5	34.6-37.4	32.3-35.1	29.4-39.9	27.2-30.9
FAIR	MALE	41.0-43.9	38.9-41.6	36.7-39.5	33.8-36.1	30.2-32.4
	FEMALE	33.8-36.1	32.1-33.9	29.5-31.6	26.9-28.7	24.5-26.5
POOR	MALE	37.1-40.3	35.4-38.1	33.0-35.6	30.2-32.5	26.5-29.4
	FEMALE	30.6-32.7	28.7-31.9	26.5-29.4	24.3-26.1	22.8-24.0
VERY POOR	MALE	27.1-36.7	26.5-34.0	24.2-32.3	22.1-29.4	18.3-25.1
	FEMALE	22.6-29.4	22.5-28.0	20.8-25.6	21.1-23.7	17.9-22.1

*$\dot{V}O_2max$ is expressed in ml O_2/kg x min.

Table **F-1**

Table F-1 lists some values for VOzmax along with their associated CR fitness levels. This table was obtained from the Institute for Aerobic Research in Dallas, Texas. These values can be used to classify a soldier's level of CR fitness based on his VO,max.

APPENDIX G

PERCEIVED EXERTION

The heart rate has traditionally been used to estimate exercise intensity. However, evidence shows that a person's own perception of the intensity of his exercise can often be just as accurate as the heart rate in gauging his exercise intensity.

The scale in Figure G-l lets a soldier rate his degree of perceived exertion (PE). This scale consists of numerical ratings for physical exercise followed by their associated descriptive ratings.

PERCEIVED EXERTION (PE) SCALE

NUMERICAL RATING	VERBAL RATING
6	
7	very, very light
8	
9	very light
10	
11	fairly light
12	
13	somewhat hard
14	
15	hard
16	
17	very hard
18	
19	very, very hard
20	

Figure G-1

To judge perceived exertion, estimate how difficult it feels to do the exercise. Do not be concerned with any one single factor such as shortness of breath or work intensity. Instead, try to concentrate on the total inner feeling of exertion.

Multiplying the rating of perceived exertion by 10 roughly approximates the heart rate during exercise. For example, a PE of 14, when multiplied by 10, equals 140.

Most soldiers with THRs between 130 and 170 BPM would exercise between a PE of 13 (somewhat hard) and 17 (very hard).

Although either percent of maximum heart rate or perceived exertion may be used during exercise, the most valid method for calculating THR is percent HRR.

THE MAJOR SKELETAL MUSCLES OF THE HUMAN BODY

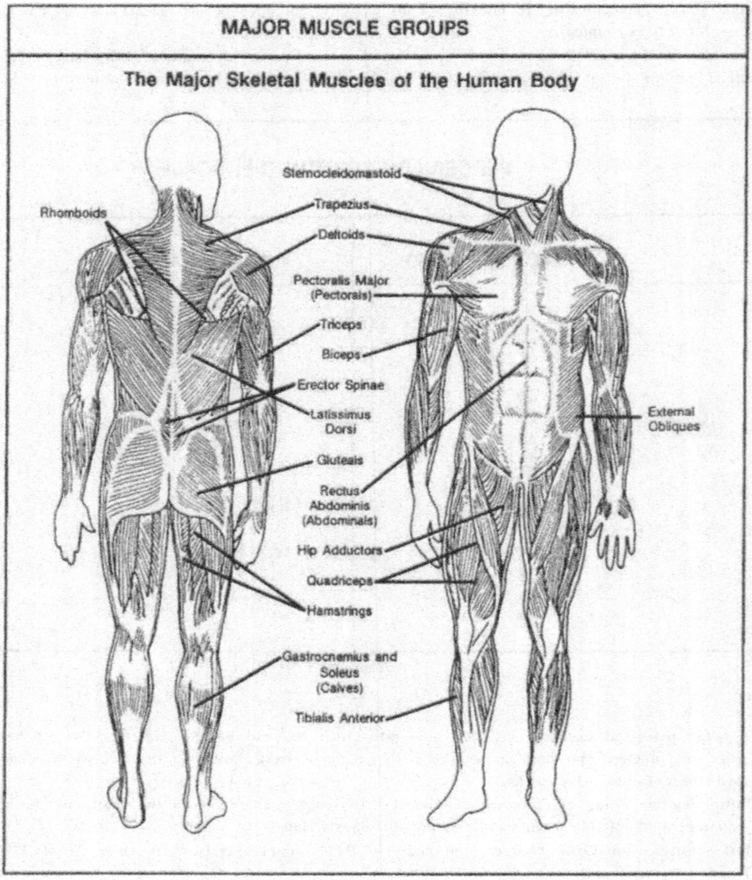

Figure H-1

The iliopoas muscle (a hip flexor) cannot be seen as it lies beneath other muscles. It attaches to the lumbar vertebrae and the femur.

GLOSSARY

Section 1: Acronyms and Abbreviations

AC	Active Component
AGR	ability group run
AIT	advanced individual training
APFT	Army Physical Fitness Test
AR	Army regulation
ARNG	Army National Guard
ARTEP	Army Training and Evaluation Program
ATP	adenosine triphosphate
BCT	basic combat training
BDU	battle dress uniform
BPM	beats per minute
BT	basic training
BTMS	Battalion Training Management System
c	centigrade
CAD	coronary artery disease
CPR	cardiopulmonary resuscitation
CPT	captain
CR	cardiorespiratory
CVSP	cardiovascular screening program
DA	Department of the Army
DOD	Department of Defense
EDRE	emergency deployment readiness exercise
EIB	Expert Infantryman Badge
EOSB	electrically operated, stationary bicycle
F	Fahrenheit
FITT	frequency, intensity, time, type
FM	field manual
FTX	field training exercise
HDL	high-density lipoprotein
HQ	headquarters
HQDA	Headquarters, Department of the Army
HRR	heart rate reserve
ID	identification
IET	initial entry training
IG	inspector general
kph	kilometers per hour

lat	latissimus dorsi
LCE	load-carrying equipment
LDL	low-density lipoprotein
MACOM	major Army command
MEDDAC	medical department activity
METL	mission-essential task list
MFT	master fitness trainer
MHR	maximum heart rate
min	minute(s)
MOS	military occupational specialty
MPH	miles per hour
MRDA	military recommended dietary allowance
MRE	meal, ready to eat
NCO	noncommissioned officer
NCOIC	noncommissioned officer in charge
NGR	National Guard regulation
No.	number
OIC	officer in charge
OST	one-station training
OSUT	one-station unit training
Pam	pamphlet
PE	perceived exertion
PNF	proprioceptive neuromuscular facilitation
PRE	partner-resisted exercise
PT	physical training
pts	points
PU	push-up
RC	Reserve Component
rep	repetition
RHR	resting heart rate
RICE	rest, ice, compression, elevation
RM	repetition maximum
ROTC	Reserve Officers' Training Corps
sec	second(s)
SCUBA	self-contained underwater breathing apparatus
SDT	self development test
SOP	standing operating procedure
SU	sit-up
TB med	technical bulletin, medical
TDA	table of distribution and allowances
THR	training heart rate
TM	technical manual
TOE	table of organization and equipment
TRADOC	U.S. Army Training and Doctrine Command
TS	timed set
TSP	training support package

U.S.	United States
USAPFS	United States Army Physical Fitness School
USAR	United States Army Reserve
VO_2max	maximum oxygen consumption per minute
WBGTI	wet bulb globe temperature index
WCF	windchill factor

Section II: Terms

extension

An increase in the angle between two bones in which a straightening movement occurs; the opposite of flexion. For example, extension of the elbow involves an increase in the angle formed by the upper and lower arm as the arm straightens at the elbow.

flexion

A decrease in the angle between two bones in which a bending movement occurs; the opposite of extension. For example, flexion of the elbow involves a decrease in the angle formed by the lower and upper arm as the arm bends at the elbow.

REFERENCES

SOURCES USED

These are the sources quoted or paraphrased in this publication.

ARMY REGULATIONS (ARs)

15-6	Procedures for Investigating Officers and Boards of Officers. May 1988.
30-1	The Army Food Service Program. January 1985.
350-15	Army Physical Fitness Program. November 1989.
385-55	Prevention of Motor Vehicle Accidents. March 1987.

OTHER ARMY PUBLICATIONS

DOD Directive 1308.1	Physical Fitness and Weight Control Program. April 1981.
FM 21-18	Foot Marches. June 1990.

DOCUMENTS NEEDED

These documents must be available to the intended users of this publication.

ARMY REGULATIONS (ARs)

40-501	Standards of Medical Fitness. July 1987.
600-8-2	Suspension of Favorable Personnel Actions (Flags). October 1987.
600-9	The Army Weight Control Program. September 1986.
600-63	Army Health Promotion. November 1987.

OTHER ARMY PUBLICATIONS

FM 25-100	Training the Force. November 1988.
NGR 40-501	Medical Examination for Members of the Army National Guard. October 1981.
TRADOC Reg 350-6	Initial Entry Training (IET) Policies and Administration. August 1989.

READINGS RECOMMENDED

These readings contain relevant supplemental information.

DEPARTMENT OF THE ARMY PAMPHLETS (DA Pares)

28-9	Unit Level Recreational Sports. June 1973.
350-15	Commander's Handbook on Physical Fitness. October 1982.
350-18	The Individual's Handbook on Physical Fitness. May 1983.
350-22	You and the Army Physical Fitness Test (APFT). September 1987.
351-4	Army Formal Schools Catalog. August 1991.

FIELD MANUALS (FMs)

21-150	Combative. December 1971.
22-5	Drill and Ceremonies. December 1986.
31-70	Basic Cold Weather Manual. April 1968.

OTHER ARMY PUBLICATIONS AND MATERIALS

AR 215-1	Administration of Army Morale, Welfare, and Recreation. February 1984.
DA Form 705	Army Physical Fitness Test Scorecard. May 1987.
DA Form 3349	Physical Profile. May 1986.
Folio No. I	"Training Facilities," Corps of Engineers Drawing No. 28-13-95. Directorate of Facilities Engineering.
SB 10-260	Master Menu. December 1989.
TB Med 507	Occupational and Environmental Health Prevention, Treatment, and Control of Heat Injury. July 1980.

TSP Physical Fitness Training - Total Fitness. July 1987.

TRAINING VIDEO TAPES (TVTS)

8-103	Standards for Determining Body Fat. 1986.
21-76	Army Physical Fitness Test (APFT). 1986.
21-191	Administration of the APFT. 1988.
21-192	Partner-Resisted Exercises (PRE). 1987.
21-218	Flexibility: The Truth About Stretching. 1989.
21-203	Push-up/Sit-up Improvement. 1988.

INDEX

This is a topical index organized alphabetically. Citations are to paragraph numbers.

FM 21-20
30 SEPTEMBER 1992

By Order of the Secretary of the Army

GORDON R. SULLIVAN
General, United States Army
Chief of Staff

Official:

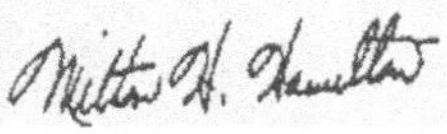

MILTON H. HAMILTON
Administrative Assistant to the
Secretary of the Army

02361

DISTRIBUTION:

Active Arm y, USAR and ARNG: To be distributed In accordance with DA Form 12-11E, requirements for FM 21-20, Physical Fitness Training (Qty rqr block no. 0165).

•U.S. Government Printing Office: 1994 – 300-421/82852

Made in the USA
Monee, IL
05 November 2026

67813164R00252